Nonlinear Targeted Energy Transfer in Mechanical and Structural Systems I

SOLID MECHANICS AND ITS APPLICATIONS
Volume 156

Series Editor: G.M.L. GLADWELL
Department of Civil Engineering
University of Waterloo
Waterloo, Ontario, Canada N2L 3GI

Aims and Scope of the Series

The fundamental questions arising in mechanics are: *Why?*, *How?*, and *How much?* The aim of this series is to provide lucid accounts written by authoritative researchers giving vision and insight in answering these questions on the subject of mechanics as it relates to solids.

The scope of the series covers the entire spectrum of solid mechanics. Thus it includes the foundation of mechanics; variational formulations; computational mechanics; statics, kinematics and dynamics of rigid and elastic bodies: vibrations of solids and structures; dynamical systems and chaos; the theories of elasticity, plasticity and viscoelasticity; composite materials; rods, beams, shells and membranes; structural control and stability; soils, rocks and geomechanics; fracture; tribology; experimental mechanics; biomechanics and machine design.

The median level of presentation is the first year graduate student. Some texts are monographs defining the current state of the field; others are accessible to final year undergraduates; but essentially the emphasis is on readability and clarity.

For other titles published in this series, go to
www.springer.com/series/6557

A.F. Vakakis • O.V. Gendelman • L.A. Bergman •
D.M. McFarland • G. Kerschen • Y.S. Lee

Nonlinear Targeted Energy Transfer in Mechanical and Structural Systems I

Alexander F. Vakakis
Department of Mechanical Science
  and Engineering
University of Illinois
Urbana, Illinois, USA
and
Mechanics Division
National Technical University of Athens
Athens, Greece

Lawrence A. Bergman
Department of Aerospace Engineering
University of Illinois
Urbana, Illinois, USA

Gaëtan Kerschen
Department of Aerospace and Mechanical
  Engineering
University of Liège
Liège, Belgium

Oleg V. Gendelman
Faculty of Mechanical Engineering
Technion – Israel Institute of Technology
Haifa, Israel

D. Michael McFarland
Department of Aerospace Engineering
University of Illinois at Urbana-Champaign
Urbana, Illinois, USA

Young Sup Lee
Department of Mechanical and Aerospace
  Engineering
New Mexico State University
Las Cruces, New Mexico, USA

ISBN-13: 978-1-4020-9125-4   e-ISBN-13: 978-1-4020-9130-8

Library of Congress Control Number: 2008940435

© 2008 Springer Science+Business Media, B.V.
No part of this work may be reproduced, stored in a retrieval system, or transmitted
in any form or by any means, electronic, mechanical, photocopying, microfilming, recording
or otherwise, without written permission from the Publisher, with the exception
of any material supplied specifically for the purpose of being entered
and executed on a computer system, for exclusive use by the purchaser of the work.

Printed on acid-free paper

9 8 7 6 5 4 3 2 1

springer.com

Contents Volume 1

Preface vii

Abbreviations xi

1 Introduction 1

2 Preliminary Concepts, Methodologies and Techniques 15
2.1 Nonlinear Normal Modes (NNMs) 16
2.2 Energy Localization in Nonlinear Systems 28
2.3 Internal Resonances, Transient and Sustained Resonance Captures 38
2.4 Averaging, Multiple Scales and Complexification 54
2.5 Methods of Advanced Signal Processing 70
    2.5.1 Numerical Wavelet Transforms 71
    2.5.2 Empirical Mode Decompositions and Hilbert Transforms 77
2.6 Perspectives on Hardware Development and Experiments 81

3 Nonlinear Targeted Energy Transfer in Discrete Linear Oscillators with Single-DOF Nonlinear Energy Sinks 93
3.1 Configurations of Single-DOF NESs 93
3.2 Numerical Evidence of TET in a SDOF Linear Oscillator with a SDOF NES 98
3.3 SDOF Linear Oscillators with SDOF NESs: Dynamics of the Underlying Hamiltonian Systems 108
    3.3.1 Numerical Study of Periodic Orbits (NNMs) 108
    3.3.2 Analytic Study of Periodic Orbits (NNMs) 124
    3.3.3 Numerical Study of Periodic Impulsive Orbits (IOs) 135
    3.3.4 Analytic Study of Periodic and Quasi-Periodic IOs 137
    3.3.5 Topological Features of the Hamiltonian Dynamics 157
3.4 SDOF Linear Oscillators with SDOF NESs: Transient Dynamics of the Damped Systems 165
    3.4.1 Nonlinear Damped Transitions Represented in the FEP 166
    3.4.2 Dynamics of TET in the Damped System 171

3.5	Multi-DOF (MDOF) Linear Oscillators with SDOF NESs: Resonance Capture Cascades and Multi-frequency TET	233
	3.5.1 Two-DOF Linear Oscillator with a SDOF NES	237
	3.5.2 Semi-Infinite Chain of Linear Oscillators with an End SDOF NES	269

4 Targeted Energy Transfer in Discrete Linear Oscillators with Multi-DOF NESs — 303

4.1	Multi-Degree-of-Freedom (MDOF) NESs	303
	4.1.1 An Alternative Way for Passive Multi-frequency Nonlinear Energy Transfers	304
	4.1.2 Numerical Evidence of TET in MDOF NESs	309
4.2	The Dynamics of the Underlying Hamiltonian System	317
	4.2.1 System I: NES with $O(1)$ Mass	320
	4.2.2 System II: NES with $O(\varepsilon)$ Mass	325
	4.2.3 Asymptotic Analysis of Nonlinear Resonant Orbits	328
	4.2.4 Analysis of Resonant Periodic Orbits	336
4.3	TRCs and TET in the Damped and Forced System	347
	4.3.1 Numerical Wavelet Transforms	347
	4.3.2 Damped Transitions on the Hamiltonian FEP	352
4.4	Concluding Remarks	365

Index — 369

Preface

This monograph evolved over a period of nine years from a series of papers and presentations addressing the subject of passive vibration control of mechanical systems subjected to broadband, transient inputs. The unifying theme is *Targeted Energy Transfer – TET*, which represents a new and unique approach to the passive control problem, in which a strongly nonlinear, fully passive, local attachment, the *Nonlinear Energy Sink – NES*, is employed to drastically alter the dynamics of the primary system to which it is attached. The intrinsic capacity of the properly designed NES to promote rapid localization of externally applied (narrowband) vibration or (broadband) shock energy to itself, where it can be captured and dissipated, provides a powerful strategy for vibration control and the opens the possibility for a wide range of applications of TET, such as, vibration and shock isolation, passive energy harvesting, aeroelastic instability (flutter) suppression, seismic mitigation, vortex shedding control, enhanced reliability designs (for example in power grids) and others. The monograph is intended to provide a thorough explanation of the analytical, computational and experimental methods needed to formulate and study TET in mechanical and structural systems. Several practical engineering applications are examined in detail, and experimental verification and validation of the theoretical predictions are provided as well. The authors also suggest a number of possible future applications where application of TET seems promising.

The authors are indebted to a number of sponsoring agencies. The Office of Naval Research – ONR (AFV, LAB), the National Science Foundation – NSF (AFV, DMM, LAB), the Air Force Office of Scientific Research – AFOSR (AFV, DMM, YSL, LAB), the Fund for Basic Research of the National Technical University of Athens (AFV), the Hellenic Secretariat for Research and Development (AFV), the Horev Fellowship Trust, the Israel Science Foundation (OG), the Belgian National Fund for Scientific Research (GK), and the Mavis Memorial Fund Fellowship of the College of Engineering of the University of Illinois (YSL) provided financial support for this work which enabled the realization of this long-standing, multinational collaborative research effort. In addition, the authors are greatly appreciative of the long-standing interest and support of Mr. Jim Lally of PCB Piezotronics, Inc., whose generosity through equipment support has enabled much of the experimental work reported in this monograph.

The authors would like to express their gratitude to Professor Leonid Manevitch of the Semenov Institute of Chemical Physics, Russian Academy of Sciences, Moscow, Russia, for his many shared insights which have proved to be critical to many of the developments included in this monograph; the authors greatly benefited from many stimulating discussions with him over the years in Moscow, Athens and Urbana. In addition, AFV will be always indebted to his late advisor and teacher, Professor Thomas K. Caughey, who through his teaching of asymptotic techniques, nonlinear dynamics and mechanical vibrations influenced much of the analytical work included in this work.

Many colleagues and former or current graduate and undergraduate students of the authors have, through their contributions, influenced various parts of this research effort. The authors are indeed indebted to (in no particular order) Dr. S. Tsakirtzis, Dr. F. Georgiadis, Dr. P. Panagopoulos, Dr. F. Nucera, Prof. Yu.V. Mikhlin, Prof. D. Quinn, Prof. R. H. Rand, Prof. T. Strganac, Prof. P. Cizmas, Prof. T. Kalmar-Nagy, Prof. V. Rothos, Prof. V. Pilipchuck, Dr. D. Gorlov, Dr. A. Musienko, Prof. X. Ma, Dr. X. Jiang, Mr. T. Sapsis, Mr. Yu. Starosvetsky, Mr. I. Karayannis, Mr. W.J. Hill, Mr. C. Nichkawde, Mr. R. Viguie, Prof. J.C. Golinval, Prof. A. Santini, Prof. D. Wang, Prof. S.C. Hong, Dr. A. Musienko, Mr. P. Kourdis, Mr. R. Viguie, Mr. J. Kowtko, Dr. Y. Wang, Mr. S. Hubbard, Ms. N. Fanouraki, Mr. C. Dumcum, Ms. C. Tripepi, Mr. F. Lo Iacono, Mr. G. Barone, Ms. M. Wise, and Ms. I. Rizou. The Departments of Aerospace Engineering and Mechanical Science and Engineering at the University of Illinois at Urbana- Champaign, the College of Engineering of the University of Illinois at Urbana-Champaign, and the Mechanics Division of the Department of Applied Sciences of the National Technical University of Athens generously provided space, resources, travel accommodation and hospitality during this nine year period, for which the authors are indeed grateful.

In addition, the authors would like to acknowledge the use of materials from their papers in archival journals and conference proceedings published by Springer Verlag, the American Society of Mechanical Engineers – ASME, Elsevier, World Scientific, the Society of Industrial and Applied Mathematics – SIAM, the American Institute of Aeronautics and Astronautics – AIAA, the American Institute of Physics – AIP, Sage Journals, and John Wiley & Sons. The original sources of these materials are referenced throughout this monograph. Moreover, the authors are appreciative of the efforts of the editorial staff at Springer Verlag, particularly of Ms. Nathalie Jacobs and Ms. Anneke Pot of Springer NL, and of the careful job of typesetting performed by Karada Publishing Services, particularly Ms. Jolanda Karada. Of course, any errors that remain are solely the responsibility of the authors.

Last, but not least, the authors would like to express their gratitude to their family members for their continued and unconditional support over the course of this long effort: Fotis Sr., Anneta, Elpida, Brian, Elias, Vasiliki, Sotiria, Marianna and Fotis Jr. (AFV), Anechka, Miriam, Sheina and Hava (OG), Jane (LAB), Karen (DMM), Carine, Julie and Maxime (GK), Ju Eun, Rose and Erica (YSL). This monograph is dedicated to them.

A.F. Vakakis
O. Gendelman
L.A. Bergman
D.M. McFarland
G. Kerschen
Y.S. Lee

NS bifurcation	Neimark–Sacker bifurcation
NS NES	Non-smooth NES
POD	Proper orthogonal decomposition
POM	Proper orthogonal mode
PZ	Propagation zone
RCC	Resonance capture cascade
r.m.s. response	Root mean square response
SDOF	Single-degree-of-freedom
SIM	Slow invariant manifold
SMR	Strongly modulated response
SN bifurcation	Saddle-node bifurcation
SP	South pole
SRC	Sustained Resonance Capture
TET	Targeted energy transfer
TMD	Tuned mass damper
VDP oscillator	Van der Pol oscillator
VI	Vibro-impact
WT	Wavelet transform

Chapter 1
Introduction

Any process in nature involves to a certain extent some type of energy transfer. From an engineering point of view, certain processes of energy transfer are undesired but still inevitable, as, for instance, energy dissipation in electromechanical systems; whereas other processes are desired and highly beneficial to the design objectives, the classical example from mechanical engineering being the addition of a vibration absorber to a machine for eliminating unwanted disturbances.

Targeted energy transfers (TETs), where energy of some form is directed from a source (donor) to a receiver (recipient) in a one-way irreversible fashion, govern a broad range of physical phenomena. One basic example of TET in nature, is resonance-driven solar energy harvesting governing photosythesis (Jenkins et al., 2004), where energy from the Sun is captured by photobiological antenna chromophores and is then transferred to reaction centers through a series of interactions between chromophore units (van Amerongen et al., 2000; Renger et al., 2001). In addition, basic problems in biopolymers concern energy self-focusing, localization and transport (Kopidakis et al., 2001), with applications in photosynthesis (Hu et al., 1998) and bioenergetic processes (Julicher et al., 1997).

From the engineering point of view, the scaling down of engineering applications from macro- to micro- and nano-scales dictates an understanding of the mechanisms governing TET and energy exchanges between components possessing different characteristic lengths with dynamics governed by different time-scales. For example, as pointed out by Wang et al. (2007) in applications such as molecular electronic devices where the scales of the dynamics are at the level of individual molecules, classical concepts of heat transport do not apply, and heat is transported by energy transfer through discrete molecular vibration excitations. Hence, understanding and analyzing energy transfer mechanisms in molecular dynamics (such as, resonance energy transfer) is key in conceiving devices or studying processes for specific macromolecular applications, such as, for example, in the area of photophysics (Andrews, 2000; Jenkins and Andrews, 2002, 2003). Moreover, molecular dynamic simulations of energy transfers (for example, through solitonic waves) in mechanistic molecular or atomistic models have been used to study thermodynamic processes, such as, melting of polymer crystals and phase transitions in polymer-clay nanocomposites (Ginzburg and Manevitch, 1991; Berlin et al., 1999; Ginzburg et al., 2001; Berlin et al., 2002; Gendelman et al., 2003). In Musumeci et al. (2003) issues related to nonlinear mechanisms for energy transfer and localization in biological macromolecules and related applications to biology are discussed. Moreover applications of nonlinear energy transfer in a broad area of applications ranging from cancer detection (Meessen, 2000; Vedruccio and Meessen, 2004) to wireless power transfer (Kurs et al., 2007) have been reported in the recent literature.

Therefore, it is not surprising that TET phenomena have received much attention in applications from diverse fields of applied mathematics, applied physics, and engineering. Representative examples are the works by Aubry and co-workers on passive targeted energy transfer (TET) between nonlinear oscillators and/or discrete breathers (Kopidakis et al., 2001; Aubry et al., 2001; Maniadis et al., 2004; Memboeuf and Aubry, 2005; Maniadis and Aubry, 2005), on breather-phonon resonances (Morgante et al., 2002), and on quantum TET between nonlinear oscillators (Maniadis et al., 2004). The dynamical mechanisms considered in these works were based on imposing conditions of nonlinear resonance between interacting dynamical systems in order to achieve TET from one to the other, and then 'breaking" this condition at the end of the energy transfer to make it irreversible. A mechanism of TET along a line or surface by means of coherent traveling solitary waves is examined in Nistazakis et al. (2002); specifically, the transfer of a solitary wave to a targeted position was studied in the nonlinear Schrödinger (NLS) equation, the underlying nonlinear dynamical mechanism being resonance energy transfer from an ac drive to the solitary wave. Applications of energy localization and TET in diverse applications, such as, biological macromolecules – proteins and DNA, arrays of Josephson junctions in superconductivity applications, and molecular crystals are given in Dauxois et al. (2004), including analytical, computational and experimental results.

In other complex phenomena, such as turbulence and chaotic dynamics, multi-scale energy transfers between different spatial and temporal scales govern the dynamics. Perhaps the best known example is turbulence, where mechanical energy is supplied to a fluid system at relatively large length scales, peculiar spatiotemporal coherent structures are formed at intermediate scales, and dissipation of energy occurs at short scales (Bohr et al., 1998). Hence, energy transfer between these scales is what makes turbulence possible. Examples of works on multi-scale energy transfers in fluids are the works by Kim et al. (1996) and Tran (2004) who studied nonlinear energy transfers in fully developed turbulence, and by Brink et al. (2004) who studied nonlinear interactions and multi-scale energy transfers among inertial modes of a rotating fluid, modeling it as a network of coupled oscillators.

All nonlinear energy transfers involve to a certain extent some type of nonlinear resonance between a donor and a receptor. Resonance energy transfer has been identified as an important mechanism for energy and electronic transports in the area of photophysics of macromolecules (Jenkins and Andrews, 2002, 2003; Andrews and Bradshaw, 2004), and has been recognized as the principal mechanism for electronic energy transport in molecular chains following initial excitations (Daniels et al., 2003). Esser and Henning (1991) analyzed energy transfer and bifurcations in a condensed molecular system. Fluorescence resonance energy transfer (FRET) where fluorescent energy from an excited fluorophore is transferred to light-absorbing molecules lying in close proximity, has been well studied; FRET has been applied as an optical microscopy technique for developing biosensors and examining physiological processes with highly temporal and spatial resolution (Cardullo and Parpura, 2003; Berland et al., 2005). Additional applications of FRET range from in vivo medical diagnosis of infections (Hwang et al., 2006), to detection of targeted DNA sequences (Xu et al., 2005), and development of biosensors and medical probes (Yesilkaya et al., 2006).

Dodaro and Herman (1998) studied analytically energy transfers in liquids through resonant vibration interactions, using a molecular dynamics approach to study the probability of vibration energy transfer between atoms. An example of a study of laser-assisted resonance energy transfer is provided in Allcock et al. (1999), and application of TET in the field of molecular motors of cochlear cells was considered in Spector (2005).

An important additional application of TET is in the area of energy harvesting, that is, of the development of efficient and reliable energy harvesters capable of efficiently capturing ambient energy from a variety of media. For example, photosynthetic organisms have developed efficient sunlight harvesting apparatus to fuel their metabolisms (Hu et al., 1998); also, dendrimeric polymers are being considered as energy harversters in nanodevices (Andrews and Bradshaw, 2004). In engineering applications energy harvesters were studied for converting ambient vibrations into usable electrical energy (Glynne-Jones et al., 2004; Lesieutre et al., 2004; Cornwell et al., 2005; Roundy, 2005; Kim et al., 2005; Stephen, 2006) but their performance is limited by the fact that ambient vibration is often low-level and broadband, occurring in random bursts. The passive nonlinear TET designs discussed in this work, result in broadband energy transfer between structural components, in some cases even at low energy levels; hence, they hold promise towards alleviating the current restrictions of current energy mechanical harvesters of ambient vibration.

Additional studies of nonlinear energy transfer in lattice models have been performed to model heat flux and test the validity of the classical Fourier law of heat conduction (Gendelman and Savin, 2004; Balakrishnan and Van den Broeck, 2005). Wang (1973) analyzed TET between nonlinearly interacting waves, and Kevrekidis et al. (2004) studied localization and resonance-induced energy transfer in mechanical lattices with geometric nonlinearity. Spire and Leon (2004) studied energy absorption due to resonance of impeding waves by discrete molecular chains, resulting in generation of solitons in these chains; it was found that both nonlinearity and discreteness effects are prerequisites for this type of nonlinear energy absorption.

If we restrict our focus to purely mechanical systems possessing no dissipation and executing vibrations, still one can point out a variety of dynamic phenomena involving strong nonlinear energy transfers. Often the process of passive nonlinear vibration energy exchange is described in terms of nonlinear interaction between different structural modes with either close or well-separated frequencies. Such exchange is not possible in linear dynamical systems since, except for the case of modes with closely spaced frequencies (giving rise to the classical beat phenomenon), modes in these systems are uncoupled and can not exchange energy between them in a passive way.

In the presence of nonlinearity, however, nonlinear energy interactions can occur due to *internal resonances*, even between structural modes with widely spaced frequencies (Guckenheimer and Holmes, 1983; Wiggins, 1990; Nayfeh and Mook, 1995; Nayfeh, 2000). In nonlinear Hamiltonian systems irreversible transfer of energy is generally precluded due to conservation of the phase volume and by virtue of the Poincaré recurrence theorem; however, in certain cases the Hamiltonian dynamics can be trapped in bounded regions of the state space for relatively long time, with subsequent release (Zaslavskii, 2005). In addition, there are special cases where complete and irreversible (targeted) transfer of energy occurs between coupled nonlinear oscillators [(Nayfeh and Mook, 1995; see also the discussion of Fermi Targeted Energy Transfer in Maniadis and Aubry (2005)]; such irreversible nonlinear energy transfers occur on heteroclinic orbits of appropriately defined slow flows of the dynamics, they occur asymptotically as time tends to infinity, and are not robust as they are realized only at specific energy levels (in fact, perturbations of these orbits destroys the irreversibility of energy transfer, and lead to excitations of quasi-periodic orbits in the slow flows). In general, nonlinear energy transfers may be realized due to symmetry-breaking as nonlinear mode bifurcations, or through spatial energy localization phenomena from the formation of localized nonlinear normal modes (NNMs) (King and Vakakis, 1995; Boivin et al., 1995; Vakakis et al., 1996; Vakakis et al., 2002; Lacarbonara et al., 2003; Jiang et al., 2005).

In the works by Nayfeh and Nayfeh (1994), Nayfeh and Mook (1995), Oh and Nayfeh (1998), Nayfeh (2000) and Malatkar and Nayfeh (2003) a new form of nonlinear energy transfer between widely spaced modes in harmonically forced structures is analyzed; this mechanism of passive energy transfer is caused by resonance interaction of the slow modulation of a higher mode (generated from a Hopf bifurcation) with a lower one. This type of energy transfer is peculiar, in the sense that the interacting modes need not satisfy conditions of internal resonance.

Kerschen et al. (2008) discuss an alternative form of nonlinear modal interaction between highly energetic NNMs. Indeed, at low energies these modes may possess incommensurate linearized natural frequencies so they do not satisfy internal resonance conditions. Due to the energy dependence of their frequencies, however, at higher energies the same NNMs may become internally resonant, as their energy-dependent frequencies may become commensurate resulting in strong nonlinear modal interactions. This underlines the fact that important, essentially nonlinear phenomena (such as this one) may be missed when resorting to perturbation techniques based on linear (harmonic) generating functions, whose range of validity is restricted to small-amplitude motions and/or weak nonlinearities [but for a perturbation technique based on strongly nonlinear yet simple (non-smooth) generating functions, valid in strongly nonlinear regimes (but not in weakly nonlinear ones!) refer to Pilipchuk (1985, 1988, 1996), Pilipchuk et al. (1997) and Pilipchuk and Vakakis (1998)].

This monograph is devoted to the study of targeted energy transfer (TET) phenomena in dissipative mechanical and structural systems possessing essentially nonlinear local attachments. We will show that the addition to a linear system of a local attachment possessing essential (nonlinearizable) stiffness nonlinearity, may significantly alter the global dynamics of the resulting integrated system. The reason lies in the lack of a preferential resonance frequency of the attachment, which, in principle, enables it to engage in nonlinear resonance with any mode of the linear system, at arbitrary frequency ranges (provided, of course, that no mode has a node in the neighborhood of the point of attachment). The actual scenario of single-mode or multi-mode nonlinear resonance interaction of the attachment with the linear system will depend on the level and spatial distribution of the instantaneous vibration energy of the integrated system. We will show that under certain conditions, passive TET from the linear system to the NES occurs, i.e., a one-way and irreversible (on the average) flow of energy from the linear system to the attachment, which acts, in effect, as a *nonlinear energy sink – NES*. Moreover, in contrast to the classical linear vibration absorber whose action is narrowband, we will show that under certain conditions the NES can resonantly interact with the linear system in a broadband fashion, and engage in a *resonance capture cascade* with a set of structural modes over a broad frequency range; then the NES, acts in essence, as a passive, adaptive, broadband boundary controller.

Hence, viewed in the context of vibration theory, the NES can be regarded as a generalization of the concept of the classical linear vibration absorber (or tuned mass damper – TMD). Viewed in the context of the theory of dynamical systems, however, the addition of the essentially nonlinear NES introduces degeneracies in the free and forced dynamics of the integrated system, opening the possibility of higher co-dimensional bifurcations and complex dynamical phenomena, certain of which might be compatible to the design objectives of the specific engineering application considered.

As a preliminary illustrative example of TET, we consider a two degree-of-freedom (DOF) dissipative unforced system described by the following equations:

$$\ddot{y}_1 + \lambda_1 \dot{y}_1 + y_1 + \lambda_2(\dot{y}_1 - \dot{y}_2) + k(y_1 - y_2)^3 = 0$$
$$\varepsilon \ddot{y}_2 + \lambda_2(\dot{y}_2 - \dot{y}_1) + k(y_2 - y_1)^3 = 0. \tag{1.1}$$

Physically, these equations describe a damped linear oscillator (LO) with mass and natural frequency normalized to unity, and viscous damping coefficient λ_1; and an essentially nonlinear attachment with normalized mass ε, normalized nonlinear stiffness coefficient k, and viscous damping coefficient λ_2. Note that system (1.1) cannot be regarded as a small perturbation of a linear system due to the strongly nonlinear coupling terms. The detailed study of this type of dynamical systems is postponed until Chapter 3, and here we will only provide a brief numerical demonstration of TET by studying its transient dynamics.

To this end, we simulate numerically system (1.1) for parameter values $\varepsilon = 0.1$, $k = 0.1$, $\lambda_1 = 0.01$ and $\lambda_2 = 0.01$. The selected initial conditions correspond to an impulse $F = A\delta(t)$ imposed to the linear oscillator [where $\delta(t)$ is Dirac's delta function – this impulsive forcing is equivalent to imposing the initial velocity $\dot{y}_1(0+) = A$] with the system being initially at rest, i.e., $y_1(0) = y_2(0) = \dot{y}_2(0) = 0$

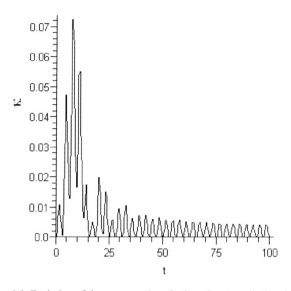

Fig. 1.1 Evolution of the energy ratio κ for impulse strength $A = 0.5$.

and $\dot{y}_1(0+) = A$. Hence, the initial energy is stored only in the LO. The instantaneous transfer of energy from the LO to the nonlinear attachment can be monitored by computing the non-dimensional energy ratio κ, which denotes the portion of instantaneous total energy stored in the nonlinear attachment,

$$\kappa = \frac{E_2}{E_1 + E_2}, \quad E_1 = \frac{1}{2}(y_1^2 + \dot{y}_1^2), \quad E_2 = \frac{\varepsilon}{2}\dot{y}_2^2 + \frac{k}{4}(y_1 - y_2)^4, \quad (1.2)$$

where E_1 and E_2 are the instantaneous energies of the LO and the attachment, respectively. Of course, all quantities in relations (1.2) are time dependent.

In Figures 1.1 and 1.2 we depict the evolution of the energy ratio κ for impulse strengths $A = 0.5$ and $A = 0.7$, respectively. From Figure 1.1 it is clear that only a small amount of energy (of the order of 7%) is transferred from the LO to the nonlinear attachment. However, for a slightly higher impulse the energy transferred climbs to almost 95% (see Figure 1.2), within a rather short time (up to $t = 15$, which is much less than the characteristic time of viscous energy dissipation in the LO). In this case, almost the entire impulsive energy is passively transferred from the LO to the nonlinear attachment, which acts as nonlinear energy sink. It should be mentioned that the mass of the attachment in this particular example is just 10% of the mass of the LO (and it will be shown that this mass can be reduced even further with similar TET results).

From this example, it appears that passive TET from the directly excited LO to the essentially nonlinear attachment in (1.1) is realized when the energy exceeds a certain critical threshold. The mathematical description of the TET process poses distinct challenges, since this phenomenon is transient (instead of steady state), and

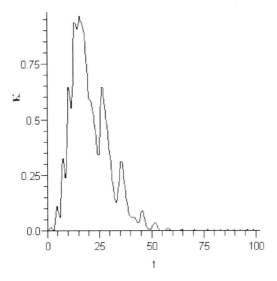

Fig. 1.2 Evolution of the energy ratio κ for impulse strength $A = 0.7$.

essentially nonlinear (instead of weakly nonlinear). Traditionally, what one does when dealing with systems of coupled oscillators like (1.1) is to consider the structure of periodic or quasi-periodic orbits of the corresponding undamped, Hamiltonian system; however, given that TET is a strongly nonlinear transient phenomenon that occurs in the dissipative system, at this point it remains unclear what its relation is to the dynamics of the underlying Hamiltonian system. In addition, it is not obvious how to analytically study the TET phenomenon, as this occurs in the strongly damped transient dynamics, where the majority of current techniques from nonlinear dynamics are inapplicable; yet an analytical study of TET is required in order for one to gain an understanding of the underlying dynamics, and apply it to practical engineering designs. Some additional obvious open questions that arise from this preliminary example concern the time scale of TET compared to the characteristic time scales of the dynamics of the LO and the NES; the realization and robustness of TET subject to other initial conditions or external excitations (such as, for example, time periodic ones); and the possible extension of TET to multi-degree-of-freedom linear oscillators or flexible structures with local essentially nonlinear attachments. These are some of the problems that we will be concerned with in this monograph.

According to the commonly accepted and perhaps correct point of view, the historical development of mechanics and, in particular, dynamics since Newton became possible because the observed motion of celestial bodies was modeled by almost conservative and nearly integrable mathematical dynamical models. Thus, the statement that the orbits of planets are ellipses allowed Newton to discover the classical laws of gravitation and motion. The solution of this problem initiated a breath of important developments in applied mathematics, which eventually grew to the general theory of integrable systems (Arnold, 1980). This type of systems possesses as many independent first integrals of motion as their number of degrees of freedom, so their n-dimensional dynamics may be reduced (at least in principle) to single-degree-of-freedom (SDOF) dynamical systems; alternatively put, the n-dimensional phase spaces of these systems are foliated by families of n-dimensional tori, so their motions may be reduced to periodic or quasi-periodic rotations on the surfaces of multi-dimensional tori. For such integrable dynamical systems no energy exchanges can occur between different modes of rotation and, of course, a process like TET is not possible at all.

Later observations demonstrated that both assertions mentioned above concerning the dynamics of celestial bodies (that is conservativity and integrability) are not exact. Indeed, these systems are not exactly conservative, primarily due to tidal phenomena; the famous manifestation of these effects is the one sidedness of the Moon. Celestial systems are also not exactly integrable, as gravitational multi-body interactions spoil the integrability; this led to the study of the celebrated three-body problem, whose proof of non-integrability led Poincaré to the development of modern geometrical dynamical systems theory and chaotic dynamics (Poincaré, 1899; Barrow-Green, 1996). Indeed, despite numerous attempts of more than two centuries, the three-body problem, that is, the dynamics of three bodies interacting via gravitational forces could not be analytically solved, until it was proven by Poincaré to be non-integrable. Until the time of Poincaré common wisdom was the Lagrangian view, that once a dynamical system is modeled by a set of differential equations its analytical solution is a matter of developing the necessary mathematical techniques; Poincaré proved that there are dynamical systems – even of simple configuration – for which no analytical solutions can exist (hence, for example, the impossibility of long-term weather prediction).

Despite their non-integrability, dynamical systems in celestial mechanics are often close to integrable ones, with the characteristic value of the perturbations from integrability being of the order of about 10^{-3} or less. This is the reason that problems of celestial mechanics provided also a major thrust to the development of regular and singular perturbation techniques in applied mathematics. With the help of these techniques, the dynamics of integrable Hamiltonian systems perturbed by small Hamiltonian perturbations were analyzed and understood rather well in the framework of the celebrated KAM (Kolmogorov–Arnold–Moser) theory (Arnold, 1963a, 1963b, 1964). If the perturbation is small enough, then for the majority of initial conditions quasi-periodic motions of the perturbed Hamiltonian system persist under the perturbation (on 'sufficiently irrational' multi-dimensional tori), whereas for special values of initial conditions (corresponding to countably infinite internal resonances of the perturbed Hamiltonian system) invariant tori are destroyed and replaced by thin layers of chaotic motions. These chaotic layers prevent the existence of a sufficient number of independent analytic independent first integrals of motion, leading to non-integrability of the perturbed Hamiltonian system.

Much less is known about the effects of non-Hamiltonian perturbations, and so in this area theoretical developments concern mainly low-dimensional systems. The main effects known include scattering by resonance and capture into the resonance

(Arnold, 1988). The former effect occurs when the orbit of the perturbed system is slightly modified due to passage through resonance [by a perturbation of $O(\varepsilon^{1/2})$, where ε is the characteristic strength of the perturbation]. Capture into the resonance occurs for a small subset of initial conditions, but the resulting variation of the perturbed trajectory is of $O(1)$.

Still, many interesting and important dynamical systems of practical significance can not be described as small perturbations of integrable ones. For this more general class of systems there are no rigorous analytical methods of solution. One approach for analyzing this class of problems is to try to apply perturbation techniques far beyond the formal boundaries of their applicability (sometimes such an approach can bring about success, for example, see the method discussed in Section 2.4). Another approach is to seek some important partial solution of the problem, for example, with the help of methods such as, harmonic balance, multiple scales, nonlinear normal modes (NNMs) (Nayfeh and Mook, 1995; Vakakis et al., 1996; Verhulst, 2005).

The systems under consideration in this work, exhibiting passive TET phenomena, belong to this latter category. Indeed, considering the dynamical system of the preliminary example (1.1), it is non-integrable, non-Hamiltonian, and besides some special cases cannot be expressed in the form of a perturbed integrable dynamical system. As mentioned previously, added challenges arise due to the type of responses that we will be interested in, namely, damped transient motions instead of steady state ones. It follows that standard perturbation techniques from the theory of dynamical systems dealing with periodic or nearly periodic motions are generally inapplicable in the problem of TET in systems of coupled discrete or continuous oscillators. Moreover, it is not clear for what types of dynamical systems is TET possible at all, and, even if it is possible, under what conditions it can be realized. Hence, in our study of TET in mechanical and structural systems certain important issues need to be addressed, including:

- The type of structural modification needed for realization of passive TET in a dynamical system, and the class of dynamical systems capable of TET.
- The robustness of the TET phenomenon to changes (and uncertainties) in system parameters, initial conditions and external excitations.
- The physical understanding of the dynamical mechanisms governing the TET, and the mathematical analysis of TET.
- The ways to enhance and optimize TET in a system according to a specific set of design objectives, and in the framework of practical applications where TET is useful.
- The comparison of TET-based designs to alternative current linear or nonlinear, passive or active designs.
- Provided that TET is theoretically proven to be beneficial according to a specific design objective, the practical implementation of these designs in engineering applications.

In this monograph we will attempt to address some of these issues, but, of course, no complete answers to all questions can be provided at this point. Instead, this work can be regarded as a first attempt towards addressing some of the above issues, and,

Gendelman, O.V., Savin, A.V., Heat conduction in a one-dimensional chain of hard disks with substrate potential, *Phys. Rev. Lett.* **92**(7), DOI: 10.1103/PhysRevLett.92.074301, 2004.

Gendelman, O.V., Manevitch, L.I., Manevitch, O.L., Solitonic mechanism of structural transition in polymer-clay nanocomposites, *J. Chem. Phys.* **119**(2), 1066–1069, 2003.

Ginzburg, V.V., Manevitch, L.I., On the theory of melting polymer crystals, *Colloid Polym. Sci.* **269**, 867–872, 1991.

Ginzburg, V.V., Gendelman, O.V., Manevitch, L.I., Simple 'kink' model of melt intercalation in polymer-clay nanocomposites, *Phys. Rev. Lett.* **86**(22), 5073–5075, 2001.

Guckenheimer, J., Holmes, P., *Nonlinear Oscillations, Dynamical System, and Bifurcation of Vector Fields*, Springer-Verlag, New York, 1983.

Hu, X., Damjanovic, A., Ritz, T., Schulten, K., Architecture and mechanism of the light-harvesting apparatus of purple bacteria, *Proc. Nat. Acad. Sciences USA* **95**(11), 5935–5941, 1998.

Hwang, Y.-C., Chen, W., Yates, M.V., Use of fluorescence resonance energy transfer for rapid detection of enteroviral infection in vivo, *Appl. Envir. Microbiology* **72**(5), 3710–3715, 2006.

Jenkins, R.D., Andrews, D.L., Four-center energy transfer and interaction pairs: Molecular quantum electrodynamics, *J. Chem. Phys.* **116**(15), 6713–6724, 2002.

Jenkins, R.D., Andrews, D.L., Multi-chromophore excitons and resonance energy transfer: Molecular quantum electrodynamics, *J. Chem. Phys.* **118**(8), 3470–3479, 2003.

Jenkins, R.D., Daniels, G.J., Andrews, D.L., Quantum pathways for resonance energy transfer, *J. Chem. Phys.* **120**(24), 11442–11448, 2004.

Jiang, D., Pierre, C., Shaw, S.W., The construction of nonlinear normal modes for systems with internal resonances, *Int. J. Nonlinear Mech.* **40**, 729–746, 2005.

Julicher, F., Ajdari, A., Prost, J., Modeling molecular motors, *Rev. Mod. Phys.* **69**(4), 1269–1281, 1997.

Kerschen, G., Peeters, M., Golinval, J.C., Vakakis, A.F., Nonlinear normal modes, Part I: A useful framework for the structural dynamicist, *Mech. Syst. Signal Proc.*, 2008 (submitted).

Kevrekidis, P.G., Dimitriev, S.V., Takeno, S., Bishop, A.R., Aifantis, E.C., Rich example of geometrically induced nonlinearity: From rotobreathers and kinks to moving localized modes and resonant energy transfer, *Phys. Rev. E* **70**, DOI: 10.1103/PhysRevE.70.066627, 2004.

Kim, J.S., Durst, R.D., Fonck, R.J., Technique for the experimental estimation of nonlinear energy transfer in fully developed turbulence, *Phys. Plasmas* **3**(11), 3998–4009, 1996.

Kim, S., Clark, W.W., Wang, Q.-M., Piezoelectric energy harvesting with a clamped circular plate: Analysis, *J. Intel. Mat. Sys. Struct.* **16**, 847–854, 2005.

King, M.E., Vakakis, A.F., An energy-based approach to computing resonant nonlinear normal modes, *J. Appl. Mech.* **63**, 810–819, 1995.

Kopidakis, G., Aubry, S., Tsironis, G.P., Targeted energy transfer through discrete breathers in nonlinear systems, *Phys. Rev. Lett.* **87**(16), DOI: 10.1103/PhysRevLett87.165501, 2001.

Kurs, A., Karalis, A., Moffatt, R., Joannopoulos, J.D., Fisher, P., Soljačić, M., Wireless power transfer via strongly coupled magnetic resonances, *Science* **317**, 83–86, 2007.

Lacarbonara, W., Rega, G., Nayfeh, A.H., Resonant nonlinear normal modes, Part I: Analytical treatment for structural one dimensional systems, *Int. J. Nonlinear Mech.* **38**, 851–872, 2003.

Lesieutre, G.A., Ottman, G.K., Hofmann, H.F., Damping as a result of piezoelectric energy harvesting, *J. Sound Vib.* **269**, 991–1001, 2004.

Malatkar, P., Nayfeh, A.H., On the transfer of energy between widely spaced modes in structures, *Nonl. Dyn.* **31**, 225–242, 2003.

Maniadis, P., Kopidakis, G., Aubry, S., Classical and quantum targeted energy transfer between nonlinear oscillators, *Physica D* **188**, 153–177, 2004.

Maniadis, P., Aubry, S., Targeted energy transfer by Fermi resonance, *Physica D* **202**(3–4), 200–217, 2005.

Meessen, A., Working principle of an EM cancer detector, Institut de Physique, Université Catholique de Louvain, Belgium, http://www.meessen.net/AMeessen/EMcancerDet2.pdf, 2000.

Memboeuf, A., Aubry, S., Targeted energy transfer between a rotor and a Morse oscillator: A model for selective chemical dissociation, *Physica D* **207**, 1–23, 2005.

Morgante, A.M., Johansson, M., Aubry, S., Breather Ű phonon resonances in finite lattices: Phantom breathers?, *J. Phys. A* **35**, 4999–5021, 2002.

Musumeci, F., Brizhik, L.S., Ho, M.-W., *Energy and Information Transfer in Biological Systems*, World Scientific, Singapore, 2003.

Nayfeh, A.H., Mook, D.T., *Nonlinear Oscillations*, John Wiley & Sons, New York, 1995.

Nayfeh, A.H., *Nonlinear Interactions: Analytical, Computational and Experimental Methods*, Wiley Interscience, New York, 2000.

Nayfeh, S.A., Nayfeh, A.H., Nonlinear interactions between two widely spaced modes – External excitation, *Int. J. Bif. Chaos* **3**, 417–427, 1993.

Nayfeh, S.A., Nayfeh, A.H., Energy transfer from high- to low-frequency modes in a flexible structure via modulation, *J. Vib. Acoust.* **116**, 203–207, 1994.

Nayfeh, A.H., Mook, D.T., Energy transfer from high-frequency to low-frequency modes in structures, *J. Vib. Acoust.* **117**, 186–195, 1995.

Nistazakis, H.E., Kevrekidis, P.G., Malomed, B.A., Frantzeskakis, D.J., Bishop, A.R., Targeted transfer of solitons in continua and lattices, *Phys. Rev. E* **66**, 015601, 2002.

Oh, K., Nayfeh, A.H., High- to low-frequency modal interactions in a cantilever composite plate, *J. Vib. Acoust.* **120**, 579–587, 1998.

Pilipchuk, V.N., The calculation of strongly nonlinear systems close to vibration-impact systems, *Prikl. Mat. Mech. (PMM)* **49**, 572–578, 1985.

Pilipchuk, V.N., A transformation for vibrating systems based on a non-smooth periodic pair of functions, *Dokl. AN Ukr. SSR Ser. A* **4**, 37–40, 1988 [in Russian].

Pilipchuk, V.N., Analytic study of vibrating systems with strong nonlinearities by employing sawtooth time transformations, *J. Sound Vib.* **192**(1), 43–64, 2006.

Pilipchuk, V.N, Vakakis A.F., Azeez, M.A.F., Study of a class of subharmonic motions using a non-smooth temporal transformation (NSTT), *Physica D* **100**, 145–164, 1997.

Pilipchuk, V.N., Vakakis A.F., Study of the oscillations of a nonlinearly supported string using non-smooth transformations, *J. Vib. Acoust.* **120**, 434–440, 1998.

Poincaré, H., *Les Methodes Nouvelles de la Mecanique Celeste*, Gauthier-Villars, Paris, 1899.

Renger, T., May, V., Kühn, O., Ultrafast excitation energy transfer dynamics in photosynthetic pigment – Protein complexes, *Phys. Rep.* **343**, 137–254, 2001.

Roundy, S., On the effectiveness of vibration-based energy harvesting, *J. Intell. Mat. Sys. Struct.* **16**, 809–823, 2005.

Spector, A.A., Effectiveness, active energy produced by molecular motors, and nonlinear capacitance of the cochlear outer hair cell, *J. Biomech. Eng.* **127**, 391–399, 2005.

Spire, A., Leon, J., Nonlinear absorption in discrete systems, *J. Phys. A* **37**, 9101–9108, 2004.

Stephen, N.G., On energy harvesting from ambient vibration, *J. Sound Vib.* **293**, 409–425, 2006.

Tran, C.V., Nonlinear transfer and spectral distribution of energy in α turbulence, *Physica D* **191**, 137–155, 2004.

Vakakis, A.F., Manevitch, L.I., Mikhlin Yu.V., Pilipchuk, V.N., Zevin, A.A., *Normal Modes and Localization in Nonlinear Systems*, Wiley Interscience, New York, 1996.

Vakakis, A.F. (Ed.), *Normal Modes and Localization in Nonlinear Systems*, Kluwer Academic Publishers, 2002 [also, Special Issue of *Nonlinear Dynamics* **25**(1–3), 2001].

Van Amerongen, H., Valkunas, L., an Grondelle, R., *Photosynthetic Excitons*, World Scientific, Singapore, 2000.

Vedruccio, C., Meessen, A., EM cancer detection by means of nonlinear resonance interaction, in *Proceedings of Progress in Electromagnetics Research Symposium (PIERS2004)*, Pisa, Italy, March 28–31, 2004.

Verhulst, F., *Methods and Applications of Singular Perturbations*, Springer-Verlag, Berlin/New York, 2005.

Wang, P.K.C., Unidirectional energy transfer in nonlinear wave-wave interactions, *J. Math. Phys.* **14**(7), 911–915, 1973.

Wang, Z., Carter, J.A., Lagutchev, A., Koh, Y.K., Seong, N.-H., Cahill, D.G., Dlott D.D., Ultrafast flash thermal conductance of molecular chains, *Science* **317**, 787–790, 2007.

Wiggins, S., *Introduction to Applied Nonlinear Dynamical Systems and Chaos*, Springer-Verlag, New York, 1990.

Xu, Q.-H., Wang S., Korystov, D., Mikhailovsky, A., Bazan, G.C., Moses, D., Heeger, A.J., The fluorescence resonance energy transfer (FRET) gate: A time-resolved study, *Proc. Nat. Acad. Sci. USA* **102**(3), 530–535, 2005.

Yesilkaya, H., Meacci, F., Niemann, S., Hillemann, D., Rüsch-Gerdes, S., LONG DRUG Study Group, Barer, M.R., Andrew, P.W., Oggioni, M.R., Evaluation of molecular-beacon, Taq-Man, and fluorescence resonance energy transfer probes for detection of antibiotic resistance-conferring single nucleotide polymorphisms in mixed Mycobacterium tuberculosis DNA extracts, *J. Clin. Microbiol.* **44**(10), 3826–3829, 2006.

Zaslavskii G.M., *Hamiltonian Chaos and Fractional Dynamics*, Oxford University Press, Oxford, 2005.

Chapter 2
Preliminary Concepts, Methodologies and Techniques

As mentioned in the Introduction (Chapter 1), the study of targeted energy transfer (TET) in strongly nonlinear and non-conservative oscillators poses some distinct technical challenges, and dictates the use of concepts, formulations, analytical methodologies and computational techniques from different fields of applied mathematics and engineering, such as dynamical systems and bifurcation theory, theory of asymptotic approximations, numerical signal processing, and experimental dynamics. Therefore, before we initiate our study of the nonlinear dynamics of TET, it is appropriate to provide first some background information related to certain key concepts and methodologies that will be applied in the work that follows.

Specifically, we will briefly discuss the concepts of *nonlinear normal mode (NNM)* and *nonlinear mode localization* in discrete and continuous oscillators, and the occurence of *nonlinear internal resonances, transient resonance captures (TRCs) and sustained resonance captures (SRCs)* in undamped or damped, forced or unforced systems of coupled oscillators. These concepts will provide us with the necessary theoretical framework to base our theoretical study of the dynamics of TET; moreover, using these concepts we will be able to identify, interpret, and place into the right context complex nonlinear dynamical phenomena related to TET.

Then, we will outline the basic elements of a special perturbation technique, namely, the *complexification-averaging (CX-A) technique* which will be one of the basic mathematical tools employed for performing the analytical derivations required for our theoretical studies. This will be followed by discussion of some selected advanced signal processing techniques, namely, wavelet transforms – WTs, empirical mode decomposition – EMD, and Hilbert transforms, which will be especially suitable for post-processing the computational nonlinear dynamical responses related to TET, and for identifying the corresponding underlying nonlinear modal interactions that govern TET or influence its effectiveness. In essence, we will work towards the formulation of an integrated post-processing methodology for analyzing strongly nonlinear transient (or steady state) modal interactions in systems with strong nonlinearities. We will end this chapter by providing some preliminary remarks on the development of the necessary hardware required for our experimental work, undertaken to validate and confirm the theoretical results related to TET.

We start our discussion by considering the concept of nonlinear normal mode (NNM), which will be central in our theoretical investigation of the dynamics of TET in systems of coupled oscillators.

2.1 Nonlinear Normal Modes (NNMs)

Engineers and physicists traditionally associate the concept of normal mode with linear vibration theory and regard it as closely related to the principle of linear superposition. Indeed, a classical result of linear vibration theory is that the normal modes of vibration of a multi-degree-of-freedom (MDOF) discrete system can be employed to decouple the equations of motion through an appropriate coordinate (modal) transformation, and to express its free or forced oscillations as superpositions of modal responses. Another result of classical linear theory is that the number of normal modes of vibration cannot exceed the number of degrees of freedom (DOF) of a discrete system, and that any forced resonances of the system under external harmonic excitation always occur in neighborhoods of frequencies of normal modes.

Although in nonlinear systems the principle of superposition does not (generally) hold, nevertheless the concept of the normal mode can still be employed. Rosenberg (1966) defined a *nonlinear normal mode (NNM)* of an undamped discrete MDOF system as a synchronous periodic oscillation where all material points of the system reach their extreme values or pass through zero simultaneously; hence, the NNM oscillation is represented by either a straight modal line (similar NNM) or a modal curve (non-similar NNM) in the configuration space of the system. NNMs are generically non-smimilar, since similarity (which is always the case in linear theory) can only be realized when special symmetries exist (Vakakis et al., 1996). Lyapunov (1947) proved the existence of n synchronous periodic solutions (NNMs) in neighborhoods of stable equilibria of n-DOF Hamiltonian systems with no internal resonances, and Weinstein (1973) and Moser (1976) extended Lyapunov's result to MDOF Hamiltonian systems with internal resonances. As discussed below, an important feature that distinguishes NNMs from linear normal modes is that they can exceed in number the degrees of freedom of an oscillator; in cases where this occurs, essentially nonlinear modes (having no analogs in linear theory) are generated through NNM bifurcations, breaking the symmetry of the dynamics and resulting in nonlinear energy localization (motion confinement) phenomena.

Similar NNMs are analogous to linear normal modes, in the sense that their modal lines do not depend on the energy of the free oscillation and space-time separation of the governing equations of motion can still be performed; however, as mentioned previously, this type of NNMs is realized only when special symmetries occur, and are not typical (generic) in nonlinear systems. More generic are non-similar NNMs, whose modal curves do depend on energy; this energy dependence prevents the direct separation of space and time in the governing equations of motion by means of non-similar NNMs, which complicates their analytical computation (Kauderer, 1958, Manevitch and Mikhlin, 1972; Vakakis et al., 1996).

In this work, we will adopt a more extended definition of NNMs, defining an NNM as a (not necessarily synchronous) time-periodic oscillation of a non-dissipative nonlinear dynamical system. This enables us to extend the NNM definition to cases of systems in internal resonance, where the resulting strongly nonlinear modal interactions render the free oscillation non-synchronous (King and Vakakis, 1996). Viewed in a different context, whereas in the absence of internal resonance a NNM can be represented by a modal line or curve in the configuration space of the system – so that functional relations of the form $y_i = \hat{y}_i(y_1)$, $y_1 \equiv \hat{y}_1(y_1)$, $i = 1, \ldots, n$ can be established between the coordinates y_j (hence, Rosenberg's original NNM definition), no such functional relations hold when internal resonances occur. Still, our extended NNM definition applies to this later case as well.

The extension of the concept of NNM to non-conservative systems with damping was studied by Shaw and Pierre (1991, 1993), who introduced the concept of *damped NNM invariant manifold* to account for the fact that the free oscillation of a damped nonlinear system is a non-synchronous, decaying motion. This NNM invariant manifold formulation is based on ideas developed by Fenichel (1971) regarding persistence and smoothness of invariant manifolds in dynamical systems, and computes damped NNM invariant manifolds of the damped dynamical flow by parametrizing the damped NNM response in terms of a reference displacement and a reference velocity. For sufficiently weak damping, the damped NNM invariant manifold can be viewed as perturbation (and analytic continuation) of the NNM of the corresponding undamped Hamiltonian system. When a motion is initiated on a damped NNM invariant manifold of a MDOF system, the response of each coordinate is in the form of a decaying oscillation with non-trivial phase difference with regard to the other coordinates. A computationally efficient extension of the invariant manifold methodology was proposed by Nayfeh and Nayfeh (1993) who reformulated the NNM invariant manifold method in a complex framework. When no resonances exist, the NNM invariant manifolds of a MDOF discrete oscillator are two-dimensional, and the NNMs are uncoupled from each other. When internal resonances exist, there occur strongly nonlinear interactions between NNMs which couple them; this causes an increase of the dimensionality of the corresponding NNM invariant manifold.

NNMs and NNM damped invariant manifolds will play a central role in our discussion of TET and related strongly nonlinear transient dynamical phenomena. Moreover, our study will indicate that a prerequisite for realization of TET from linear systems to strongly nonlinear boundary attachments is the existence of some form of energy dissipation in the system; although in this study the main energy dissipation mechanism considered is (weak) viscous damping, other forms of energy dissipation may also qualify for TET, such as, for example, energy transmission to the far field of unbounded media by traveling waves (see Section 3.5.2). A paradoxical fact, however, is that although TET is realized only in the (weakly) dissipative system, in essence its dynamics is governed by the dynamics, and, especially, the

NNMs of the underlying non-dissipative system. Indeed, it will be shown, that the properties and bifurcations of NNMs of the non-dissipative system determine the conditions (i.e., the ranges of system parameters, external excitations and initial conditions) for the realization of TET in the dissipative system. The topological structure and bifurcations of the NNMs of the underlying non-dissipative systems will be carefully studied in this work – especially in the frequency-energy domain, since the energy dependencies of NNMs (and damped NNM invariant manifolds) play a key role regarding TET; this holds especially for NNMs whose spatial distributions change from non-localized to localized with decreasing energy.

But there are additional benefits to be gained by adopting a NNM-based framework for our study. It will be shown that NNM bifurcations govern complex nonlinear transitions occurring in the damped dynamics with decreasing energy. This becomes clear when one considers that in a weakly damped system the NNMs and NNM bifurcations are preserved as weakly damped NNM invariant manifolds and as bifurcations of these manifolds, respectively, which lie in neighborhoods of the corresponding undamped NNMs. It follows that the weakly damped, transient, nonlinear dynamics follow approximately paths along NNM invariant manifolds, and that bifurcations of NNM invariant manifolds appear as sudden transitions (jumps) in the damped transient dynamics. These may lead to complex, multi-modal and multi-frequency complex transitions in the dynamics, which, however, may be fully interpreted, modeled and analytically studied by adopting a theoretical framework based on NNMs and damped NNM invariant manifolds. More importantly, using such a framework TET can be analyzed and optimized according to a set of design criteria, which is needed for the implementation of TET in practical applications. Indeed a NNM-based approach seems to be natural for the study of TET and the associated strongly nonlinear phenomena discussed in this work.

Returning now to our brief review of NNM-related works, constructive methods for computing NNMs in discrete oscillators with no internal resonances have been developed (see, for example, Rand, 1971, 1974; Manevitch and Miklhin, 1972; Mikhlin, 1985; Bellizzi and Bouc, 2005), and NNMs in systems with internal resonances (where strong nonlinear modal interactions take place) have also been studied (see, for example, Boivin et al., 1995; King and Vakakis, 1996; Nayfeh et al., 1996; Jiang et al., 2005a). In an additional series of works (King and Vakakis, 1993, 1994, 1995a; Vakakis and King, 1995; Andrianov, 2008) methodologies for analysing the NNMs (and their bifurcations) of nonlinear elastic and continuous systems have been developed. In King and Vakakis (1994) stationary and traveling solitary waves (breathers) in a class of nonlinear partial differential equations are regarded as localized NNMs over domains of infinite spatial extent and are studied analytically. These methods and some additional ones for analyzing and computing NNMs in discrete and continuous oscillators are reviewed in Manevitch et al. (1989), Vakakis et al. (1996), Vakakis (1996, 1997, 2002), Pierre et al. (2006), Kerschen et al. (2008a) and Peeters et al. (2008).

An additional interesting feature of NNMs, which clearly distinguishes them from classical linear normal modes, is that they can exceed in number the degrees of freedom of a dynamical system. This is due to NNM bifurcations which may also

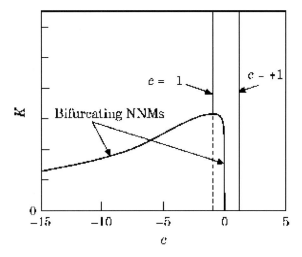

Fig. 2.1 NNMs of system (2.1); —: stable, - - -: unstable NNMs.

lead to NNM instability (a feature which, again, is distinct from what predicted by linear theory). We illustrate this by a simple example. To this end, we consider the following two-DOF Hamiltonian system with cubic stiffness nonlinearities (Vakakis and Rand, 1992a, 1992b):

$$\ddot{y}_1 + y_1 + y_1^3 + K(y_1 - y_2)^3 = 0$$
$$\ddot{y}_2 + y_2 + y_2^3 + K(y_2 - y_1)^3 = 0 \qquad (2.1)$$

Due to its symmetry, this system possesses only similar NNMs which are computed by imposing the following functional relationship:

$$y_2 = \hat{y}_2(y_1) \equiv c\, y_1 \qquad (2.2)$$

where $c \in R$ is a real modal constant. Substituting (2.2) into (2.1), we derive the following algebraic equation satisfied by the modal constant:

$$K(1+c)(c-1)^3 = c(1-c^2) \qquad (2.3)$$

In Figure 2.1 the real values of the modal constant c are depicted for varying coupling stiffness coefficient K, from which we infer that a pitchfork bifurcation (Wiggins, 1990) of NNMs occurs in the Hamiltonian system. This type of bifurcation is realized due to the symmetry of system (2.1) and is expected to 'break' into saddle node (SN) bifurcation(s) when this symmetry is perturbed. Referring to Figure 2.1, we note that system (2.1) always possesses the NNMs $y_2 = \pm y_1$ corresponding to solutions $c = \pm 1$ of (2.3), irrespectively of the coupling strength K; these correspond to in-phase and out-of-phase similar NNMs, respectively, which can be regarded as continuations of the two normal modes of the corresponding linear system. However, as noted from the bifurcation diagram of Figure 2.1, the nonlinear system possesses two additional NNMs which bifurcate from the out-of-phase NNM at $K = 1/4$. The bifurcating NNMs are out-of-phase, essentially nonlinear, time-periodic motions of system (2.1) having no analogs in linear theory; both of these NNMs become *strongly localized* as $K \to 0$ (i.e., for sufficiently weak coupling) to one of the two SDOF oscillators of system (2.1). Hence, in the limit of weak coupling *nonlinear mode localization* occurs in the symmetric system. In the next section nonlinear localization in dynamical systems is discussed in more detail.

This simple example demonstrates that the NNMs of a dynamical system may exceed in number its degrees of freedom. In this particular case, the NNM bifurcation is due to 1:1 internal resonance between the two SDOF nonlinear oscillators of system (2.1). An additional interesting conclusion drawn from this specific example is that NNM bifurcations may result in mode instability; indeed, for $K < 1/4$ the out-of-phase NNM $x_2 = -x_1$ becomes unstable (Vakakis et al., 1996), a result which, as shown below, has implications on the global Hamiltonian dynamics of system (2.1). We mention that the instability of the out-of-phase NNM is manifested in the form of modulated (instead of a periodic) oscillation, and not in the form of an exponentially growing motion; in other words, in system (2.1) only *orbital stability* (Nayfeh and Mook, 1995) has meaning, as *Lyapunov asymptotic stability* is not possible in the nonlinear Hamiltonian oscillator (2.1) due to the dependence of the frequency of oscillation on the energy.

To show this, we construct numerical Poincaré maps of the global dynamics. First, we reduce the dynamical flow of system (2.1) on its three-dimensional isoenergetic manifold, defined by the relation

$$H(y_1, \dot{y}_1, y_2, \dot{y}_2) \equiv \frac{\dot{y}_1^2 + \dot{y}_2^2}{2} + \frac{y_1^2 + y_2^2}{2} + \frac{y_1^4 + y_2^4 + K(y_1 - y_2)^4}{4} = h \quad (2.4)$$

where h is the (conserved) level of energy. Then we intersect the isoenergetic flow by the two-dimensional cut section

$$\Sigma = \{y_1 = 0, \dot{y}_1 > 0\} \cap \{H = h\} \quad (2.5)$$

which is everywhere transverse to the flow. Moreover, the resulting two-dimensional Poincaré map is orientation-preserving due to the restriction imposed on the sign of the velocity $\dot{y}_1$ at the cut section. In Figure 2.2 we depict the Poincaré maps of system (2.1) for the low energy level $h = 0.4$, and two values of K corresponding to relatively strong (Figure 2.2a) and weak (Figure 2.2b) coupling.

We note that the in-phase NNM normal mode (appearing as the upper equilibrium point in both maps) is orbitally stable, as it appears as a center surrounded by closed orbits (which are intersections of invariant tori of the Hamiltonian with the cut section Σ). Considering the out-of-phase NNM, above the bifurcation it is stable, whereas below it is unstable and possesses a double homoclinic loop, as inferred from Figure 2.2b. The (seemingly smooth) homoclinic orbits (loops) are formed by the coalescence of the stable and unstable invariant manifolds of the unstable out-

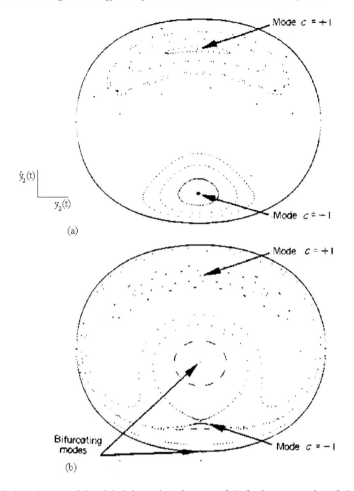

Fig. 2.2 Poincaré maps of the global dynamics of system (2.1) for low energy $h = 0.4$: (a) $K = 0.4 > 1/4$, (b) $K = 0.1 < 1/4$.

of-phase NNM, and represent the boundaries between trajectories that encircle only one of the bifurcating NNMs and those that enclose both. The Poincaré plots of Figure 2.2 (which correspond to a relatively low value of energy) are rather deceiving, however, since they give the impression that the global dynamics of the oscillator (2.1) is regular and completely predictable [in fact, for low energies the dynamics can be asymptotically approximated by the method of multiple scales (Vakakis and Rand, 1992a)].

In fact, since the oscillator (2.1) is non-integrable, 'rational' and some 'irrational' invariant tori of the flow are expected to 'break' according to the KAM theorem (MacKay and Meiss, 1987), giving rise to random-like chaotic motions; *local (small-scale) chaos* then results in layers of stochasticity surrounding countable infinities of stable subharmonic periodic orbits (Guckenheimer and Holmes, 1983; Veerman and Holmes, 1985; Wiggins, 1990) that result from the 'breakdown' of invariant tori. This local chaos is due to exponentially small splittings of the stable and unstable manifolds of unstable subharmonic orbits, producing transverse intersections of these manifolds close to resonance bands of the dynamics (Veerman and Holmes, 1986).

Apart from causing global qualitative changes in the dynamics, the NNM bifurcation depicted in Figure 2.1 gives rise to *global (large-scale) chaos* in the Hamiltonian system (2.1), and, hence, to large-scale instability. This is a consequence of the splitting and transverse intersections of the stable and unstable invariant manifolds that form the seemingly smooth (at low energies) homoclinic loops of the unstable NNM in the Poincaré map of Figure 2.2b; this results in large-scale homoclinic tangles and chaotic Smale horseshoe maps (Wiggins, 1990) leading to large-scale chaos in system (2.1). This is demonstrated in the Poincaré maps of Figure 2.3, corresponding to relatively high energy levels $h = 50.0$ and $h = 150.0$, and weak coupling (i.e., after the NNM bifurcation has taken place). We note that there is a large region [a *sea of stochasticity* (Lichtenberg and Lieberrman, 1983)] in each of these maps, inside which the orbits of the oscillator seem to wander in an erratic fashion. These regions contain chaotic motions, i.e., motions with extreme sensitivity on initial conditions. In each Poincaré map the region of large-scale chaos occupies a neighborhood of the unstable out-of-phase NNM and the domain where transverse intersections of the invariant manifolds of that NNM occur.

The occurrence of large-scale chaotic motions in the Hamiltonian system (2.1) is a direct consequence of the pitchfork bifurcation of NNMs, since they appear only after the NNM bifurcation has occurred (i.e., only for $K < 1/4$). Therefore, a necessary condition for large-scale chaos in system (2.1) is the orbital instability of the out-of-phase NNM (since only then can large-scale transverse homoclinic intersections of invariant manifolds occur). As a result, in this case *the bifurcation of NNMs increases the complexity of the global dynamics and adds global instability into the system*. This is a first indication of the global effects on the dynamics that a NNM bifurcation can introduce. In the course of this work we will show that NNM bifurcations can affect in a critical way the dynamics of TET, and that they play an important role when optimizing for robust, fast-scale and strong passive TET from a directly forced linear system to an essentially nonlinear boundary attachment.

To illustrate the frequency-energy dependence and some additional interesting features of NNMs we consider another example of a two-DOF system, consisting of a nonlinear oscillator linearly coupled to a linear one (Kerschen et al., 2008a):

$$\ddot{y}_1 + 2y_1 - y_2 + 0.5y_1^3 = 0$$
$$\ddot{y}_2 + 2y_2 - y_1 = 0 \qquad (2.6)$$

In contrast to (2.1) this system is not symmetric so it can possess only non-similar NNMs. As mentioned previously, this is the generic type of NNMs encountered in dynamical systems, so this example aims to demonstrate certain features of non-similar NNMs that are typical for a broad class of nonlinear coupled oscillators.

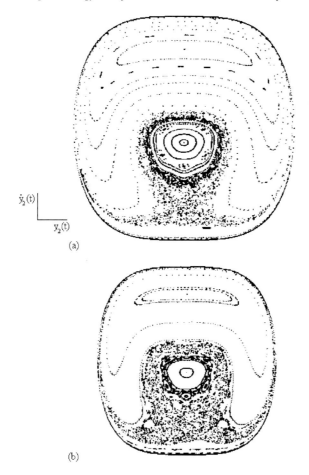

Fig. 2.3 Poincaré maps of the global dynamics of system (2.1) for high energies and $K = 0.1 < 1/4$: (a) $h = 50.0$, (b) $h = 150.0$.

The non-similar NNMs of this system are approximately computed by the method of harmonic balance (Nayfeh and Mook, 1995), i.e., by seeking time-periodic responses in the form

$$y_1(t) \approx A\cos\omega t, \quad y_2(t) \approx B\cos\omega t \qquad (2.7)$$

Note that the computation of non-similar NNMs is approximate, in contrast to the exact expressions derived for the similar NNMs in the previous example. When the *ansatz* (2.7) is substituted into (2.6), and a matching of coefficients of the various harmonic functions is performed, we obtain the following expressions for the amplitudes:

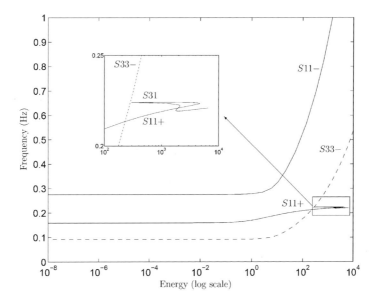

Fig. 2.4 NNMs of system (2.6) depicted in a frequency-energy plot (FEP); the corresponding modal curves in the configuration plane are inset, horizontal and vertical axes in these plots depict the displacements of the nonlinear and linear oscillators, respectively.

$$A = \pm \left\{ \frac{8(\omega^2 - \omega_2^2)(\omega^2 - \omega_1^2)}{3(\omega^2 - 2)} \right\}^{1/2}, \quad B = \frac{A}{2 - \omega^2} \tag{2.8}$$

with the natural frequencies of the linearized system given by $\omega_1 = 1$ and $\omega_2 = \sqrt{3}$. This result demonstrates the frequency dependence of the amplitudes of the NNMs of system (2.6).

The appropriate graphical depiction of NNMs is key to their exploitation. In this work extensive use will be made of *frequency-energy plots* (FEPs) where the amplitude of a NNM is plotted as function of its (conserved) energy. The NNMs of system (2.6) were computed numerically (Peeters et al., 2008) and are depicted in Figure 2.4. There exist two main backbone branches of NNMs, an in-phase branch, $S11+$, originating (for low energies) from the first linearized natural frequency and an out-of-phase one, $S11-$, originating from the second linearized natural frequency. The notation 'S' used for these NNMs refer to the symmetric character of these solutions [i.e., both oscillators of (2.6) execute synchronous motions], whereas the indices indicate that the two oscillators of system (2.6) vibrate with the same dominant frequency. A detailed discussion of FEPs and the corresponding notations of branches of NNMs depicted on them, is postponed until Section 3.3.

The FEP of Figure 2.4 clearly shows that the nonlinear modal parameters have a strong dependence on the (conserved) energy of the oscillation. Specifically, the frequencies of the in-phase and out-of-phase NNMs increase with energy, which reveals the hardening characteristic of system (2.6). Moreover, the modal curves change with increasing energy, since the in-phase NNM tends to localize to the linear oscillator (i.e., its modal curve tends to become vertical in the corresponding configuration plane with increasing energy), whereas the out-of-phase NNM tends to localize to the nonlinear oscillator (its modal curve tends to become horizontal with increasing energy). This tendency of NNMs to localize with varying energy is key for the realization of TET in the corresponding weakly dissipative system, as discussed in Chapter 3 [see also Pilipchuck (2008) for an additional study of nonlinear mode localization and TET due to the dependence of the shapes of NNMs on energy]. In this work we will make extensive use of the FEP, and show that it is a valuable tool not only for examining NNMs of Hamiltonian systems, but also for investigating nonlinear transitions leading to TET in weakly dissipative ones.

Another salient feature of NNMs is that they may nonlinearly interact without their linearized natural frequencies necessarily satisfying conditions of internal resonance. These strongly nonlinear modal interactions [which differ from nonlinear modal interactions considered in the current literature, see (Nayfeh, 2000) for example] occur at relatively high energy levels (so that nonlinear effects are dominant in the motion), and can be clearly studied by representing the NNMs in the FEP. Such *internally resonant NNMs* have no counterparts in linear theory and are generated through NNM bifurcations. The FEP of system (2.6) depicts internally resonant NNMs at high energies (see Figure 2.4). In particular, we note an additional branch of NNMs lying on a *subharmonic tongue* emanating from the in-phase backbone branch $S11+$. This tongue is denoted by $S31$, since it corresponds to a 3:1 internal resonance of the in-phase and out-of-phase NNMs at those energy levels. Surprisingly, the ratio of the linearized natural frequencies of system (2.6) is equal to $\sqrt{3}$, but due to the energy dependence of the frequencies of the NNMs, a 3:1 ratio between the two frequencies of the NNMs can still be realized; hence, conditions of 3:1 internal resonance are realized at high energies, although no such conditions are possible at lower energies. This result clearly demonstrates that *NNMs can be internally resonant without necessarily having commensurate linearised natural frequencies*, a feature that is rarely discussed in the literature. This also underlines that important features of nonlinear dynamics can be missed when resorting exclusively to perturbation techniques based on linearized generating solutions, and, thus, being limited to small-amplitude motions (Kerschen et al., 2008a).

To better illustrate this interesting high-energy nonlinear resonance mechanism, the branch $S11-$ is represented by dashed line as $S33-$ in the FEP of Figure 2.4, at a third of its frequency. This is permissible, because a periodic solution of period T is also periodic with period $3T$, so the branch $S33-$ can be considered as identical to $S11-$. Using this notation it is clear that at the energy range of 3:1 internal resonance there occurs a smooth transition from branch $S11-$ to branch $S33-$ through the subharmonic tongue $S31$. In Figure 2.5 we present a closeup of the FEP in the energy range of existence of the 3:1 internally resonance NNMs; the subharmonic tongue is more clearly depicted, and the stability of the various branches of internally resonance is examined.

This discussion indicates that additional nonlinear resonance scenarios are realized when we further increase the energy of the seemingly simple system (2.6).

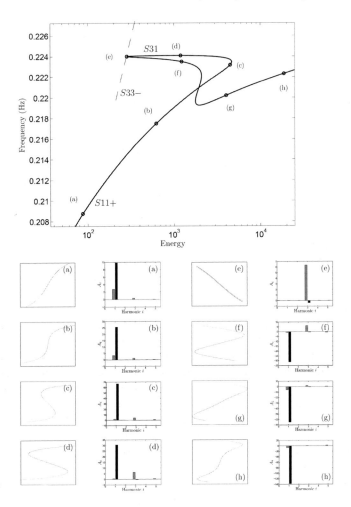

Fig. 2.5 Energy range of the FEP of existence of 3:1 internally resonant NNMs in system (2.6), —•—•—• unstable NNMs; the corresponding modal representations in the configuration plane are depicted at selected energies.

This is supported by the fact that for increasing energy the frequencies of the out-of-phase NNMs on branch $S11-$ increase steadily, whereas the frequencies of the in-phase NNMs on $S11+$ tend to the asymptotic limit $\omega_2 = \sqrt{3}$. Following this reasoning, we expect the existence of a *countable infinity of internal resonances between the in-phase and out-of-phase NNMs at specific higher energy ranges*. This is confirmed by the numerical results presented in Kerschen et al. (2008a). In this work we will investigate in detail FEPs similar to those depicted in Figures 2.4 and 2.5, and show that the energy dependencies of the NNM backbone branches and subharmonic tongues of NNMs dictate the different forms of possible targeted energy transfers in coupled oscillators with essentially nonlinear attachments.

The previous examples highlight the advantages of adopting nonlinear theoretical frameworks (instead of linearized ones) for analyzing nonlinear dynamical responses. As shown previously, there are cases where a nonlinear dynamical system may possess essentially nonlinear modes or can exhibit essentially nonlinear dynamical responses that have no counterparts in linear theory. Applying linear concepts such as modal analysis and frequency response plots to such a nonlinear system may model only partially the dynamics, so alternative approaches that take into full account the effects of the nonlinearity must be applied instead. In that context, the concepts of NNM and damped NNM invariant manifold provide a solid theoretical framework for analyzing, interpreting and modeling strongly nonlinear responses of dynamical systems.

An additional important characteristic of NNMs relates to forced resonances, since in analogy to linear theory, forced resonances of nonlinear systems excited by periodic excitations occur in neighborhoods of NNMs (Mikhlin, 1974) [this may lead to quite complex structures of forced resonances as discussed by King and Vakakis (1995b)]. Hence, knowledge of the structure of NNMs of a nonlinear oscillator can provide valuable insight on its fundamental or secondary (subharmonic, superharmonic or combinantion) resonances (Nayfeh and Mook, 1995), a feature which is of considerable engineering importance. The structure of forced resonances of nonlinear oscillators is determined, in essence, from the structure and bifurcations of their NNMs, so performing forced response analysis based on linear eigenspaces and not taking into account the possibility that essentially nonlinear modes might exist, may lead to inadequate modeling of the dynamics.

Moreover, it was shown in recent studies (Pesheck, 2000; Pesheck et al., 2002; Jiang et al., 2005b; Touzé et al., 2004, 2007a, 2007b; Touzé and Amabili, 2006) that NNMs can provide effective bases for constructing reduced-order models of the dynamics of discrete and continuous nonlinear oscillators. Indeed, NNM-based Galerkin projections for discretizing the dynamics were proven to be more accurate in predicting the nonlinear dynamics of these systems compared to linear mode-based Galerkin projections. These results demonstrate one additional application of NNMs; that is, even though NNMs do not satisfy orthogonality properties (as classical linear normal modes do) they can still be used as bases for accurate, low-order Galerkin projections of the dynamics of discrete and continuous weakly or strongly nonlinear oscillators. The resulting low-order reduced models are expected to be much more accurate compared to linear mode-based ones (especially in systems with strong or even nonlinearizable nonlinearities). The reason for the enhanced accuracy of NNM-based reduced-order models lies on the invariance properties of NNMs, and on the fact that they represent exact solutions of the free or forced nonlinear dynamics of the oscillators considered. Hence, free or forced oscillations of a nonlinear structure in the neighborhoods of NNMs can be accurately captured by either isolated NNMs (in the absence of multi-modal nonlinear interactions), or by a small subset of NNMs (when internal resonances between NNMs occur). Hence, NNMs hold promise as bases for efficient and accurate low-order reduction of the dynamics of systems with many degrees-of-freedom, for example, of finite-element (FE) computational models; this holds, in spite the fact that NNMs do not satisfy any form of orthogonality conditions.

NNMs can be applied in additional areas of vibration theory, as in the area of modal analysis and system identification. Traditional techniques for modeling the dynamics of nonlinear structures are based on the assumptions of weak nonlinearities and of a nonlinear modal structure similar to that of the underlying linearised system. As shown in the previous examples even a simple two-DOF system can have more normal modes than its degrees of freedom; hence, in performing nonlinear modal analysis one should consider the possibility that certain of the modes might be essentially nonlinear, having no counterparts in linear theory. The bifurcating NNMs of the previous examples represent precisely this type of essentially nonlinear modes; they change qualitatively the modal structure of the dynamical system by adding essentially nonlinear components that do not exist in the context of linear theory. It follows that the concept of the NNM can provide the necessary framework for developing nonlinear modal analysis techniques, capable of modeling essentially nonlinear dynamics (Kerschen et al., 2005).

It was mentioned previously that the bifurcating similar NNMs of the symmetric system (2.1) become localized to either one of the two oscillators of the system as coupling between them becomes weak, so that nonlinear mode localization occurs in the weakly coupled system. Moreover, we showed that the non-similar NNMs of the non-symmetric system (2.6) become localized with varying energy, even in the absence of NNM bifurcations. Hence, it becomes clear that nonlinear localization is an important feature of the dynamics of coupled nonlinear dynamical systems. *Nonlinear mode localization* and its applications are discussed in the next section.

2.2 Energy Localization in Nonlinear Systems

One of the most interesting features of NNMs is that they may induce nonlinear mode localization in dynamical systems, i.e., a subset of NNMs may be spatially localized to subcomponents of dynamical systems. Mode localization may occur also in linear systems composed of multiple coupled subsystems (Anderson, 1958; Pierre and Dowell, 1987; Hodges, 1982), however, it only results due to the interplay between break of symmetry (structural disorder) and weak coupling between subsystems. In nonlinear systems, structural disorder is not a prerequisite for mode localization, since the dependence of the frequency of oscillation on the amplitude (energy) provides an 'effective disorder' (or 'mistuning') in the dynamics (Vakakis et al., 1993; Vakakis, 1994; King et al., 1995; Vakakis et al., 1996).

Nonlinear mode localization was realized in both examples of systems of unforced coupled oscillators examined in Section 2.1, either due to a bifurcation of similar NNMs in the symmetric system (2.1), or due to the energy dependence of the nonlinear mode shape of non-similar NNMs in system (2.6). Moreover, forced nonlinear localization in systems under harmonic excitation has been studied (Vakakis,

1992; Vakakis et al., 1994), and nonlinear mode localization in flexible systems with smooth (Vakakis, 1994; Aubrecht and Vakakis, 1996; Aubrecht et al., 1996) and non-smooth nonlinearities (Emaci et al., 1997) has been investigated. A review of mode localization in systems governed by nonlinear partial differential equations was given in Vakakis (1996).

We will demonstrate some aspects of nonlinear mode localization in coupled oscillators by considering two examples, one involving a low-dimensional cyclic system, and a second one with a nonlinear medium of infinite spatial extent. The latter example will underline the theoretical link between NNMs and solitary waves. We start by considering a cyclic assembly of coupled oscillators, governed by the following set of ordinary differential equations (Vakakis et al., 1993):

$$\ddot{y}_i + y_i + \varepsilon\mu y_i^3 + \varepsilon k(y_i - y_{i+1}) + \varepsilon k(y_i - y_{i-1}) = 0, \quad i = 0, \ldots, N$$

$$y_0 \equiv y_N, \quad y_{N+1} \equiv y_1 \tag{2.9}$$

This symmetric system possesses similar NNMs, which can be approximately computed by the method of multiple scales (Nayfeh and Mook, 1995) as follows:

$$\left.\begin{aligned}
y_1(t) &= a_1 \cos[(1+\varepsilon\alpha)t + \beta_1] + O(\varepsilon) \\
y_2(t) &= y_{N-1}(t) = -a_2 \cos[(1+\varepsilon\alpha)t + \beta_1] + O(\epsilon) \\
y_3(t) &= y_{N-2}(t) = -a_3 \cos[(1+\varepsilon\alpha)t + \beta_1] + O(\epsilon) \\
&\quad \bullet \ \bullet \ \bullet
\end{aligned}\right\} (N = 2p+1)$$

or

$$\left.\begin{aligned}
y_1(t) &= a_1 \cos[(1+\varepsilon\alpha)t + \beta_1] + O(\varepsilon) \\
y_2(t) &= y_{N-1}(t) = -a_2 \cos[(1+\varepsilon\alpha)t + \beta_1] + O(\varepsilon) \\
y_3(t) &= y_{N-2}(t) = a_3 \cos[(1+\varepsilon\alpha)t + \beta_1] + O(\varepsilon) \\
&\quad \bullet \ \bullet \ \bullet \\
y_{p-1}(t) &= y_{p+1}(t) = (-1)^p a_{p-1} \cos[(1+\varepsilon\alpha)t + \beta_1] + O(\varepsilon) \\
y_p(t) &= (-1)^{p+1} a_p \cos[(1+\varepsilon\alpha)t + \beta_1] + O(\varepsilon)
\end{aligned}\right\} (N = 2p+1)$$

(2.10)

where p is an integer and $a_k \geq 0$; the phase β_1 depends on the initial conditions, and $\varepsilon\alpha = \varepsilon\alpha(a_1)$ is the small amplitude-dependent nonlinear correction to the frequency of the NNM. The ratios (a_n/a_m) in (2.10) were determined in (Vakakis et al., 1993) for systems with even and odd degrees of freedom.

In Figure 2.6 we present a subset of NNMs for systems (2.9) with $N = 4$ and $N = 5$ degrees of freedom. It can be shown that for fixed (conserved) energy level, the parameter that conrols nonlinear mode localization is the ratio (k/μ), i.e., the relative magnitude of coupling with respect to stiffness nonlinearity. For low values of this ratio the NNMs depicted in Figures 2.6a, b become localized to the first oscillator, i.e., the amplitude a_1 becomes much larger than the corresponding amplitudes of the other oscillators; this occurs in spite of direct coupling between oscillators. As the coupling to nonlinearity ratio (k/μ) increases from relatively small

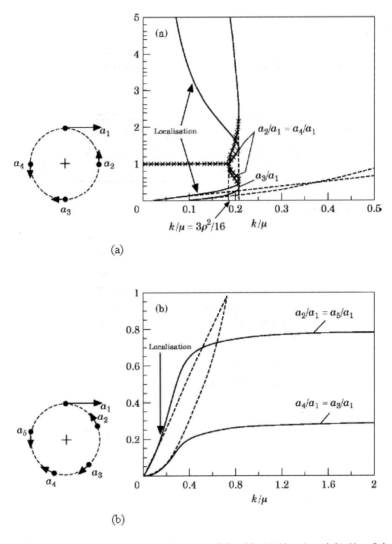

Fig. 2.6 Nonlinear localization in the cyclic system (2.9) with, (a) $N = 4$, and (b) $N = 5$ degrees of freedom; —: stable NNMs, — — —: unstable NNMs; - - - -: asymptotic approximations.

to relatively large values, there occur two distinct scenarios of delocalization, as the energy of the NNM gradually becomes spatially extended. Specifically, for the system with even DOF ($N = 4$, see Figure 2.6a), the localized NNM branches become delocalized through a bifurcation with the out-of-phase (spatially extended) NNM $a_1 = a_2 = a_3 = a_4$; this bifurcation signifies the end of localization in this system. A different scenario of delocalization occurs in the system with odd DOF ($N = 5$, see Figure 2.6b), since as the ratio (k/μ) increases the localized branches of NNMs become delocalized through smooth transitions to spatially extended NNMs; the absence of NNM bifurcations in this case is a reflection of the symmetry group of this system which differs from that of the system with even DOF (for group theoretic approaches to problems in dynamical systems, see Manevitch and Pinsky, 1972a, 1972b, and also Manevitch et al., 1970; Vakakis et al., 1996).

We note that although the previous results prove that for weak coupling to nonlinearity ratios (strong) localization of motion occurs in the first oscillator of system (2.9), due to cyclic symmetry this result can be extended to each of the other oscillators. Hence, we can prove that system (2.9) possesses N (strongly) localized NNMs with the vibration being passively confined mainly to one of the oscillators. Moreover, these localized NNMs are stable, and, hence, physically realizable (Vakakis et al., 1993); in the same reference, it is proven that additional (weakly) localized NNMs occur, with motion passively confined mainly to a subset of oscillators.

Nonlinear localization can greatly infuence the transient structural response since it can lead to *passive motion confinement of disturbances generated by external forces*. When localized NNMs of such structures are excited by external impulsive forces, the oscillations remain passively confined close to the point where they are initially generated instead of 'spreading' through the entire structure. Such passive confinement can also occur in linear systems but only in the presence of disorder and weak substructure coupling (Pierre and Dowell, 1987). To demonstrate the passive nonlinear motion confinement phenomenon, we consider the impulsive response of the cyclic system (2.9). As discussed previously, as $k/\mu \to 0+$ branches of similar NNMs localize (strongly) to a single oscillator.

A system with $N = 50$ oscillators is considered in the numerical simulations and the numerical results are obtained by finite element (FE) computations (Vakakis et al., 1993). A force with unit magnitude and duration $\Delta t = 0.2$ is applied to the first oscillator, and the transient response of the system is depicted in Figure 2.7 for parameters $\varepsilon\mu = 0.3$, $\varepsilon k = 0.05$ and $k/\mu = 0.166$; at this energy level and for the chosen system parameters the cyclic system possesses strongly localized NNMs, so passive nonlinear motion confinement of the impulsive response is expected. Indeed, as shown in Figure 2.7 the nonlinear response remains confined to the directly forced oscillator, instead of 'leaking' to the entire system. For comparison purposes the responses of the corresponding linear system with $\varepsilon\mu = 0$ are also shown in the plots of Figure 2.7, from which we conclude that in the linear case there is a gradual 'spreading' of the impulsive energy to all oscillators; moreover, the spreading of energy in the linear system becomes increasingly more profound as time increases. The motion confinement of disturbances in the nonlinear system can only be attributed to the excitation of strongly localized NNMs by the external impulse, yielding motion confinement due to their invariance properties. In fact, it is the invariance property of the stable strongly localized NNM that yields transient motion confinement of disturbances in the system under consideration.

The second example will demonstrate that there is a theoretical link between NNMs and spatially localized solitary waves in nonlinear media of infinite spatial extent. For this we consider spatially localized NNMs in the following nonlinear partial differential equation (King and Vakakis, 1994):

where again the short-hand notation for partial differentiation is used. The amplitude $A > 0$ of the NNM is the maximum amplitude attained by the response $u_0(t)$ of the reference point, when the system reaches the position of maximum potential energy. Equation (2.15) has to be solved simultaneously with the following two additional conditions:

$$\lim_{x \to \pm\infty} U[x, u_0(t)] = 0 \qquad (2.16)$$

$$\{-U(x_0, A) - \varepsilon\lambda U_{xx}(x_0, A) - \varepsilon\alpha U^3(x_0, A)\}U_{u_0}(x, A) =$$
$$-U(x, A) - \varepsilon\lambda U_{xx}(x, A) - \varepsilon\alpha U^3(x, A) \qquad (2.17)$$

Condition (2.16) is self-explanatory [it corresponds to the first of relations (2.14a)], whereas condition (2.17) needs further justification. A careful examination of the functional relation (2.15) reveals that it becomes singular when the system reaches the position of maximum potential energy $u_0 = A$; indeed, the coefficient of the highest-order partial derivative $U_{u_0 u_0}$ becomes zero when $u_0 = A$, so this represents a regular singular point of the mathematical problem. Therefore, the solution for $U(x, u_0)$ must be, (i) first asymptotically approximated in semi-open intervals $0 \leq u_0(t) < A$, and then, (ii) analytically continued up to the maximum potential energy level $u_0 = A$; this analytical continuation is accomplished by imposing the condition (2.17) which guarantees that the solution of the functional equation (2.15) is extended up to the point of maximum potential energy.

The non-similar NNM governed by relations (2.15)–(2.17) was solved asymptotically in King and Vakakis (1994), leading to the following analytical approximation for the modal function

$$U[x, u_0(t)] = [a_1^{(0)}(x) + \varepsilon a_1^{(1)}(x) + O(\varepsilon^2)]u_0(t)$$
$$+ [\varepsilon a_3^{(1)}(x) + O(\varepsilon^3)]u_0^3(t) + O[\varepsilon u_0^5(t)] \qquad (2.18)$$

where

$$a_1^{(0)}(x) = \mathrm{sec}\, hz,$$

$$a_1^{(1)}(x) = [(1/24)(\alpha A^2 + K_1 + AK_2)z \cosh z - (\alpha A^2/48) \sinh z] \tanh z \,\mathrm{sec}\, h^2 z,$$

$$a_3^{(1)}(x) = -(\alpha/8)(1 - \mathrm{sec}\, h^2 z)\, \mathrm{sec}\, hz, \quad z = A(3\alpha/8\lambda)^{1/2}(x - x_0),$$

$$K_1 = \left[-\int_{-\infty}^{+\infty} \lambda a_1^{(0)'2}(x)\, dx\right] \Big/ \left[\int_{-\infty}^{+\infty} a_1^{(0)2}(x)\, dx\right],$$

$$K_2 = \left\{-\int_{-\infty}^{+\infty} \left[2a_1^{(0)}(x)a_3^{(1)}(x) + (\alpha/2)a_1^{(0)4}(x)\right] dx\right\} \Big/ \left[\int_{-\infty}^{+\infty} a_1^{(0)2}(x)\, dx\right]$$

and prime denotes differentiation with respect to x. Note that although no space-time separation is possible for this problem (since the sought NNM is non-similar), the derived asymptotic approximation is based on solving an hierarchy of subproblems at increasing orders of ε which are separable, so they can be solved analytically (King and Vakakis, 1994).

After computing the approximation for the modal function (2.18), the reference response $u_0(t)$ is computed by substituting (2.18) into (2.11) and evaluating the resulting expression at the reference point $x = x_0$. This results in the following nonlinear *modal oscillator*:

$$\ddot{u}_0(t) + [1 + \varepsilon \lambda a_1^{(0)''}(x_0) + \varepsilon^2 \lambda a_1^{(1)''}(x_0)]u_0(t)$$
$$+ [\varepsilon \alpha + \varepsilon^2 \lambda a_3^{(1)''}(x_0)]u_0^3(t) + O[\varepsilon^2 u_0^5(t), \varepsilon^3] = 0 \quad (2.19)$$

For specific initial conditions the response $u_0(t)$ of the modal oscillator can be computed in closed form in terms of Jacobian elliptic functions. This computes also the frequency of the oscillation of the non-similar NNM, and reveals its dependence on energy (the initial condition). For example, for initial conditions $u_0(0) = A, \dot{u}_0(0) = 0$ (i.e., for initiation of the NNM oscillation at the point of maximum potential energy) the solution of (2.19) is expressed as

$$u_0(t) = A \operatorname{cn}(pt, k^2),$$
$$p = \{1 + \varepsilon \lambda a_1^{(0)''}(x_0) + \varepsilon^2 \lambda a_1^{(1)''}(x_0) + [\varepsilon \alpha + \varepsilon^2 \lambda a_3^{(1)''}(x_0)]A^2\}^{1/2} \quad (2.20)$$

where $k^2 = [\varepsilon \alpha + \varepsilon^2 \lambda a_3^{(1)'''}(x_0)]A/2p^2$ is the elliptic modulus (Byrd and Friedman, 1954). The frequency of the NNM coincides with the frequency of the periodic response (2.20),

$$\omega = \omega(A) = \frac{\pi p}{2K(k)} \quad (2.21)$$

where $K(\bullet)$ is the complete elliptic integral of the first kind (Byrd and Friedman, 1954). This completes the analytic approximation of the NNM of system (2.11).

The solution $u(x, t) = U[x, u_0(t)]$ given by expressions (2.18)–(2.21) represents a stationary, spatially localized, time-periodic response of the nonlinear medium, i.e., a *stationary breather* or stationary solitary wave. Since this stationary wave represents synchronous (in-unison) oscillations of all points of the nonlinear medium, it can be regarded as a localized NNM of the medium of infinite spatial extent. Hence, the previous results provide a theoretical link between NNMs and stationary solitary waves (breathers) in nonlinear partial differential equations. In Figure 2.8 we depict snapshots of the stationary breather for parameters $\alpha = 1.2, \lambda = 0.9, A = 0.25, x_0 = 0$ and $\varepsilon = 0.01$. As discussed in King and Vakakis (1994), based on the stationary solution (2.18)–(2.21) a family of *travelling breathers* of system (2.11) can be computed by imposing the following Lorentz coordinate transformation:

$$\tilde{u}(x, t) = U\left[\frac{x + vt\sqrt{\varepsilon \lambda}}{\sqrt{1 + v^2}}, u_0\left(\frac{t - vx/\sqrt{\varepsilon \lambda}}{\sqrt{1 + v^2}}\right)\right] \quad (2.22)$$

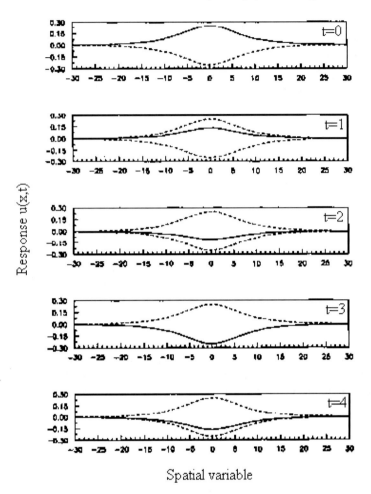

Fig. 2.8 Localized NNM of system (2.11) – stationary breather.

The traveling wave velocity (group velocity) v is related to the frequency ω by modifying the frequency-energy relation as follows:

$$\omega = \omega(v, A) = \frac{\pi p \sqrt{1+v^2}}{2K(k)} \tag{2.23}$$

We note that the NNM (2.18) can be considered as special case of the traveling breather solution (2.22) with zero group velocity, $v = 0$.

From a practical point of view, nonlinear mode localisation phenomena can be implemented in active or passive vibration isolation designs, where unwanted disturbances generated by external forces are initially spatially confined to predetermined, specially designed subcomponents of the structure, and then passively or actively dissipated locally. Indeed, inducing localized NNMs in flexible structures of large spatial extent is expected to enhance the controllability of these structures, since in designing for active control one would need to consider only local structural components where the unwanted disturbances are to be confined, instead of considering the structures in their entireties; of course, issues of observability, controllability, spill-over effects, and possible instabilities by excitation of unwanted or unmodelled modes should be addressed in these control designs.

In addition, the study of motion confinement phenomena due to nonlinear effects can prove to be beneficial in applications where such localization phenomena are unwanted. For example, localization of vibration energy in rotating turbine blade assemblies can be catastrophic since it may lead to failure of high-speed rotating blades. Understanding the interplay between (and effects of) structural disorders, coupling forces and stiffness or damping nonlinearities on localization can prevent such failures and prolong the operational life of mechanical or structural components.

We end this section by providing a remark concerning the relation between nonlinear mode localization and nonlinear targeted energy transfer (TET) phenomena considered in this work. Simply stated, nonlinear mode localization can be regarded as a *static* way of passive energy confinement: in structures with localized NNMs, energy confinement can be achieved only as long as stable localized NNMs are excited either by the external excitations and/or the initial conditions of the problem; it follows that energy confinement through nonlinear mode localization relies mainly on passive confinement of disturbances at the points of their generation through direct excitation of stable localized NNMs. It follows that no passive energy transfer is possible in this case. On the other hand, TET can be regarded as a *dynamic* way of passive energy confinement: indeed, TET relies of the passive, directed transfer of unwanted vibration energy from the point of its generation to isolated or sets of nonlinear energy sinks (NESs) where this energy is confined and dissipated locally; moreover, we will show that TET can be realized for a broad range of external excitations and/or initial conditions, and can result in broadband energy transfer between different parts of a system. As mentioned in the previous section (and as discussed in detail in the following chapters), nonlinear mode localization plays a key role in TET, as TET critically depends on the variation of the shapes of excited NNMs, from being non-localized to becoming localized with varying energy.

In the next section we continue our discussion of introductory concepts by discussing the nonlinear phenomena or internal resonances, transient resonance captures (TRCs) and sustained resonance captures (SRCs) in nonlinear dynamical systems.

2.3 Internal Resonances, Transient and Sustained Resonance Captures

A general n-DOF time-invariant, linear Hamiltonian vibrating system with n distinct natural frequencies possesses n linear normal modes which form a complete orthogonal basis in R^n; if a natural frequency has multiplicity p – for example, due to special symmetries of the system – the set of $(n - p + 1)$ independent normal modes can be complemented by $(p - 1)$ generalized modes (Meirovitch, 1980) to form again a complete and orthogonal basis in R^n. This can be extended to infinite dimensions in the case of bounded, time-invariant, unforced linear continuous systems [since unbounded elastic media possess continuous spectra of eignevalues and support waves instead of vibration modes (Courant and Hilbert, 1989)].

Viewed from a geometric perspective, the $2n$-dimensional phase space of the n-DOF linear time-invariant Hamiltonian system is foliated by an infinite family of invariant n-tori, parametrized by the Hamiltonian (which in most cases coincides with the total conserved energy of the motion); this is due to the fact that linear systems are always *integrable*. To give an example, consider the following two-DOF linear system of coupled oscillators:

$$\ddot{y}_1 + y_1 + K(y_1 - y_2) = 0$$
$$\ddot{y}_2 + y_2 + K(y_2 - y_1) = 0 \qquad (2.24)$$

This system possesses an in-phase mode with natural frequency $\omega_1 = 1$, and an out-of-phase mode with natural frequency $\omega_2 = \sqrt{1 + 2K}$. To get a geometric picture of the dynamics in phase space, we introduce the action-angle variable transformation, $(y_1, \dot{y}_1, y_2, \dot{y}_2) \in R^4 \to (I_1, I_2, \phi_1, \phi_2) \in (R^+ \times R^+ \times S^1 \times S^1)$, which can be regarded as a form of nonlinear polar transformation (Persival and Richards, 1982), and defined by the relations

$$y_1 = \sqrt{2I_1/\omega_1} \sin \phi_1, \quad \dot{y}_1 = \sqrt{2I_1\omega_1} \cos \phi_1$$
$$y_2 = \sqrt{2I_2/\omega_2} \sin \phi_2, \quad \dot{y}_2 = \sqrt{2I_2\omega_2} \cos \phi_2 \qquad (2.25)$$

In terms of the new variables, the system of coupled oscillators (2.24) can be expressed as follows:

$$\dot{I}_1 = 0 \Rightarrow I_1 = I_{10}$$
$$\dot{I}_2 = 0 \Rightarrow I_2 = I_{20}$$
$$\dot{\phi}_1 = \omega_1$$
$$\dot{\phi}_2 = \omega_2 \qquad (2.26)$$

The leading two equations in (2.26) are trivially solved, and represent conservation of energy for each of the two normal modes of system (2.24); actually, these

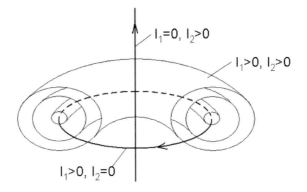

Fig. 2.9 Foliation of the phase space of two-DOF linear Hamiltonian system (2.24) by an infinite family of invariant two-tori parametrized by energy.

additional first integrals of motion render the two-DOF linear system (2.24) fully integrable (for linear systems the integrability property can be extended to R^n). It follows that at a given energy level the dynamics of (2.24) is reduced to the dynamics of the angles ϕ_1 and ϕ_2 on an isoenergetic two-torus T^2, with the resulting motion being either periodic (if the frequency ratio ω_1/ω_2 is a rational number) or quasi-periodic (if ω_1/ω_2 is irrational). By varying the energy of the motion (through changes in initial conditions) the dynamics in phase space takes place on different isoenergetic two-tori, so the entire phase space of system (2.24) is foliated by an infinite family of invariant two-tori parametrized by energy. In Figure 2.9 we present a schematic depiction of this family of isoenergetic two-tori which are invariant for the dynamical flow of (2.24). We note that the limiting cases where only one of the two modes of the system is excited by the initial conditions (i.e., $I_1 = 0$ or $I_2 = 0$) correspond to degeneracies of the family of tori and are represented by one-dimensional manifolds (lines) as shown in Figure 2.9.

Returning to our discussion of the general n-DOF time-invariant linear Hamiltonian system, the energy imparted at $t = 0$ in the system by the initial conditions is partitioned among the linear modes (i.e., the motion takes place on a specific n-torus T^n in phase space), and no further energy exchanges between modes is possible for $t > 0$. Each linear mode conserves its own energy and participates accordingly in the (periodic or quasi-periodic) response of the system through linear superposition with the responses of the other modes.

This nice structure of the linear phase space in terms of the foliation by the infinite family of invariant tori is not expected to be preserved when the Hamiltonian system is perturbed by nonlinear terms. For example, considering perturbations of the integrable Hamiltonian system (2.24) by nonlinear non-Hamiltonian perturbations, no tori survive the perturbation. For Hamiltonian perturbations, however, the KAM (Kolmogorov–Arnold–Moser) theorem (MacKay and Meiss, 1987) guarantees that 'sufficiently irrational' tori (i.e., those for which the ratio ω_1/ω_2 is 'poorly' approximated by rational numbers in a number-theoretic setting) are preserved, filled with quasi-periodic orbits. Generically, the remaining invariant n-tori of the infinite foliation 'break-up' by the perturbation, leading to an infinite number of stable-unstable pairs of periodic orbits of arbitrarily large periods and to chaotic trajectories located in local chaotic layers; this renders the perturbed nonlinear system non-integrable (this scenario was demonstrated in the example with the two-DOF Hamiltonian system and the corresponding Poincaré maps of Figure 2.3 in Section 2.1). There are, however, cases of integrable nonlinear Hamiltonian systems where the foliation of phase space by invariant tori is still preserved (in similarity to the linear case) (Moser, 2003); it is conjectured, however, that full intergability is not a generic property of nonlinear Hamiltonian systems.

The previously described scenario of 'break-up' of rational and 'insufficiently irrational' tori in nonlinear Hamiltonian systems underlines a nonlinear dynamical mechanism that enables energy exchanges between modes, even if they are well separated in frequency (clearly, this would not be possible in linear theory). This mechanism is the phenomenon of *internal resonance* which results in nonlinear coupling between modes, and gives rise to mode bifurcations and nonlinear beat phenomena during which strong energy exchanges between modes occur. This is not possible in linear theory, since, as discussed above, there is no mechanism for exchanging energy between well separated modes (although, it is well-known that beat phenomena can occur when linear modes are closely spaced in frequency).

Internal resonances in nonlinear Hamiltonian systems are associated with the failure of the averaging theorem with respect to certain *'slow angles'* of the problem in neighborhoods of *resonance manifolds*. We show this in the following brief exposition which follows Arnold (1988), Lochak and Meunier (1988) and Verhulst (2005). Consider the following $2n$-dimensional nonlinear Hamiltonian system in action-angle variables,

$$\left. \begin{array}{l} \dot{I} = \varepsilon F(\phi, I) \\ \dot{\phi} = \omega(I) + \varepsilon G(\phi, I) \end{array} \right\} (I, \phi) \in (R^{+n} \times T^n) \qquad (2.27)$$

which, for $|\varepsilon| \ll 1$ is a perturbation of the $2n$-dimensional integrable Hamiltonian system, $\dot{I} = 0$, $\dot{\phi} = \Omega(I)$. We consider the general case where the n frequencies $\omega = [\omega_1 \ldots \omega_n]^T$ depend on the n-vector of actions I [this is typical in nonlinear Hamiltonian systems, but it does not hold for the linear system (2.24)–(2.26)]. In (2.27) we assume that F and G are sufficiently smooth functions which are 2π-periodic in the n-vector of angles $\phi = [\phi_1, \ldots, \phi_n]^T$; moreover, as in previous examples, by T^n we denote the n-torus.

It follows that the n-vector of functions F can be expanded in complex Fourier series in terms of the n angles as follows:

$$F(\phi, I) = \sum_{k_1, \ldots, k_n = -\infty}^{+\infty} c_{k_1 \ldots k_n}(I) e^{j(k_1 \phi_1 + \ldots + k_n \phi_n)} \qquad (2.28)$$

where $j = (-1)^{1/2}$ and $c_{k_1 \ldots k_n}(I)$ is an n-vector of complex coefficients of the harmonic characterized by the indices $(k_1, \ldots, k_n) \in Z^n$. A resonance manifold of the dynamics of (2.28) is defined by the relation,

$$\hat{k}_1\omega_1(I) + \hat{k}_2\omega_1(I) + \ldots + \hat{k}_n\omega_n(I) = 0 \Rightarrow I = I^{(\hat{k}_1:\hat{k}_2:\ldots:\hat{k}_n)} \quad (2.29)$$

for some $(\hat{k}_1, \hat{k}_2, \ldots, \hat{k}_n) \in Z^n$, provided that the corresponding vector of Fourier coefficients in (2.28) does not vanish, $\|c_{\hat{k}_1\hat{k}_2\ldots\hat{k}_n}(I^{(\hat{k}_1:\hat{k}_2:\ldots:\hat{k}_n)})\| \neq 0$. If the resonance manifold is a low-dimensional submanifold of R^n, in its neighborhood we can average out the angles that do not participate in the internal resonance (these angles possess time-like behavior and are regarded as *'fast'* angles), and reduce accordingly the dimensionality of the dynamics. This is performed by defining appropriate *'slow' angles* (which are not time-like and cannot be averaged out of the dynamics) as combinations of the angles that participate in the resonance condition (2.29). In essence, the internal resonance provides nonlinear coupling between all participating modes, and results in energy exchanges between these modes.

In the absence of internal resonance, all angles in (2.27) possess time-like behavior (and, hence, are 'fast' angles) so they can be averaged out of the problem to reduce it to the following n-dimensional averaged dynamical system,

$$\dot{I}_a = \varepsilon c_{0\ldots 0}(I_a) \quad (2.30)$$

i.e., in terms of the vector of coefficients of the Fourier term in (2.28) not depending on ϕ. Given an initial condition $I(0) = I_a(0)$, it can be proven that $I(t) - I_a(t) = O(\varepsilon)$ on the timescale $1/\varepsilon$ (Verhulst, 2005). In the absence of internal resonances no nonlinear modal interactions occur, and each mode retains its energy, in similarity to the linear case [at least correct to $O(1)$ – small modal energy exchanges occur at higher orders of ε so they are insignificant]. It follows that no significant energy exchanges between modes can occur in the absence of internal resonances.

The effect of an internal resonance on the dynamics of a nonlinear system is illustrated by the following example. We consider a two-DOF system composed of a linear oscillator weakly coupled to a strongly nonlinear attachment (Vakakis and Gendelman, 2001),

$$\ddot{y}_1 + Cy_1^3 + \varepsilon(y_1 - y_2) = 0$$
$$\ddot{y}_2 + \omega_2^2 y_2 + \varepsilon(y_2 - y_1) = 0 \quad (2.31)$$

where the stiffness characteristic of the weak coupling, $0 < \varepsilon \ll 1$, is the small parameter of the problem. For $\varepsilon = 0$ the two oscillators become uncoupled, and the nonlinear system is integrable. We wish to study the effects of internal resonance on the dynamics of this system when we perturb it by weak coupling terms. In terms of the terminology introduced in Chapter 3, this system represents a linear oscillator (LO) with an attached grounded nonlinear energy sink (NES) (see Section 3.1).

First, we bring this system in the form (2.27) by transforming in terms of the action-angle variables $(I_1, I_2, \phi_1, \phi_2) \in (R^+ \times R^+ \times T^2)$ of the unperturbed system,

$$y_1 = \Lambda I_1^{1/3} \operatorname{cn}[2K(1/2)\phi_1/\pi, 1/2]$$

$$\dot{y}_1 = -[\Lambda I_1^{1/3} \omega_1(I_1) 2K(1/2)/\pi] \operatorname{sn}[2K(1/2)\phi_1/\pi, 1/2] \times$$
$$\operatorname{dn}[2K(1/2)\phi_1/\pi, 1/2]$$
$$y_2 = \sqrt{2I_2/\omega_2} \sin \phi_2$$
$$\dot{y}_2 = \sqrt{2I_2\omega_2} \cos \phi_2 \qquad (2.32)$$

where $\omega_1(I_1) = \Xi I_1^{1/3}$ is the frequency of oscillation of the uncoupled nonlinear oscillator, $K(1/2)$ is the complete elliptic integral of the first kind (Byrd and Friedman, 1954), and $\Lambda = (4C)^{-1/6}[3\pi/K(1/2)]^{1/3}$, $\Xi = [3\pi^4 C/8K^4]^{1/3}$. Introducing these transformations into the perturbed system (2.31), we express it in the form (2.27):

$$\dot{I}_1 = \varepsilon F_1(I_1, I_2, \phi_1, \phi_2)$$
$$\dot{I}_2 = \varepsilon F_2(I_1, I_2, \phi_1, \phi_2)$$
$$\dot{\phi}_1 = \omega_1(I_1) + \varepsilon G_1(I_1, I_2, \phi_1, \phi_2)$$
$$\dot{\phi}_2 = \omega_2 + \varepsilon G_2(I_1, I_2, \phi_1, \phi_2) \qquad (2.33)$$

By construction, the functions F_1, F_2, G_1 and G_2 are 2π-periodic in ϕ_1 and ϕ_2 and are listed explicitly later in this section, and also in Vakakis and Gendelman (2001).

Equations (2.33) represent a two-frequency dynamical system in $(R^+ \times R^+ \times T^2)$, and are in a form directly amenable to two-frequency averaging (Lochak and Meunier, 1988). Indeed, by applying straightforward averaging with respect to the two angles ϕ_1 and ϕ_2 we obtain the following simplified averaged system:

$$\dot{I}_{1a} = \varepsilon(1/4\pi^2) \int_0^{2\pi} \int_0^{2\pi} F_1(I_{1a}, I_{2a}, \phi_1, \phi_2) d\phi_1 d\phi_2 = 0$$
$$\dot{I}_{2a} = \varepsilon(1/4\pi^2) \int_0^{2\pi} \int_0^{2\pi} F_2(I_{1a}, I_{2a}, \phi_1, \phi_2) d\phi_1 d\phi_2 = 0 \qquad (2.34)$$

which is of the general form (2.30). Hence, in the averaged system the two oscillators [conserve to $O(1)$] their initial energies, inspite of the weak coupling. Clearly, this is will not be case when internal resonances occur in the dynamics.

The condition under which the dynamics of the averaged system (2.33) accurately describes the dynamics of the full system (2.34) has been addressed in previous works (Neishtadt, 1975; Morozov and Shilnikov, 1984; Arnold, 1988). Arnold's theorem (1988) answers this question. If the condition

$$\frac{d}{dt}\left[\frac{\omega_1(I_1)}{\omega_2}\right] \neq 0$$

is satisfied along the trajectories of the dynamical flow of (2.33), then the full dynamics is close to the averaged dynamics up to time of $O(1/\varepsilon)$. That is, if $I(0) = I_a(0)$, then $\|I(t) - I_a(t)\| \leq \kappa\sqrt{\varepsilon}$ for $0 < t < 1/\varepsilon$.

The condition of the theorem precludes any trajectory of (2.33) from being captured on a resonance manifold. According to our previous discussion, the conditions for the existence of an $(m:n)$ resonance manifold of (2.33) are as follows:

$$m\omega_1(I_1) - n\omega_2 = 0,$$

$$\int_0^{2\pi}\int_0^{2\pi} F_p(I_1, I_2, \phi_1, \phi_2)e^{-j(m\phi_1 - n\phi_2)}d\phi_1 d\phi_2 \neq 0, \quad p = 1, 2 \quad (2.35)$$

where m and n are integers.

In what follows we study in detail the 1:1 internal resonance in the dynamics of the Hamiltonian system (2.31) or (2.33), corresponding to the following level of the action of the nonlinear oscillator:

$$\omega_1(I_1) - \omega_2 = 0 \Rightarrow I_1 \equiv I_1^{(1:1)} = (\omega_2/\Xi)^3 \quad (2.36)$$

We restrict our analysis to an $O(\sqrt{\varepsilon})$ boundary layer of the 1:1 resonance manifold by defining the 'slow' angle $\psi = \phi_1 - \phi_2$, and introducing the angle transformation $(\phi_1, \phi_2) \to (\psi, \phi_2)$ and the action transformation $I_1 = I_1^{(1:1)} + \sqrt{\varepsilon}\xi$. Introducing these transformations into the last of equations (2.33), we express the independent variable as $t = (\phi_2/\omega_2) + O(\varepsilon)$, which shows that ϕ_2 is time-like, and, hence, a 'fast' angle. It follows that we can replace t by ϕ_2 as the independent variable of the remaining three equations of (2.33), and obtain the following reduced *local* dynamical system in the neighborhood of the (1:1) resonance manifold,

$$\xi' = \sqrt{\varepsilon}\omega_2^{-1}\tilde{F}_1(I_1^{(1:1)}, I_2, \psi, \phi_2) + \varepsilon\omega_2^{-1}\frac{\partial \tilde{F}_1}{\partial I_1}(I_1^{(1:1)}, I_2, \psi, \phi_2) + O(\varepsilon^{3/2})$$

$$I_2' = \varepsilon\omega_2^{-1}\tilde{F}_2(I_1^{(1:1)}, I_2, \psi, \phi_2) + O(\varepsilon^{3/2})$$

$$\psi' = \sqrt{\varepsilon}\omega_1'(I_1^{(1:1)})\omega_2^{-1}\xi + \varepsilon\omega_2^{-1}[\omega_1''(I_1^{(1:1)})\xi^2/2$$

$$+ \tilde{G}_1(I_1^{(1:1)}, I_2, \psi, \phi_2) - \tilde{G}_2(I_1^{(1:1)}, I_2, \psi, \phi_2)] + O(\varepsilon^{3/2}) \quad (2.37)$$

where primes denote differentiation with respect to ϕ_2, and the notation $F_p(I_1, I_2, \phi_1 = \psi + \phi_2, \phi_2) \equiv \tilde{F}_p(I_1, I_2, \psi, \phi_2)$, $p = 1, 2$, and a similar notation for $\tilde{G}_p$, $p = 1, 2$ are adopted. We emphasize that due to 1:1 internal resonance only one 'fast' (time-like) angle remains in the dynamics of the reduced averaged system (the angle ϕ_2), and the new 'slow' angle ψ appears. Moreover, although averaging with respect to the 'fast' angle ϕ_2 can still be performed, this cannot be done with respect to the 'slow' angle ψ since the conditions of the averaging theorem do not apply with respect to that angle; hence, *1:1 internal resonance is associated with failure of the averaging theorem in the neighborhood of the corresponding resonance manifold.*

The dynamics of the local model (2.37) which describes the nonlinear interaction between the two oscillators in the $O(\sqrt{\varepsilon})$ neighborhood of the 1:1 resonance manifold can be analyzed by asymptotic techniques such as the method of multiple

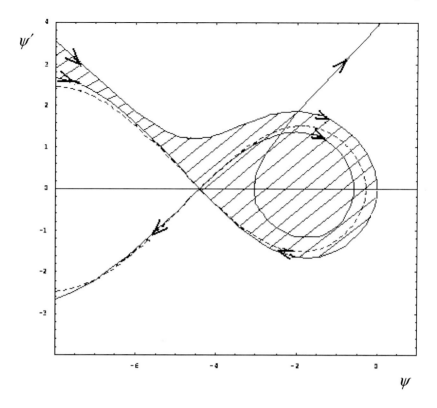

Fig. 2.13 Perturbation of the homoclinic loop of the phase portrait of the 'slow' angle by $O(\sqrt{\varepsilon})$ terms, showing the domain of attraction of 1:1 resonance capture; the unperturbed homoclinic loop is indicated by dashed lines (Panagopoulos et al., 2004).

2.4 Averaging, Multiple Scales and Complexification

Different versions and combinations of multiple scales and averaging techniques are widely used for analyzing the responses of nonlinear dynamical systems (Kevorkian and Cole, 1996), and a general review of asymptotic methods in mechanics is provided in Andrianov et al. (2003). In this work we will make extensive use of a special technique, the so called *complexification-averaging (CX-A) technique*, based on complexification of the dynamics and then averaging over 'fast' time-scales. We will show that the CX-A technique is especially suitable for analyzing strongly nonlinear transient responses of the type that we will be concerned with in our study of TET. Indeed, the employment of this special technique is dictated by the fact that the majority of TET problems considered in this work will be formulated in the transient domain (although in Chapter 6 we will examine steady state TET as well), so conventional perturbation or asymptotic techniques such as the methods of averaging, multiple scales and Lindtstead–Poincaré which are more suitable for analyzing steady state motions (such as periodic orbits) are not directly applicable in the majority of problems that we will be concerned with in this work (however, these methods will be employed after application of CX-A in our further analysis of the resulting slow flows). Moreover, the systems considered in this work possess strong (and even nonlinearizable) nonlinearities, so perturbation techniques based on linear generating functions and based on the assumption of weak nonlinearity, again are not directly applicable for the TET-related problems examined herein.

The complex representation of a nonlinear oscillatory system was initially considered as a phenomenological model that provides enhanced possibility for analyzing nonlinear effects (Scott et al., 1985; Kosevitch and Kovalyov, 1989). Moreover, the use of complexification techniques is widely used in the applied physics literature for studying nonlinear dynamics and wave phenomena. It has been shown recently (Manevitch, 1999, 2001) that this type of complexified models can be formally obtained for anharmonic oscillators and nonlinear oscillatory chains, through the CX-A technique, in order to replace the classical equations of motion by a set of first-order complex (modulation) equations. The method is based on an initial transformation of real coordinates to complex ones, and subsequent use of averaging or multiple scale expansions with further selection of resonance terms for obtaining the main nonlinear approximations (Manevitch 1999, 2001; Gendelman and Manevitch, 2003).

We illustrate the different approaches for applying the CX-A method by means of two examples. The first deals with a common and simple model widely used in the nonlinear dynamics literature, namely, the weakly nonlinear Duffing oscillator forced by weak harmonic excitation. The second example concerns the application of the CX-A method to the analysis of the strongly nonlinear transient response of the system of damped oscillators (2.41) undergoing 1:1 transient resonance capture; in this way, we will demonstrate an alternative approach for analyzing transient resonance captures in that system.

The first example provides a formal introduction to the CX-A method considering a system that has been analyzed in the literature with a breath of analytical techniques; hence, it will help us relate and compare the application of the CX-A technique with other analytical methods of nonlinear dynamics. The second example demonstrates the application of the CX-A technique to a more complicated, strongly nonlinear transient problem, which lies beyond the formal range of validity of weakly nonlinear conventional methods. Additional and more complicated applications of the CX-A technique will be presented throughout this work to a variety of problems; indeed, this technique will serve as our main theoretical tool for obtaining analytic approximations of the transient damped dynamics of coupled, essentially nonlinear oscillators leading to TET.

We start by analyzing the dynamics of the following harmonically forced oscillator,

$$\ddot{y} + y + \varepsilon(8y^3 - 2\cos t) = 0 \qquad (2.49)$$

where $0 < \varepsilon \ll 1$ is a dimensionless formal small parameter.

The first step for applying the method is complexification of the dynamics, which is performed by introducing the new complex variable,

$$\psi = \dot{y} + jy \qquad (2.50)$$

where $j = (-1)^{1/2}$. The transformation of variables $(y, \dot{y}) \to \psi$ corresponds physically to studying the dynamics from a fixed to a rotating coordinate frame. Transforming the original equation (2.49) in terms of the complex variable (2.50), and recognizing that $\cos t = (e^{jt} + e^{-jt})/2$, we obtain the following alternative complex differential equation of motion:

$$\dot{\psi} - j\psi + \varepsilon[j(\psi - \psi^*)^3 - (e^{jt} + e^{-jt})] = 0 \qquad (2.51)$$

where $*$ denotes complex conjugate. This equation is exact, as it is derived from the original real equation of motion without omitting any terms in the process.

At this point we make an assumption regarding the dynamics. In particular, we aim to study the dynamics of (2.51) under the assumption of 1:1 (fundamental) resonance, i.e., under condition that the response of the oscillator has a dominant harmonic component with frequency equal to the frequency of the harmonic excitation. Hence, we express the complex variable in the following polar form:

$$\psi(t) = \varphi(t) e^{jt} \qquad (2.52)$$

As shown in later chapters, under proper modifications – i.e., multi-fast frequency partitions – the CX-A method can be extended to systems whose responses possess more than one dominant (fast) frequency components. Substituting the representation (2.52) into (2.51) yields the following alternative (still exact) equation of motion:

$$\dot{\varphi} + j\varepsilon[\varphi^3 \exp(2jt) - 3|\varphi|^2\varphi + 3|\varphi|^2\varphi^* \exp(-2jt) \\ - \varphi^{*3} \exp(-4jt)] - \varepsilon[1 + \exp(-2jt)] = 0 \qquad (2.53)$$

At this point there are two ways of proceeding with the analysis, both of which involve approximations of a certain extent. The first way to proceed is to apply a multiple scales analysis to system (2.53) and reduce the problem to solving an hierarchy of linear subproblems at orders of increasing powers of the formal small parameter, ε^k, $k = 0, 1, 2, \ldots$. In what follows we will demonstrate the application of this approach by analyzing (2.53). The second way of approximately analyzing (2.53) is to average out terms possessing (fast) frequencies higher than unity; this amounts to assuming that (2.52) represents a slow-fast partition of the dynamics (note that no such slow-fast partition is imposed in the first way to solving the problem), with $e^{j\omega t}$ representing the fast oscillation, and $\varphi(t)$ its (complex) slow modulation. This alternative approach, which is especially suitable in neighborhoods of resonances of strongly nonlinear transient problems [where the nonlinear terms are not scaled by a formal small parameter as in (2.53)], will be demonstrated in the second example considered later in this section.

Since no formal assumption regarding the fast frequencies of the system (2.53) was imposed, the multiple scales singlular perturbation technique is applied to analyze its dynamics. To this end, the following asymptotic decomposition of the dependent variable, and the corresponding transformation of the independent variable are introduced:

$$\varphi(t) = \varphi_0(\tau_0, \tau_1, \ldots) + \varepsilon \varphi_1(\tau_0, \tau_1, \ldots) + \varepsilon^2 \varphi_2(\tau_0, \tau_1, \ldots) + O(\varepsilon^3)$$
$$\frac{d}{dt} = \frac{\partial}{\partial \tau_0} + \varepsilon[1 + \varepsilon f_2(\tau_1)]\frac{\partial}{\partial \tau_1} + O(\varepsilon^2) \tag{2.54}$$

where $\tau_0 = t$ is the fast time scale, and $\tau_1 = \varepsilon t$ is the leading-order slow time scale; the higher-order, slower time scales τ_k, $k = 2, 3, \ldots$ are obtained by proper inversion of the second of equations (2.54) once the slow functions $f_2(\tau_1), \ldots$ are determined (see discussion below). We emphasize the point that the second of expansions (2.54) is slightly different than those used in conventional multiple scales expansions, and the necessity for introducing slow multiplicative factors such as $f_2(\tau_1)$ in the $O(\varepsilon^2)$ terms will be explained below. Apparently this type of decomposition has been used for the first time by Lighthill (1960), but in the rather different context of problems in aerodynamics.

Transforming the slow flow (2.53) by (2.54), we obtain the following hierarchy of linear subproblems at different orders of approximation. The subproblem at $O(1)$ yields the following solution:

$$\frac{\partial \varphi_0}{\partial \tau_0} = 0 \Rightarrow \varphi_0 = \varphi_0(\tau_1) \tag{2.55}$$

which indicates that the main approximation for φ is slowly-varying (at time scale $\tau_1 = \varepsilon t$); this indicates that under the assumptions of this analysis the *ansatz* (2.52) indeed represents a slow-fast partition of the dynamics (although this was not assumed *a priori*).

Proceeding to the next order of approximation, we obtain the following linear subproblem governing φ_1:

$$\frac{\partial \varphi_0}{\partial \tau_1} + \frac{\partial \varphi_1}{\partial \tau_0} + j[\varphi_0^3 \exp(2j\tau_0) - 3|\varphi_0|^2 \varphi_0 + 3|\varphi_0|^2 \varphi_0^* \exp(-2j\tau_0)$$
$$- \varphi_0^{*3} \exp(-4j\tau_0)] - 1 - \exp(-2j\tau_0) = 0 \tag{2.56}$$

This equation represents the $O(\varepsilon)$ approximation of the *slow flow dynamics* of the system, i.e., it governs approximately the slow evolution of the complex amplitude φ with time. In order to avoid the secular growth of φ_1 with respect to the fast time scale, i.e., to avoid a response that will not be uniformly valid with increasing time, we need to eliminate from (2.56) non-oscillating terms. Hence, the following condition must be imposed:

$$\frac{\partial \varphi_0}{\partial \tau_1} - 3j|\varphi_0|^2 \varphi_0 - 1 = 0 \tag{2.57}$$

Equation (2.57) is integrable, yielding the following first integral of motion for the $O(\varepsilon)$ approximation [but not for the original equation of motion (2.49) or (2.53)]:

$$h = \frac{3j}{2}|\varphi_0|^4 + \varphi_0^* - \varphi_0 \tag{2.58}$$

This means that the $O(\varepsilon)$ approximation can be analytically computed in closed form. It should be mentioned that the appearance of a first integral of motion is a common feature of CX-A calculations for Hamiltonian systems. Indeed, the exact system (2.49) has a time-dependent Hamiltonian, and by applying averaging, it can be shown that (2.58) is a first integral of the corresponding slow flow.

After introducing a polar decomposition of φ_0 in terms of a real amplitude and a real phase, $\varphi_0(\tau_1) = N(\tau_1)\exp[j\delta(\tau_1)]$, equations (2.57) and (2.58) are rewritten as:

$$\frac{\partial N}{\partial \tau_1} = \cos\delta, \quad \frac{\partial \delta}{\partial \tau_1} = 3N^2 - \frac{1}{N}\sin\delta$$

$$h = \frac{3}{2}N^4 - 2N\sin\delta = \text{const} \tag{2.59}$$

Introducing the notation $Z = N^2$, combining the first of equations (2.59) with the first integral of motion h into a single equation in terms of Z, and integrating it by quadratures we obtain the following explicit solution for the amplitude N:

$$N(\tau_1) = -\left\{\frac{aq\,\text{sn}^2\left(\frac{3}{2}\sqrt{pq}\tau_1, k\right) + bp\left[1 + \text{cn}\left(\frac{3}{2}\sqrt{pq}\tau_1, k\right)\right]^2}{q\,\text{sn}^2\left(\frac{3}{2}\sqrt{pq}\tau_1, k\right) + p\left[1 + \text{cn}\left(\frac{3}{2}\sqrt{pq}\tau_1, k\right)\right]^2}\right\}^{1/2} \tag{2.60}$$

where a and b are the two real roots of the algebraic equation

$$4Z - \left[(3/2)Z^2 - h\right]^2 = 0$$

(with the other two roots being complex and expressed as $m \pm jn$) and the remaining parameters are defined according to:

$$p = \sqrt{(m-a)^2 + n^2}, \quad q = \sqrt{(m-b)^2 + n^2}, \quad k = \frac{1}{2}\sqrt{\frac{-(p-q)^2 + (a-b)^2}{pq}}$$

In the above expressions k is the modulus of the Jacobi elliptic functions sn($\bullet$) and cn($\bullet$). From (2.60) the real phase $\delta(\tau_1)$ is evaluated directly from the first of equations (2.59).

It should be mentioned that the expression for the $O(1)$ approximation φ_0 has been computed by considering $O(\varepsilon)$ terms and applying the method of multiple scales. The same result could be obtained without formal resort to the method of multiple scales by merely omitting all non-resonant terms from the initial (exact) equation (2.53) – i.e., by performing 'naive averaging' with respect to the fast time scale τ_0 (actually this approach will be used in the second example of CX-A technique that follows). This observation means that *the seemingly voluntary trick of omitting non-resonant terms from the orgininal exact equation (2.53) may be substantiated by formal use of multiple scales, and thus the efficiency of the CX-A approach may be explained formally, at least for the case of weak nonlinearity.*

The computation of the next approximation constitutes a somewhat non-trivial problem. To this end, the explicit expression for the first approximation is obtained by solving equation (2.56) after eliminating secular terms through (2.57),

$$\varphi_1(\tau_0, \tau_1) = -\frac{1}{2}\varphi_0^3 \exp(2j\tau_0) + \frac{3}{2}|\varphi_0|^2\varphi_0^* \exp(-2j\tau_0)$$

$$-\frac{1}{4}\varphi_0^{*3} \exp(-4j\tau_0) + \frac{j}{2}\exp(-2j\tau_0) + C_1(\tau_1) \quad (2.61)$$

where the slow-varying function $C_1(\tau_1)$ is a constant of integration with respect to the fast time scale τ_0, and is computed by considering the equation governing the $O(\varepsilon^2)$ approximation:

$$\frac{\partial \varphi_2}{\partial \tau_0} + f_2(\tau_1)\frac{\partial \varphi_0}{\partial \tau_1} - \frac{3j}{4}\varphi_0^5 \exp(4j\tau_0)$$

$$+ \left[-(15j/2)|\varphi_0|^2\varphi_0^3 + 3j\varphi_0^2 C_1(\tau_1) - 3\varphi_0^2\right]\exp(2j\tau_0)$$

$$+ \left\{\frac{\partial C_1}{\partial \tau_1} + \frac{51j}{4}|\varphi_0|^4\varphi_0 + 3|\varphi_0|^2 - \frac{3}{2}\varphi_0^2 - 3j[2|\varphi_0|^2 C_1(\tau_1) + \varphi_0^2 C_1^*(\tau_1)]\right\}$$

$$+ \left[-(69j/4)|\varphi_0|^4\varphi_0^* + 3j\varphi_0^2 C_1(\tau_1) + 6j|\varphi_0|^2 C_1^*(\tau_1) + 6|\varphi_0|^2\right]\exp(-2j\tau_0)$$

$$+ \left[(21j/4)|\varphi_0|^2\varphi_0^{*3} - (9/4)\varphi_0^{*2} - 3j\varphi_0^{*2} C_1^*(\tau_1)\right]\exp(-4j\tau_0)$$

$$+ (3j/4)\varphi_0^{*5}\exp(-6j\tau_0) = 0 \quad (2.62)$$

Secular terms in (2.62) are eliminated by imposing the following condition:

$$f_2(\tau_1)\frac{\partial \varphi_0}{\partial \tau_1} + \frac{dC_1(\tau_1)}{d\tau_1} + \frac{51j}{4}|\varphi_0|^4\varphi_0 + 3|\varphi_0|^2 - \frac{3}{2}\varphi_0^2$$

$$- 3j[2|\varphi_0|^2 C_1(\tau_1) + \varphi_0^2 C_1^*(\tau_1)] = 0 \quad (2.63)$$

Now it is possible to demonstrate that the term containing the unknown function $f_2(\tau_1)$ is unavoidable, so it is necessary to be included in the initial multiple scale expansions (2.54). Indeed, if we set $f_2 \equiv 0$ equation (2.63) has the following solution,

$$C_1(\tau_1) = \frac{17}{24}|\varphi_0|^2\varphi_0 - \frac{19j}{72} + \left(D - \frac{17}{6}\int_0^{\tau_1}|\varphi_0(u)|^2 du\right)(3j|\varphi_0|^2\varphi_0 + 1)$$

where D is a real constant of integration. The integral term leads to global divergence of the solution, although at time scales of order higher than $1/\varepsilon$. It should be mentioned that normal averaging procedures guarantee the accuracy at similar time scale, but *the approach developed here enables the extension of the analytical solution to even larger time scales*. In other words, in order to avoid weak secularity of $C_1(\tau_1)$ we need to introduce an additional function $f_2(\tau_1)$ through the definition (2.54).

The first way to compute $f_2(\tau_1)$ is to set the function $C_1(\tau_1)$ equal to zero, and to compensate for the secular terms in (2.63) by appropriate selection of $f_2(\tau_1)$, as follows:

$$f_2 = -\frac{(51j/4)|\varphi_0|^4\varphi_0 + 3|\varphi_0|^2 - (3/2)\varphi_0^2}{3j|\varphi_0|^2\varphi_0 + 1}, \quad C_1 \equiv 0 \qquad (2.64)$$

The approximate solution for this choice of f_2 is computed by combining the previous results (2.59)–(2.63). The dependence of the slow time scale on the original temporal variable is obtained by appropriate inversion of (2.54) with account of the explicit expression (2.64). These expressions may be trivially computed but are not presented here due to their awkwardness.

This way of computing $f_2(\tau_1)$ and $C_1(\tau_1)$ has two shortcomings. First, it is inapplicable in the vicinity of stationary points of equation (2.57) because of divergence of f_2 there. Second, the slow time variable becomes complex, and additional divergence problems may occur in neighborhoods of the poles of the elliptic functions in (2.60). Despite these shortcomings, the previously outlined procedure may be performed at any order of approximation. However, it is possible to derive an analytic approximation free from the above shortcomings. To this end, one can demonstrate that the requirements of non-diverging $C_1(\tau_1)$, and of real and non-diverging $f_2(\tau_1)$ may be satisfied by a *unique* choice of these functions as follows:

$$f_2 = \frac{17}{6}|\varphi_0|^2, \quad C_1 = \frac{17}{24}|\varphi_0|^2\varphi_0 - \frac{19j}{72} \qquad (2.65)$$

Then, the corresponding approximation for the solution is given by

$$\psi = \varphi_0 \exp(jt) + \varepsilon \left[-\frac{1}{2}\varphi_0^3 \exp(3jt) + \frac{3}{2}|\varphi_0|^2\varphi_0^* \exp(-jt) \right.$$
$$\left. - \frac{1}{4}\varphi_0^{*3} \exp(-3jt) + \frac{j}{2}\exp(-jt) + \left(\frac{17}{24}|\varphi_0|^2\varphi_0 - \frac{19j}{72}\right)\exp(jt) \right]$$
$$(2.66)$$

where $\varphi_0 = N(\varepsilon t)\exp[j\delta(\varepsilon t)]$. The shortcoming of this approach is that in order to compute the second-order approximation we have to solve the equation that eliminates secular terms of the equation at the third degree of the small parameter, which is a rather cumbersome task.

We now compare the results obtained by CX-A approach with direct numerical simulations of equation (2.49) for different values of the small parameter and various initial conditions. The numerical parameters used for these simulations are

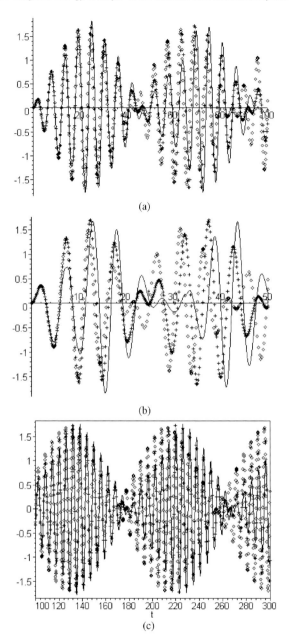

Fig. 2.14 CX-A solution of (2.49) for initial conditions $y(0) = \dot{y}(0) = 0$: (a) $\varepsilon = 0.065$, (b) $\varepsilon = 0.13$, (c) $\varepsilon = 0.03$; exact solution is represented by crosses (+ + +), the analytical approximation based on (2.64) by a solid line (—), and the analytical approximation based on (2.65) by diamonds ($\Diamond\Diamond\Diamond$).

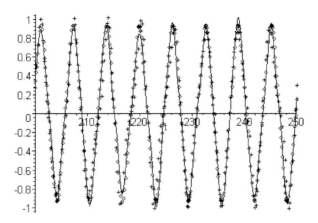

Fig. 2.15 CX-A solution of (2.49) for initial conditions $y(0) = 0.7$, $\dot{y}(0) = 0$ (close to fundamental resonance) and $\varepsilon = 0.5$; exact solution is represented by crosses (+ + +), the analytical approximation based on (2.64) by a solid line (—), and the analytical approximation based on (2.65) by diamonds ($\Diamond\Diamond\Diamond$).

listed in the corresponding figure captions. The results depicted in Figure 2.14 indicate that the analytical approximation including terms up to $O(\varepsilon^2)$ and based on (2.65) provides a better approximation to the solution, compared to the corresponding analytical approximation based on (2.64) with $C_1 \equiv 0$. Besides, the accuracy of the analytical approximation decreases with increasing values of the small parameter ε, at least in the range considered in the simulations.

It should be stressed that large values of ε do not necessarily imply that the derived analytical approximations will be poor. The numerical simulation depicted in Figure 2.15 demonstrates that close to fundamental resonance the analytical solution is close to the exact solution despite the relatively large value of ε used in this particular simulation. Both analytical approximations based on (2.64) and (2.65) provide good approximations to the exact solution, even at relatively large times.

The results presented in Figures 2.16–2.18 provide comparisons of exact solutions with the analytical approximation (2.66) based on conditions (2.65), for various values of ε and initial conditions. We note that the accuracy of the analytical approximation decreases with increasing ε. In general, for these simulations the analytical approximation based on conditions (2.64) provides accuracy comparable to the othe analytical approximations depicted in these figures.

From the analysis of the dynamical system (2.49) we conclude that the CX-A technique, when applied together with a modified multiple scales procedure, provides good analytical approximations for the forced nonlinear response. Moreover, in regions of resonance the CX-A approach provides good approximations even for relatively large values of the small parameter of the problem (i.e., beyond the formal range of applicability of the multiple-scales approach). Two different approaches were proposed for computing the higher-order approximations, both providing rather reliable predictions in their corresponding regions of applicability. It

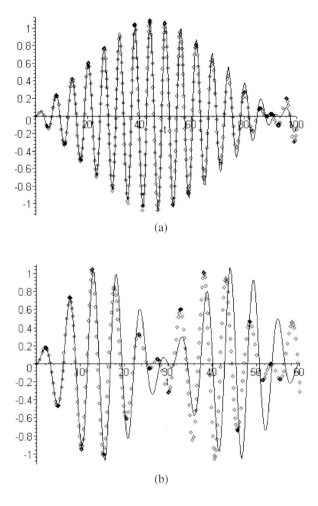

Fig. 2.16 CX-A solution of (2.49) for initial conditions $y(0) = 0$, $\dot{y}(0) = 0$, (a) $\varepsilon = 0.03$, (b) $\varepsilon = 0.1$; exact solution is represented by a solid line (—), and the analytical approximation based on (2.65) by diamonds ($\Diamond\Diamond\Diamond$).

should be mentioned that the dimensionless formal parameter ε in (2.49), commonly regarded as the small parameter in conventional asymptotic analyses of this problem, turns out not to be a 'true' perturbation parameter.

Specifically, the analysis of the previous example demonstrates that the accuracy of the derived asymptotic approximations depends on the relationship between the frequency of the slow modulation φ and the (fast) frequency of the main (fundamental) resonance of the problem; however, it is not yet clear how one could select appropriate perturbation parameters to scale this relationship in the analysis, so this issues remains an open problem.

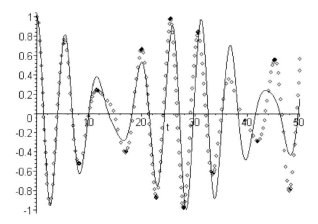

Fig. 2.17 CX-A solution of (2.49) for initial conditions $y(0) = 1$, $\dot{y}(0) = 0$ (close to fundamental resonance) and $\varepsilon = 0.1$; exact solution is represented by solid line (—), and the analytical approximation based on (2.65) by diamonds ($\lozenge\lozenge\lozenge$).

Another possibility for accurate asymptotic expansions using the CX-A technique arises in cases when the initial conditions of the response are in the neighborhood of the stationary point of equation (2.57) (or, in other terms, close to the regime of fundamental resonance of the problem). The small parameter in this case measures the deviation of the response from the stationary point. This case is of major importance in applications of the CX-A technique when systems with strong nonlinearity are studied (where the nonlinear terms are not scaled by a formal small parameter – this is the case in the next example of application of the CX-A technique).

Thus, it is justified to apply the CX-A technique even in dynamical systems that do not formally satisfy the conditions of the averaging theorem (see, for example, Kevorkian and Cole, 1996), but only in response regimes that are either close to exact resonance, or in the domains of attraction of the corresponding resonance manifolds. This observation paves the way for the application of the CX-A technique to TET-related problems, where transient or sustained resonance captures on fundamental or subharmonic resonance manifolds are dominant in the corresponding damped, nonlinear transient resposes.

It should be mentioned that the very presentation of the equations of motion in complex form [i.e., equation (2.53)], and the elimination of non-resonant terms for the modulation equations much resembles the well-known method of normal forms (Guckenheimer and Holmes, 1983; Wiggins, 1990; Nayfeh, 1993; Kahn and Zarmi, 1997). Still, the CX-A technique outlined above is based on different ideas of multiple scales and averaging and seems to lead to essential simplifications of these well-known methods.

In the second example considered in this section we demonstrate the application of the CX-A technique to a strongly nonlinear transient problem, and show that the method is capable of analytically modeling the regime of 1:1 transient resonance

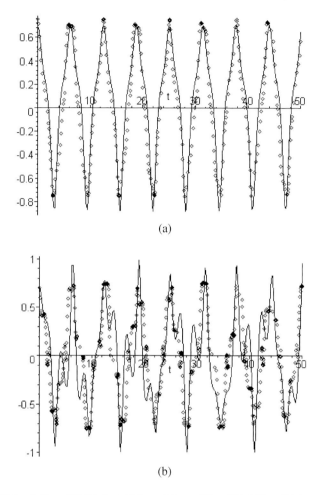

Fig. 2.18 CX-A solution of (2.49) for initial conditions $y(0) = 0.7$, $\dot{y}(0) = 0$ (close to fundamental resonance), (a) $\varepsilon = 1.0$, (b) $\varepsilon = 1.3$; exact solution is represented by a solid line (—), and the analytical approximation based on (2.65) by diamonds ($\Diamond\Diamond\Diamond$).

capture (i.e., of 1:1 transient resonance) in a system of coupled oscillators, in accordance with the previous discussion. To this end, we reconsider the two-DOF system of coupled damped oscillators (2.41) examined in the previous section, and apply the CX-A technique to study the regime of 1:1 TRC (Vakakis and Gendelman, 2001); this response regime was studied in the previous section using an alternative methodology, i.e., by resorting to action-angle transformations and analyzing the corresponding local model in the neighborhood of the 1:1 resonance manifold by the method of multiple scales.

Rewriting the system (2.41) in the form

$$\ddot{y}_1 + \varepsilon\lambda\dot{y}_1 + Cy_1^3 + \varepsilon(y_1 - y_2) = 0$$
$$\ddot{y}_2 + \varepsilon\lambda\dot{y}_2 + \omega^2 y_2 - \varepsilon y_1 = 0 \tag{2.67}$$

where $\omega^2 = \omega_2^2 + \varepsilon$, and introducing the new complex variables,

$$\psi_1 = \dot{y}_1 + j\omega y_1, \quad \psi_2 = \dot{y}_2 + j\omega y_2 \tag{2.68}$$

we express (2.67) as the following set of first-order complex differential equations:

$$\dot{\psi}_1 - \frac{(j\omega + \varepsilon\lambda)}{2}(\psi_1 + \psi_1^*) - \frac{j\varepsilon}{2\omega}(\psi_1 - \psi_1^*)$$
$$+ \frac{jC}{8\omega^3}(\psi_1 + \psi_1^*)^3 + \frac{j\varepsilon}{2\omega}(\psi_2 - \psi_2^*) = 0$$
$$\dot{\psi}_2 - j\omega\psi_2 + \frac{\varepsilon\lambda}{2}(\psi_2 + \psi_2^*) + \frac{j\varepsilon}{2\omega}(\psi_1 - \psi_1^*) = 0 \tag{2.69}$$

The set of equations (2.69) is exact, and, in contrast to the previous example, it represents a strongly nonlinear system since the nonlinear terms are not scaled by a small parameter and the initial conditions are assumed to be $O(1)$ quantities.

We now seek an approximate solution of (2.69) based on the assumption of 1:1 resonance, i.e., by assuming that both oscillators execute slowly-modulated oscillations with identical 'fast' frequencies equal to ω:

$$\psi_1 = \varphi_1 e^{j\omega t}, \quad \psi_2 = \varphi_2 e^{j\omega t} \tag{2.70}$$

In essence, in the regime of 1:1 TRC we partition the dynamics in terms of the 'slow' complex amplitudes φ_i, $i = 1, 2$ modulating the 'fast' oscillatory terms $e^{j\omega t}$. Hence, in contrast to the previous example, and in the absence of a formal small parameter scaling the nonlinear terms, we make the basic assumption that there exists a single fast frequency ω in the dynamics as a means of simplifying the analysis. This is needed in view of the fact that formal application of the method of multiple scales [at least with linear trigonometric generating functions – but see Belhaq and Lakrad (2000) and Lakrad and Belhaq (2002) for extensions of the multiple scales method with Jacobi elliptic functions and Yang et al. (2004) and Chen and Cheung (1996) for extension of averaging and other perturbation schemes based on elliptic generating functions] is not justified in this strongly nonlinear problem.

Substituting (2.70) into (2.69) and averaging out terms that contain fast frequencies higher than ω (such as terms multipled by $e^{2j\omega t}$, $e^{3j\omega t}$, ...) we obtain the following approximate *slow flow* valid in the regime of 1:1 TRC:

$$\dot{\varphi}_1 + \frac{j}{2}\left(\omega - \frac{\omega}{2} + \frac{\varepsilon\lambda}{2}\right)\varphi_1 - \frac{3jC}{8\omega^3}|\varphi_1|^2\varphi_1 + \frac{j\varepsilon}{2\omega}\varphi_2 = 0$$
$$\dot{\varphi}_2 + \frac{\varepsilon\lambda}{2}\varphi_2 + \frac{j\varepsilon}{2\omega}\varphi_1 = 0 \tag{2.71}$$

The fact that (2.71) is an averaged system, among other approximations, poses certain restrictions concerning the time domain of its validity. As mentioned earlier, when first-order averaging is performed in systems in standard form containing a small parameter ε [for example, see relations (2.33) and (2.42)], the validity of the results is only up to times of $O(1/\varepsilon)$. In the CX-A approach described above there is no formal small parameter to describe the slowly-varying character of the complex modulations φ_1 and φ_2, so we cannot provide a formal result regarding its range of validity. In this regard, we can only state that the averaged slow flow (2.71) is valid only up to finite times, as long as the basic assumptions outlined above (regarding the slow-fast partition and the existence of a single 'fast' frequency in the dynamics) are satisfied.

Returning to the analysis of the slow flow (2.71), in order to account for the amplitude decays of the two oscillators due to damping dissipation we introduce the new variables, σ_1 and σ_2 defined by the relations, $\varphi_i = \sigma_i e^{-\varepsilon\lambda t/2}$, $i = 1, 2$, and express the averaged slow flow in the following form:

$$\dot{\sigma}_1 + \frac{j}{2}\left(\omega - \frac{\omega}{2}\right)\sigma_1 - \frac{3jCe^{-\varepsilon\lambda t}}{8\omega^3}|\sigma_1|^2\sigma_1 + \frac{j\varepsilon}{2\omega}\sigma_2 = 0$$

$$\dot{\sigma}_2 + \frac{j\varepsilon}{2\omega}\sigma_1 = 0 \quad (2.72)$$

We now show that the above dynamical system is fully integrable. To this end, we multiply the first of equations (2.72) by the complex conjugate σ_1^*; then we take the complex conjugate of the same first equation and multiply it by σ_1. We perform similar operations on the second of equations (2.72), i.e., we first multiply it by σ_2^* and then multiply its complex conjugate by σ_2. By adding the so derived four complex expressions we show that the averaged system (2.72) possesses the following first integral of motion:

$$\dot{\sigma}_1\sigma_1^* + \dot{\sigma}_1^*\sigma_1 + \dot{\sigma}_2\sigma_2^* + \dot{\sigma}_2^*\sigma_2 = 0 \Rightarrow$$

$$\frac{d(|\sigma_1|^2 + |\sigma_2|^2)}{dt} = 0 \Rightarrow |\sigma_1|^2 + |\sigma_2|^2 = \rho^2 \quad (2.73)$$

This first integral is a conservation-of-energy-like integral of the averaged system when expressed in terms of the σ-variables. This enables us to express the complex amplitudes in the following polar representations:

$$\sigma_1 = \rho \sin\theta e^{j\delta_1}, \quad \sigma_2 = \rho \cos\theta e^{j\delta_2} \quad (2.74)$$

which, when substituted into (2.72) and following certain algebraic manipulations, reduce the isoenergetic averaged dynamics (i.e., for $\rho = $ const) to the following dynamical system on the two-torus $(\delta, \theta) \in T^2$:

$$\dot{\delta} + \frac{\omega}{2} - \frac{3C\rho e^{-\varepsilon\lambda t}}{8\omega^3}\sin^2\theta + \frac{\varepsilon}{\omega}\cot 2\theta \cos\delta = 0$$

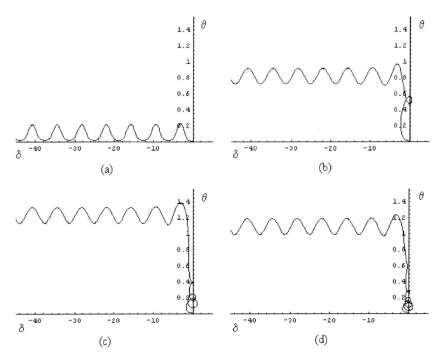

Fig. 2.19 Phase plots of the reduced slow flow (2.75): (a) case of no resonance capture, $\rho = 7.84$; and cases of 1:1 resonance capture, (b) $\rho = 16.0$, (c) $\rho = 100.0$ and (d) $\rho = 225.0$.

$$\dot{\theta} + \frac{\varepsilon}{2\omega} \sin \delta = 0 \qquad (2.75)$$

In (2.75) we introduced the phase difference $\delta = \delta_1 - \delta_2$, which denotes the relative phase between the two oscillators during 1:1 resonance, and the angle θ which determines their corresponding amplitudes ($\theta \approx 0$ denotes localization of the oscillation to the linear oscillator, whereas $\theta \approx \pi/2$ denotes localization to the nonlinear oscillator). Moreover, in the averaged slow flow there occurs a slow 'drift' of the 'instantaneous equilibrium points' of the reduced flow (2.75) due to the previously introduced exponentially decaying coordinate transformation that relates the complex amplitudes φ_i and σ_i. Hence, the present analysis accurately captures the $O(\varepsilon)$ slow 'drift' of the equilibrium points of the slow flow [as discussed in the analysis of system (2.41) in Section 2.3].

The numerical integrations of system (2.75) for varying values of the initial first integral ρ reveal clearly the 1:1 resonance capture in the system. These results are presented in the (δ, θ) phase plots of Figure 2.19 for parameters $\omega = 1.0$, $C = 2.0$, $\varepsilon = 0.1$, $\lambda = 1.0$ and initial conditions $\delta(0) = 0.0$ and $\theta(0) = 0.01$ (i.e., for motion initially localized to the linear oscillator). For $\rho = 7.84$ (see Figure 2.19a) the initial energy localized in the linear oscillator remains confined to that oscillator (indicated by the fact that θ is in the neighborhood of zero for the entire duration of

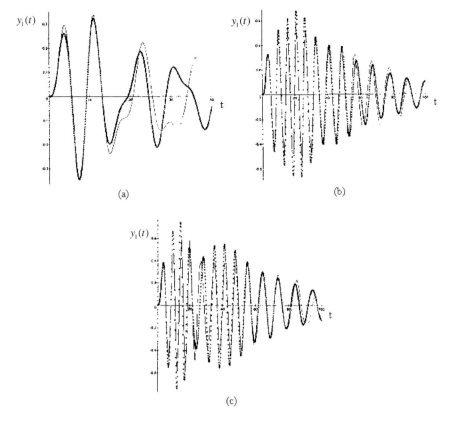

Fig. 2.20 Transient response of the nonlinear oscillator of system (2.67) for, (a) $h = 0.5$ (no resonance capture), (b) $h = 0.8$ and (c) $h = 1.125$ (cases of 1:1 TRC); — exact numerical simulation, ◊◊◊ CX-A analysis.

the motion). At higher values of ρ (see Figures 2.19b–d) we note targeted energy transfer from the linear to the nonlinear oscillator; indeed, orbits that start initially with $\theta \approx 0$, after some transients settle to damped oscillations with $\theta \approx \pi/2$, i.e., localize to the nonlinear oscillator. Of particular interest is the fact that the analytical results capture accurately not only the 1:1 resonance capture of the dynamics, but also the transition to resonance capture as the dynamics is attracted towards the neighborhood of the 1:1 resonance manifold.

To assess the accuracy of the analytical predictions obtained by the CX-A technique, in Figure 2.20 we compare the theoretically predicted response $y_1(t)$ of the nonlinear oscillator through application of the previous CX-A technique, to the corresponding numerical response derived by direct numerical simulation of the original equations of motion (2.67). For these results we used the system parameters $\omega_2^2 = 0.9$, $C = 5.0$, $\varepsilon = 0.1$ and $\lambda = 0.5$, and set all initial conditions to zero except for the initial velocity of the linear oscillator, $\dot{y}_2(0) = \sqrt{2h}$. In Figure 2.20a we depict the low-energy damped response for $h = 0.5$; in this case no resonance capture occurs in the dynamics, and there is poor agreement between the analytical and numerical results. This is justified by the fact that the basic assumption of the CX-A analysis (i.e., that both oscillators possess a single dominant fast frequency nearly equal to ω_2) does not hold in this low-energy regime. In Figures 2.20b, c where 1:1 TRC (and TET) takes place there is satisfactory agreement between the predicted and numerical transient responses, although some overshooting or undershooting can be noted in certain time intervals. These errors can be attributed to the averaging approximations introduced in the CX-A analysis, and to the strong nonlinearities of the system considered. These results demonstrate the potential of the CX-A technique to accurately model strongly nonlinear transient responses under conditions of resonance capture.

In this work the CX-A technique will be applied to various problems involving TRCs in coupled oscillators whose responses possess single or multiple fast frequencies. It will be shown that this method is a valuable analytical tool for studying strongly nonlinear damped responses resulting in single- or multi-frequency TET. Moreover, coupled with advanced signal processing algorithms, the CX-A technique can be also applied to studies of identification of strongly nonlinear modal interactions governing TET in practical applications, such as aeroelastic instability suppression, shock isolation of flexible structures, and passive seismic mitigation. In the next section we provide a brief discussion of some advanced signal processing techniques that will be used throughout this work to analyze strongly nonlinear transient responses related to TET.

2.5 Methods of Advanced Signal Processing

The strongly nonlinear dynamics governing TET require the use of special techniques for their analysis and post-processing. In this work we will make extensive use of advanced signal processing techniques that are especially suited for post-processing non-stationary nonlinear time series. In this section we provide a brief introduction to these techniques. Specifically, one way to carry out the study of strongly nonlinear weakly damped dynamics coupled oscillators considered in this work will be to superimpose the wavelet transform (WT) spectra of the transient responses in frequency-energy plots (FEPs) of the corresponding Hamiltonian dynamics, as discussed in Section 2.1. In performing this procedure we recognize that the effect of weak damping on the transient dynamics is rather parasitic (in the sense that it does not generate 'new dynamics,' but rather acts as perturbation of the underlying Hamiltonian responses), so that the damped transient responses are expected to occur in neighborhoods of periodic (or quasi-periodic) Hamiltonian motions. Once this is recognized, the interpretation of the damped nonlinear dynamics and the full understanding of the associated multi-frequency modal interactions become possible. In addition, analysis of strongly nonlinear damped transitions will be performed by applying Empirical Mode Decomposition (EMD) to the measured time series,

Hilbert-transforming the resulting Intrinsic Model Functions (IMFs) and then comparing the results to Wavelet Transform spectra. We will show that this process can help us identify and classify the different strongly nonlinear transitions that take place in multi-frequency transient data of the type considered in TET applications.

Since these methodologies will be applied throughout this work, in what follows we give a brief exposition of their basic elements and provide some preliminary examples of their applications to nonlinear time series analysis.

2.5.1 Numerical Wavelet Transforms

The WT can be viewed as a basis for functional representation, but is at the same time a relevant technique for time-frequency analysis. In contrast to the Fast Fourier Transform (FFT) which assumes signal stationarity, the WT involves a windowing technique with variable-sized regions. Small time intervals are considered for high frequency components, whereas the size of the interval is increased for lower frequency components thereby providing better time and frequency resolutions than the corresponding FFTs.

Hence, the Wavelet Transform (WT) can be viewed as the 'dynamic' extension of the 'static' Fourier Transform (FT), in the sense that instead of decomposing a time series (signal) in the frequency domain using the cosine and sine trigonometric functions (as in the FT), in the WT alternative families of orthogonal functions are employed which are localized in frequency and time. These families of orthogonal functions, the so-called *wavelets* can be adapted in time and frequency to provide details of the frequency components of the signal during the time interval analyzed. These wavelets result from a *mother wavelet* function through successive iterations. As a result, the WT provides the transient evolution of the main frequency components of the time series, in contrast to the FT that provides a 'static' description of the frequency of the signal.

In this work, the results of applying the numerical WT are presented in terms of *WT spectra*. These contour plots depict the amplitude of the WT as a function of frequency (vertical axis) and time (horizontal axis). Heavy shaded regions correspond to regions where the amplitude of the WT is high, whereas lightly shaded ones correspond to low amplitudes. Such plots enable one to deduce the temporal evolutions of the dominant frequency components of the signals analyzed. The Matlab® program used for the WT computations reported in this work was developed at the University of Liège by Dr. V. Lenaerts in collaboration with Dr. P. Argoul from the Ecole Nationale des Ponts et Chaussées (Paris, France). Two types of mother wavelets $\psi_M(t)$ are considered: (a) The Morlet wavelet which is a Gaussian-windowed complex sinusoid of frequency ω_0, $\psi_M(t) = e^{-t^2/2} e^{j\omega_0 t}$; and (b) the Cauchy wavelet of order n, $\psi_M(t) = [j/(t+j)]^{n+1}$, where $j = (-1)^{1/2}$. The frequency ω_0 for the Morlet WT and the order n for the Cauchy WT are user-specified parameters which allow one to tune the frequency and time resolutions of the results. It should be noted that these two mother wavelets provide similar results when applied to the signals considered in the present work. In recent works by Argoul and co-workers (Argoul and Le, 2003; Le and Argoul, 2004; Yin et al., 2004; Erlicher and Argoul, 2007), the continuous Cauchy Wavelet transform was applied to system identification of linear dynamical systems.

We demonstrate the application of the numerical WT by an example taken from the dissertation thesis by Tsakirtzis (2006). Specifically, we consider a two-DOF linear system weakly coupled to a three-DOF attachment composed of strongly nonlinear coupled oscillators (this system will be studied in detail in Chapter 4, Section 4.1.2, where TET from linear systems to strongly nonlinear MDOF attachments will be analyzed):

$$\ddot{u}_1 + (\omega_0^2 + \alpha)u_1 - \alpha u_2 + \varepsilon\lambda\dot{u}_1 = F_1(t)$$

$$\ddot{u}_2 + (\omega_0^2 + \alpha + \varepsilon)u_2 - \alpha u_1 - \varepsilon v_1 + \varepsilon\lambda\dot{u}_2 = F_2(t)$$

$$\mu\ddot{v}_1 + C_1(v_1 - v_2)^3 + \varepsilon(v_1 - u_2) + \varepsilon\lambda(\dot{v}_1 - \dot{v}_2) = 0$$

$$\mu\ddot{v}_2 + C_1(v_2 - v_1)^3 + C_2(v_2 - v_3)^3 + \varepsilon\lambda(2\dot{v}_2 - \dot{v}_1 - \dot{v}_3) = 0$$

$$\mu\ddot{v}_3 + C_2(v_3 - v_2)^3 + \varepsilon\lambda(\dot{v}_3 - \dot{v}_2) = 0 \qquad (2.76)$$

In Figures 2.21–2.23 we present the WT spectra of the relative responses $v_2 - v_1$ and $v_3 - v_2$ of the strongly nonlinear attachment for parameters $\varepsilon = 0.2$, $\alpha = 1.0$, $C_1 = 4.0$, $C_2 = 0.05$, $\varepsilon\lambda = 0.01$, $\mu = 0.08$, and $\omega_0^2 = 1.0$; in the simulations out-of-phase impulsive excitations are considered, $F_1(t) = -F_2(t) = Y\delta(t)$ with zero initial conditions.

First, we consider the WT spectra of the weakly forced responses depicted in Figure 2.21. In this case there occurs strong targeted energy transfer (TET) from the directly forced linear system to the nonlinear attachement (amounting to nearly 90% of input energy transferred and dissipated by the attachment). Examination of the WT spectra reveals certain interesting features of the dynamics. Indeed, we note that there occurs a transient resonance capture (TRC) of the dynamics of the relative response $v_1 - v_2$ by a strongly nonlinear mode whose frequency varies in time and lies in between the two natural frequencies of the uncoupled and undamped linear system; that this is a strongly nonlinear mode is signified by the fact that it does not lie close to either one of the linear natural frequencies of the system, which implies that this mode localizes predominantly to the nonlinear attachment. The strong nonlinearity of the response is further signified by the occurrence of an initial multi-frequency beat oscillation (subharmonic or quasi-periodic), as evidenced by the existence of an initial high frequency component in the spectrum of $v_1 - v_2$. In addition, the second nonlinear stiffness-damper pair of the attachment (corresponding to the relative response $v_2 - v_3$) absorbs (and dissipates) broadband energy from the both modes of the linear system; this is evidenced by the fact that the corresponding WT spectrum of Figure 2.21b possesses a broad range of frequency components that includes both natural frequencies of the linear system.

These results indicate that *strong TET in this case is associated with TRCs of the dynamics of the nonlinear attachment by strongly nonlinear modes that predom-*

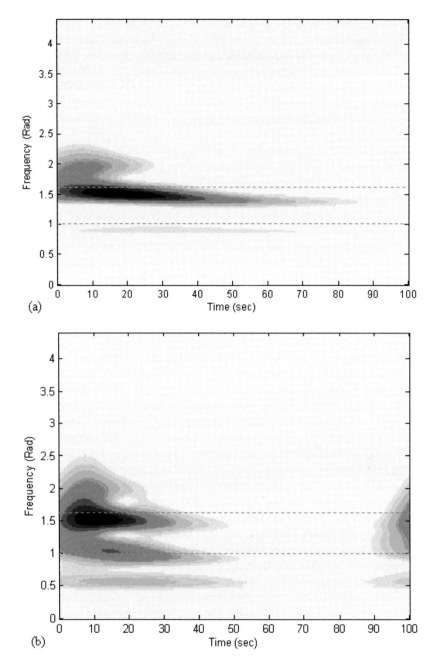

Fig. 2.21 WT spectra of the relative responses, (a) $v_1 - v_2$, and (b) $v_2 - v_3$ of system (2.76) for out-of-phase impulse excitation of magnitude $Y = 0.1$; the linear natural frequencies of the uncoupled and undamped linear system ($\varepsilon = 0$) are indicated by dashed lines.

inantly localize to the attachment; moreover these TRCs take place over a broad frequency range, resulting in broadband TET. Hence, it becomes clear that the numerical WT spectra provide important information not only regarding the frequency contents of the nonlinear responses, but also on the temporal evolution of each individual frequency component as the interaction between the linear and nonlinear subsystems progresses in time. This underlines the usefulness of the WT as a tool to analyze essentially nonlinear dynamical interactions of the type considered in this work.

By increasing the magnitude of the impulse to $Y = 1.0$, there occurs a marked deterioration of TET from the linear system to the nonlinear attachment. In Figures 2.22a, b we depict the corresponding WT spectra of the relative responses of the nonlinear attachment in this case, which reveal the reason for poor TET. Indeed, the dynamics of the nonlinear attachment appears to engage in sustained resonance capture (SRC) predominantly with two weakly nonlinear modes lying in the corresponding neighborhoods of the in-phase and out-of-phase modes of the unforced and undamped linear system. Moreover, the fact that the weakly nonlinear in-phase and out-of-phase modes localize predominantly to the linear system, prevents significant localization of the vibration to the NES, a feature that contributes to weaker TET. We conclude that *weak TET in this case is associated with SRC of the dynamics with weakly nonlinear modes that are predominantly localized to the linear subsystem*.

Finally, in Figures 2.23a, b we depict the corresponding WT spectra for the system with stong out-of-phase excitation $Y = 1.5$. Similarly to the case depicted in Figures 2.21a, b, we note the occurrence of a strong TRC of the dynamics on a strongly nonlinear mode localized predominantly to the nonlinear attachment; this TRC leads to strong TET from the linear system to the attachment. Comparing the WT spectra of Figures 2.23a, b to those of the case of weak TET (depicted in Figures 2.22a, b), we note that in the later case the transient responses are dominated by sustained frequency components (i.e., by SRCs), indicating excitation of weakly nonlinear modes which are mere analytic continuations of linearized modes of the system. On the contrary, in cases where strong TET occurs, the frequencies of the nonlinear modes involved in the TRCs are not close to linearized natural frequencies, indicating that these are strongly nonlinear modes having no linear analogs; as a result, these modes localize predominantly to the NES.

A general conclusion drawn from the examination of these WT spectra is that *the TET efficiency of system (2.76) may be explained by the examination of the resonance captures depicted in the WTs of the transient responses*. Indeed, strong TET in the system is associated with TRCs of the dynamics with essentially (strongly) nonlinear modes localized predominantly to the nonlinear attachment; whereas weak TET involves SRCs, i.e., sustained excitation of weakly nonlinear modes (i.e., modes that are analytic continuations of linearized modes of the system) localized predominantly to the linear system.

This application demonstrates clearly the potential of the numerical WT as a tool for analyzing and interpreting strongly nonlinear transient dynamics in terms of transient or sustained resonance captures. Moreover, when combined with *Em-*

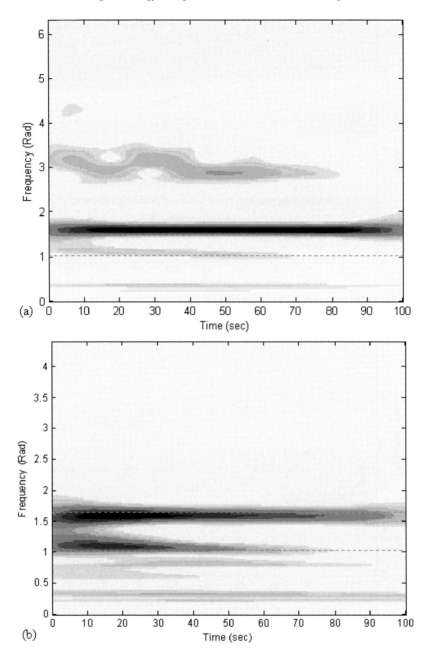

Fig. 2.22 WT spectra of the relative responses, (a) $v_1 - v_2$, and (b) $v_2 - v_3$ of system (2.76) for out-of-phase impulse excitation of magnitude $Y = 1.0$; the linear natural frequencies of the uncoupled and undamped linear system ($\varepsilon = 0$) are indicated by dashed lines.

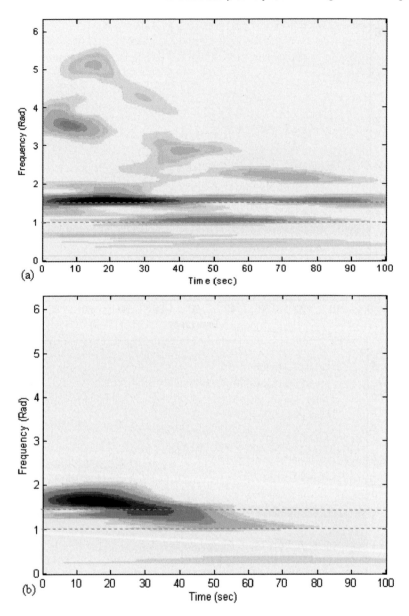

Fig. 2.23 WT spectra of the relative responses, (a)$v_1 - v_2$, and (b) $v_2 - v_3$ of system (2.76) for out-of-phase impulse excitation of magnitude $Y = 1.5$; the linear natural frequencies of the uncoupled and undamped linear system ($\varepsilon = 0$) are indicated by dashed lines.

pirical Mode Decomposition and the Hilbert transform it can form the basis of an integrated nonlinear approach for identifying the transient dynamics as well as the modal interactions that occur in the dynamics of systems with strongly nonlinear substructures.

2.5.2 Empirical Mode Decompositions and Hilbert Transforms

The *Empirical Mode Decomposition (EMD)* is a technique for decomposing a signal in terms of intrinsic oscillatory modes that are termed *intrinsic mode functions (IMFs)*. The IMFs satisfy the following three main conditions, which are imposed in *an ad* hoc fashion: (a) For the duration of the entire time series, the number of extrema and of zero crossings of each IMF should either be equal or differ at most by one; (b) at any given time instant, the mean value (moving average) of the local envelopes of the IMFs defined by their local maxima and minima should be zero; and (c) the linear superposition of all IMFs should reconstruct the original time series.

The EMD algorithm for computing the intrinsic mode functions (IMFs) of a signal (time series), say $x(t)$, is called *sifting process* and involves the following steps (Huang et al., 1998a, 1998b, 2003):

(a) Consider separately the envelopes defined by the local maxima and minima of $x(t)$, and interpolate the locus of all local maxima of $x(t)$ through a spline approximation, thus constructing an upper envelope of the signal $e^1_{max}(t)$; similarly interpolate the locus of all local minima of $x(t)$ thus creating a lower envelope of the signal, $e^1_{min}(t)$.
(b) Compute the moving average $R_1(t)$ between the lower and the upper envelopes, and define the modified, zero-mean signal $h_1(t) = x(t) - R_1(t)$.
(c) Repeat this procedure k times starting from $h_1(t)$ until the signal computed at the k-th iteration, say $h_{1k}(t) \equiv c_1(t)$, satisfies the properties of an IMF therefore one stop criterion must be applied. The stop criteria of the repeatable procedure can be various; one of them is being applied in each case. In our applications we use either the standard deviation between the $(k-1)$-th and k-th steps or the number of successive repetitions of the sifting process. This process yields the first IMF of the signal $x(t)$, namely, $c_1(t)$.
(d) The second-order remainder of the signal, $x_2(t)$, is defined by the relation $x_2(t) = x(t) - c_1(t)$, on which the previous procedure is repeated to extract the second IMF, $c_2(t)$.
(e) The outlined procedure is repeated until the n-th order remainder, $x_n(t)$, becomes a monotonic function of time.

As discussed above, one can employ alternative convergence criteria for completing the outlined iterative algorithm; two of them are extracted directly from the afore-mentioned properties of the IMFs. The first convergence criterion determines convergence when the following standard deviation between the $(k-1)$-th and k-th steps,

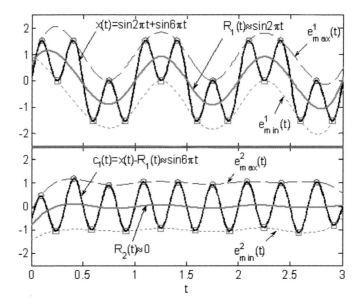

Fig. 2.24 Schematic presentation of application of the empirical mode decomposition to the signal $x(t) = \sin \omega_0 t + \sin 3\omega_0 t$ where $\omega_0 = 2\pi$.

$$SD = \sum_{t=0}^{T} \left[\frac{|h_{1(k-1)}(t) - h_{1(k)}(t)|^2}{h_{1(k-1)}^2(t)} \right]$$

is reduced below a preset tolerance, and T is the signal duration; in this work this tolerance was chosen in the range [0.2, 0.3]. Practically, this criterion implies that the k-th iteration $h_{1k}(t) \equiv c_1(t)$ is approximately (within the specified tolerance) zero-mean. A second convergence criterion consistent with the properties of the IMFs, is to determine convergence by computing the successive repetitions of the sifting process, and determining if the number of zero crossings and the number of extrema are equal or differ by one for S repetitions; in this work S was chosen to be equal to either 2 or 3. In this study, we utilize Matlab® codes developed by Rilling et al. (2003) to perform numerical EMD.

Figure 2.24 depicts schematically the extraction of IMFs from the signal $x(t) = \sin 2\pi t + \sin 6\pi t$. Since there is no control of the sifting process, end effects appear in the results. Following the previous notation the two IMFs of this signal are computed as $c_1(t) \approx \sin 6\pi t$ (i.e., the high-frequency component is extracted first), and $c_2(t) = x(t) - c_1(t) \approx \sin 2\pi t$.

By the construction algorithm outlined above, *the lowest-order IMFs contain the oscillatory components (IMFs) of the signal with the highest frequency components.* As the order of the IMFs increases, their corresponding frequency contents decrease accordingly. Hence, EMD analysis extracts oscillating modulations or modes imbedded in the data, which could be regarded as the 'oscillatory building blocks' of the signal. It follows that *the essence of the EMD method is to empirically identify the intrinsic oscillatory modes in the data (time series), and to categorize them in terms of their characteristic time scales, by considering the successive extreme values of the signal.* Hence, the result of the analysis is a multi-scale separation of the time series in terms of its oscillating components, with the different time scales being extracted automatically by the algorithm itself. As discussed below, the EMD algorithm, when combined with the Hilbert transform can provide further insightful information on the decomposition of the signal.

After applying the EMD analysis to the time series, the extracted IMFs are Hilbert-transformed in order to compute their approximate transient amplitudes and phases. The Hilbert transform $H[c(t)] \equiv \hat{c}(t)$ of a signal (time series) $c(t)$ is defined as follows:

$$\hat{c}(t) = \left(\frac{1}{\pi}\right) \int_{-\infty}^{+\infty} \frac{c(\tau)}{t-\tau} d\tau \equiv \frac{1}{\pi t} * c(t) \qquad (2.77)$$

where (*) denotes the convolution operator:

$$f(t) * g(t) = \int_{-\infty}^{t} f(\tau)g(t-\tau)d\tau = \int_{-\infty}^{t} f(t-\tau)g(\tau)d\tau$$

Hence, the Hilbert transform does not change the domain of the signal, as it transforms the signal from the time domain to the time domain.

In the context of the following analysis, the Hilbert transform of the signal $c(t)$ can be regarded as the 'imaginary' part of the signal, enabling one to perform a complexification of that signal. Indeed, defining the complexified analytical signal

$$\psi(t) = c(t) + j\hat{c}(t) \qquad (2.78)$$

where $j = (-1)^{1/2}$, we compute its amplitude $A(t)$ and phase $\varphi(t)$ by expressing the complexification in polar form:

$$\psi(t) = A(t)e^{j\varphi(t)} = A(t)\cos\varphi(t) + jA(t)\sin\varphi(t) \qquad (2.79)$$

It follows that the signal can be represented in the form

$$c(t) = A(t)\cos\varphi(t) \qquad (2.80)$$

with amplitude and phase given by

$$A(t) = \sqrt{c(t)^2 + \hat{c}(t)^2}, \quad \varphi(t) = \tan^{-1}\left[\frac{\hat{c}(t)}{c(t)}\right] \qquad (2.81)$$

These decompositions enable one to compute the *instantaneous frequency of the signal* $c(t)$ according to the following definition:

$$f(t) = \frac{\dot{\varphi}(t)}{2\pi} = \frac{c(t)\dot{\hat{c}}(t) - \hat{c}(t)\dot{c}(t)}{2\pi[c(t)^2 + \hat{c}(t)^2]} \qquad (2.82)$$

Therefore, by applying the Hilbert transform to each IMF component resulting from EMD of a signal, we can determine the variation of the instantaneous frequency of each IMF; this, in turn, enables us to get valuable insight into the dominant frequency components that are contained in each IMF and to study resonant modal interactions between IMFs of responses of different components of a system.

It is precisely these results that make the combined EMD-Hilbert transform useful for the TET problems considered in this work. Indeed, the decomposition of the transient responses of different components of a system in terms of their oscillatory components (IMFs), and the subsequent computation of the instantaneous frequencies of these IMFs, provides a useful tool for studying nonlinear resonant interactions between these components. To this end, we say that a (k:m) *transient resonance capture* (TRC) occurs between two IMFs $c_1(t)$ and $c_2(t)$ with phases $\varphi_1(t)$ and $\varphi_2(t)$, respectively, whenever their instantaneous frequencies satisfy the following approximate relation,

$$k\varphi_1(t) - m\varphi_2(t) \approx \text{const} \Rightarrow k\dot{\varphi}_1(t) \approx m\dot{\varphi}_2(t), \quad t \in [T_1, T_2] \quad (2.83)$$

The time interval $[T_1, T_2]$ defines the duration of the TRC between the two IMFs.

A more complete picture for the TRC between two IMFs can be gained by constructing appropriate phase plots of the dynamics of the phase difference $\Delta\varphi_{12}(t) = \varphi_1(t) - \varphi_2(t)$. More specifically, a resonance capture is signified by the existence of a loop in the phase plot of $\Delta\varphi_{12}(t)$ when plotted against $\Delta\dot{\varphi}_{12}(t)$, whereas absence of (or escape from) TRC is signified by time-like (that is, monotonically varying) behavior of $\Delta\varphi_{12}(t)$ and $\Delta\dot{\varphi}_{12}(t)$. In addition, the ratio of instantaneous frequencies of the IMFs, $\dot{\varphi}_1(t)/\dot{\varphi}_2(t)$, provides an estimate of the order of the resonance capture.

Ending this brief exposition we mention that the dominant (see discussion below) IMFs of a signal have usually a physical interpretation as far as their characteristic scales are concerned; indeed, certain IMFs may possess instantaneous frequencies that are nearly identical to resonance frequencies of components of the system examined, but this need not always be the case. This implies that certain IMFs may represent artificial (non-physical) oscillating modes of the data. As shown in Kerschen et al. (2006, 2008b), the leading-order (dominant) IMFs coincide with the responses of the slow flow generated by the set of modulation equations of the system; this interesting observation, paves the way for a physics-based interpretation of the IMFs, in terms of the slow flow dynamics (which represent the 'essential' dynamics of the system).

EMD when combined with the WT enables one to determine the *dominant IMFs* of a nonlinear time series. This is achieved by superimposing the plots of instantaneous frequencies of the IMFs to the corresponding WT spectra of the time series. The instantaneous frequencies of the dominant IMFs should coincide with the main (dominant) harmonic components of the corresponding WT spectra in the corresponding time windows of the response. It follows that by combining EMD and the WT one is able to determine the main dominant oscillating components in a measured time series and, hence, to perform order reduction and low-order modeling of measured transient signals. In this work EMDs and numerical WTs are implemented in Matlab®. Focusing in the specific applications examined in this work, *this integrated approach provides the characteristic time scales of the dominant nonlinear dynamics and the modal interactions occurring between components of a system.* Moreover, by adopting this analysis one can identify and analyze the most important nonlinear resonance interactions that are responsible for nonlinear energy exchanges and TET between these components.

2.6 Perspectives on Hardware Development and Experiments

We conclude this chapter by discussing certain issues related to the experimental validation of the theoretical results related to TET derived in this work. Experimental studies of TET will be performed by considering SDOF nonlinear oscillators attached to SDOF or MDOF linear systems. As discussed in the theoretical derivations of Chapter 3, important prerequisites for the realization of passive TET in these systems is that the nonlinear attachments possess *essential (nonlinearizable) stiffness nonlinearities*, and that there exists *weak damping dissipation* in the integrated linear system – nonlinear attachment configuration. The later is easily implementable, since to a certain extent all practical experimental fixtures possess some degree of damping (inherent damping, or damping added at joints or supports); so the main concern in the experiments is with regard to the accurate measurement and estimation of damping in the exterimental fixtures. The former requirement of essential stiffness nonlinearity, however, is more difficult to implement, so in the experimental work special care was paid towards the design and practical implementation of essentially nonlinear stiffness elements and the accurate measurement of their stiffness characteristics.

Passive stiffness nonlinearity in practical settings can be implemented by taking advantage of geometric nonlinearity realized during oscillations of elastic elements. Following this approach, recent works employed different linear spring combinantions to develop geometrically nonlinear stiffness designs. Virgin et al. (2007) considered absorbers with geometrically nonlinear stiffnesses and studied their vibration isolation capacities. Carella et al. (2007a, 2007b) considered vertical linear springs acting in parallel with oblique linear springs, and showed that this configuration could be designed to possess zero dynamic stiffness at their static equilibrium positions. DeSalvo (2007) combined horizontal and vertical linear springs in an arrangement yielding a geometrically nonlinear overall stiffness characteristic, and applied this design to the problem of passive seismic mitigation. Lee et al. (2007) designed spring mechanisms with 'negative stiffness in the large' and applied them to vehicle suspension designs; their approach was based on the large-amplitude post-buckling behavior of elastic 'springing' (thin shell) elements.

In our approach essential (nonlinearizable) stiffness nonlinearity of the third degree was realized experimentally by adopting the simple configuration of Figure 2.25. A thin rod (piano wire) *with no pretension* was clamped at both ends, and was restricted to perform transverse vibrations at its center. Assuming that the

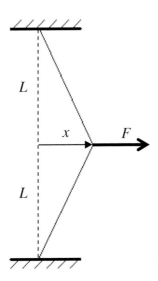

Fig. 2.25 Realization of essential stiffness nonlinearity of the third degree.

wire is composed of linearly elastic material, a static force F will cause a transverse displacement x, which from geometry can be expressed as:

$$F = kx[1 - L(L^2 + x^2)^{-1/2}] \tag{2.84}$$

The stiffness characteristic $k = 2EA/L$ represents the stiffness constant of the wire in axial displacement, E and A are the modulus of elasticity and cross sectional area of the wire, respectively, and L the half-length of the wire. The nonlinear force-displacement relationship (2.84) is a consequence of the geometric nonlinearity of this system, eventhough the wire itself is linearly elastic.

For small displacements x we Taylor-expand the expression in the bracket of (2.84) about $x = 0$, yielding

$$(L^2 + x^2)^{-1/2} = \frac{1}{L} - \frac{x^2}{2L^3} + \frac{3x^4}{8L^5} + O(x^6) \tag{2.85}$$

so that the force displacement relation (2.84) is approximated as follows:

$$F = \frac{k}{2L^2}x^3 + O(x^5) = \frac{EA}{L^3}x^3 + O(x^5) \tag{2.86}$$

Hence, the geometric nonlinearity of the system considered produces, to the leading order of approximation, a cubic stiffness nonlinearity with coefficient $C = EA/L^3$. Moreover, the corrective terms for increasing displacement are of higher order in x, and do not add a linear term in the stiffness characteristic (2.86). If, however, the thin wire is preloaded, a highly undesirable linear term, proportional to the initial

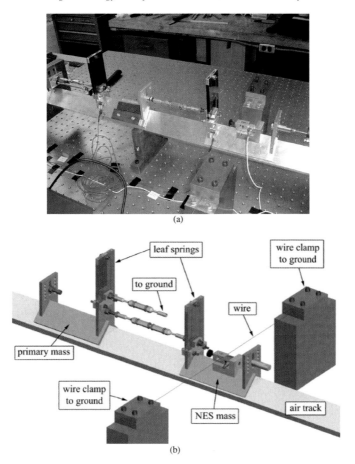

Fig. 2.26 Experimental realization of Configuration I of nonlinear attachment (grounded attachment with essential cubic stiffness nonlinearity): (a) experimental fixture, (b) schematic describing the various components of the fixture.

preload tension, appears, and the resulting stiffness becomes linearizable. Hence, special care in the experimental setups was given to minimize pretension in the wire; in practical realizations of (2.86) a small linear term (due to unavoidable small pretension) always appears, however, this does not affect the TET results.

In the experiments three different configurations of essentially nonlinear attachments were considered. The first configuration (labeled Configuration I) consists of a grounded, essentially nonlinear attachment (termed *nonlinear energy sink – NES*, see Chapter 3), and its practical implementation is depicted in the experimental fixture of Figure 2.26. The fixture consists of two single-degree-of-freedom oscillators connected by means of a linear coupling stiffness. The left oscillator (the linear system) is grounded by means of a linear spring, whereas the right one (the NES) is grounded by means of a nonlinear spring with essential cubic nonlinearity (the

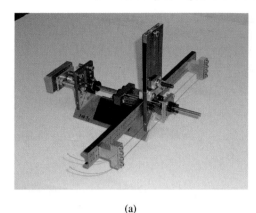

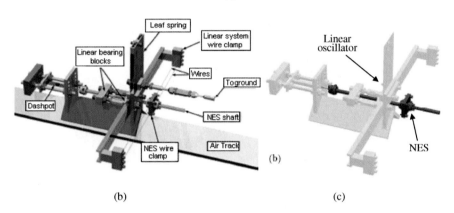

Fig. 2.27 Experimental realization of Configuration II of nonlinear attachment (ungrounded attachment with essential cubic stiffness nonlinearity): (a) experimental fixture, (b) schematic describing the various components of the fixture, (c) schematic indicating the NES portioned from the linear oscillator.

clamped wire design presented in Figure 2.25); an additional viscous damper exists in the NES.

The second configuration of essentially nonlinear attachment (NES) (labeled Configuration II) consists of an ungrounded nonlinear attachment, that is coupled to the linear system through an essential stiffness element. In Figure 2.27 we depict this Configuration. The advantage of this design compared to Configuration I is its versatility, since it can be connected to ungrounded structures (such as moving ones); moreover, it will be shown that even lightweight ungrounded NESs can be effective passive absorbers and local energy dissipators, making them primary candidates for realizing TET in practical applications. Experimental results with fixtures implementing Configurations I and II will be reported in Chapters 3 and 8 of this work (for example, an experimental fixture depicting an ungrounded NES

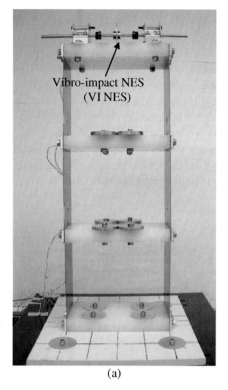

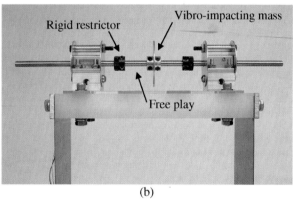

Fig. 2.28 Experimental realization of a vibro-impact attachment: (a) experimental fixture, (b) detail of VI NES.

configuration attached to a two-DOF linear system of coupled oscillators is depicted in Figure 3.96).

A third experimental configuration with a vibro-impact attachment will be considered in our study of passive seismic mitigation by means of TET. The vibro-impact configuration is depicted in Figure 2.28. In this design, the essential stiffness nonlinearity of the attachment is realized by vibro-impacts, which, as argued in Chapter 7, can be viewed as a limiting case of a family of 'smooth' essentially nonlinear stiffnesses; in that context, the vibro-impact nonlinearity can be regarded as the 'strongest possible' stiffness nonlinearity of this family of essentially nonlinear stiffnesses. In the experimental fixture considered in this work, the vibro-impact nonlinearity of the attachment is realized by imposing rigid restrictors to the free motion of the mass of the attachment (see Figure 2.28b). We will demonstrate that, apart from their relative simplicity, properly designed vibro-impact attachments can act as strong passive absorbers and energy dissipators of broadband vibrations from the structures to which they are attached. Vibro-impact TET will concern us in Chapters 7 and 10.

References

Anderson, P.W., Absence of diffusion in certain random lattices, *Phys. Rev.* **109**, 1492–1505, 1958.
Andrianov, I.V., Asymptotic construction of nonlinear normal modes for continuous systems, *Nonl. Dyn.* **51**, 99–109, 2008.
Andrianov, I.V., Awrejcewicz, J., Barantsev, R.G., Asymptotic approaches in mechanics: New parameters and procedures, *Appl. Mech. Rev.* **56**(1), 87–110, 2003.
Argoul, P., Le, T.P., Instantaneous indicators of structural behavior based on the continuous Cauchy wavelet analysis, *Mech. Syst. Signal Proces.* **17**(1), 243–250, 2003.
Arnold, V.I. (Ed.), *Dynamical Systems III*, Encyclopaedia of Mathematical Sciences, Vol. 3, Springer-Verlag, Berlin, 1988.
Aubrecht, J., Vakakis, A.F., Localized and non-localized nonlinear normal modes in a multi-span beam with geometric nonlinearities, *J. Vib. Acoust.* **118**(4), 533–542, 1996.
Aubrecht, J., Vakakis, A.F., Tsao, T.C., Bentsman, J., Experimental study of nonlinear transient motion confinement in a system of coupled beams, *J. Sound Vib.* **195**(4), 629–648, 1996.
Bakhtin, V.I., Averaging in multi-frequency systems, *Funct. An. Appl.* **20** 83–88 (English translation from *Funkts. Anal. Prilozh.* **20**, 1–7, 1986 [in Russian]).
Belhaq, M., Lakrad, F., The elliptic multiple scales method for a class of autonomous strongly nonlinear oscillators, *J. Sound Vib.* **234**(3), 547–553, 2000.
Bellizzi, S., Bouc, R., A new formulation for the existence and calculation of nonlinear normal modes, *J. Sound Vib.* **287**, 545–569, 2005.
Belokonov, V., Zabolotnov, M., Estimation of the probability of capture into a resonance mode of motion for a spacecraft during its descent in the atmosphere. *Cosmic Res.* **40**, 467–478, 2002 (translated from *Kosmicheskie Issledovaniya* **40**, 503–514, 2002).
Boivin, N., Pierre, C., Shaw, S.W., Nonlinear modal analysis of structural systems featuring internal resonances, *J. Sound Vib.* **182**, 336–341, 1995.
Bosley, D., Kevorkian, J., Adiabatic invariance and transient resonance in very slowly varying oscillatory Hamiltonian systems, *SIAM J. Appl. Math.* **52**, 494–527, 1992.
Burns, T., Jones, C., A mechanism for capture into resonance, *Physica D* **69**, 85–106, 1993.
Byrd, P.F., Friedman, M.D., *Handbook of Elliptic Integrals for Engineers and Physicists*, Springer-Verlag, Berlin/New York, 1954.
Carrella, A., Brennan, M.J., Waters, T.P., Static analysis of a passive vibration isolator with quasi-zero-stiffness characteristic, *J. Sound Vib.* **301**, 678–689, 2007a.
Carrella, A., Brennan, M.J., Waters, T.P., Optimization of a quasi-zero-stiffness isolator, *J. Mech. Sc. Tech.* **21**(6), 946–949, 2007b.

Chen, S.H., Cheung, Y.K., An elliptic perturbation method for certain strongly nonlinear oscillators, *J. Sound Vib.* **192**(2), 453–464, 1996.

Courant, R., Hilbert, D., *Methods of Mathematical Physics, I and II*, Wiley Interscience, New York, 1989.

Dermott, S.F., Murray, C.D., Nature of the Kirkwood gaps in the asteroid belt, *Nature* **301**, 201–205, 1983.

DeSalvo, R., Passive, nonlinear, mechanical structures for seismic attenuation, *J. Comp. Nonl. Dyn.* **2**, 290–298, 2007.

Dodson, M.M., Rynne, B.P., Vickers, J.A.G., Averaging in multi-frequency systems, *Nonlinearity* **2**, 137–148, 1989.

Emaci, E., Nayfeh, T.A., Vakakis, A.F., Numerical and experimental study of nonlinear localization in a flexible structure with vibro-impacts, *ZAMM* **77**(7), 527–541, 1997.

Erlicher, S., Argoul, P., Modal identification of linear non-proportionally damped systems by wavelet transforms, *Mech. Syst. Signal Proces.* **21**(3), 1386–1421, 2007.

Fenichel, N., Persistence and smoothness of invariant manifolds for flows, *Indiana Univ. Math. J.* **21**, 193–225, 1971.

Gendelman, O.V., Manevitch, L.I., Method of complex amplitudes: Harmonically excited Oscillator with Strong Cubic Nonlinearity, in *Proceedings of the ASME DETC03 Design Engineering Technical Conferences and Computers and Information in Engineering Conference*, Chicago, Illinois, September 2–6, 2003.

Gildenburg, V.B., Semenov, V.E., Vvedenskii, N.V., Self-similar sharpening structures and traveling resonance fronts in nonlocal HF ionization processes, *Physica D* **152–153**, 714–722, 2001.

Guckenheimer, J., Holmes, P., *Nonlinear Oscillations, Dynamical System, and Bifurcation of Vector Fields*, Springer-Verlag, New York, 1983.

Guevara, M.R., Glass, L., Shrier, A., Phase locking, period-doubling bifurcations, and irregular dynamics in periodically stimulated cardiac cells, *Science* **214**(4527), 1350–1353, 1981.

Hodges, C.H., Confinement of vibration by structural irregularity, *J. Sound Vib.* **82**(3), 411–424, 1982.

Huang, N.E., Shen, Z., Long, S.R., Wu, M.C., Shih, H.H., Zheng, Q., Yen, N.C., Tung, C.C., Liu, H.H., The empirical mode decomposition and the Hilbert spectrum for nonlinear and nonstationary time series analysis, *Proc. Royal Soc. London, Ser. A* **454**, 903–995, 1998a.

Huang, W., Shen, Z., Huang, N.E., Fung, Y.C., Engineering analysis of biological variables: an example of blood pressure over 1 day, *Proc. Nat. Acad. Sci.* **95**, 4816–4821, 1998b.

Huang, N.E., Wu, M.C., Long, S.R., Shen, S.S.P., Qu, W., Gloersen, P., Fan, K.L., A confidence limit for the empirical mode decomposition and Hilbert spectral analysis, *Proc. Royal Soc. London, Ser. A* **459**, 2317–2345, 2003.

Itin, A., Neishtadt, A., Vasiliev, A., Captures into resonance and scattering on resonance in the dynamics of a charged relativistic particle in magnetic field and electrostatic wave, *Physica D* **141**, 281–296, 2000.

Jiang, D., Pierre, C., Shaw, S.W., The construction of nonlinear normal modes for systems with internal resonance, *Int. J. Nonlinear Mech.* **40**, 729–746, 2005a.

Jiang, D., Pierre, C., Shaw, S.W., Nonlinear normal modes for vibratory systems under harmonic excitation, *J. Sound Vib.* **288**(4–5), 791–812, 2005b.

Kahn, P.B., Zarmi, Y., *Nonlinear Dynamics: Exploration through Normal Forms*, J. Wiley & Sons, 1997.

Kath, W., Necessary conditions for sustained reentry roll resonance, *SIAM J. Appl. Math.* **43**, 314–324, 1983a.

Kath, W., Conditions for sustained resonance II, *SIAM J. Appl. Math.* **43**, 579–583, 1983b.

Kauderer, H., *Nichtlineare Mechanik*, Springer-Verlag, Berlin/New York, 1958.

Kerschen, G., Worden, K., Vakakis, A.F., Golinval, J.C., Past, present and future of nonlinear system identification in structural dynamics, *Mech. Syst. Signal Proces.* **20**, 505–592, 2006.

Kerschen, G., Vakakis, A.F., Lee, Y.S., McFarland, D.M., Bergman, L.A., Toward a fundamental understanding of the Hilbert–Huang transform in nonlinear structural dynamics, in *Proceedings of the 24th International Modal Analysis Conference (IMAC)*, St-Louis, MO, 2006.

Kerschen, G., Peeters, M., Golinval, J.C., Vakakis, A.F., Nonlinear normal modes, Part I: A useful framework for the structural dynamicist, *Mech. Syst. Sign. Proc.*, 2008a (submitted).

Kerschen, G., Vakakis, A.F., Lee, Y.S., McFarland, D.M., Bergman, L.A., Toward a fundamental understanding of the Hilbert-Huang transform in nonlinear structural dynamics, *J. Vib. Control* **14**, 77–105, 2008b.

Kevorkian, J., Cole, J.D., *Multiple Scale and Singular Perturbation Methods*, Springer-Verlag, Berlin/New York, 1996.

King, M.E., Vakakis, A.F., An energy-based approach to computing nonlinear normal modes in undamped continuous systems, *J. Vib. Acoust.* **116**, 332–340, 1993.

King, M.E., Vakakis, A.F., A method for studyng waves with spatially localized envelopes in a class of weakly nonlinear partial differential equations, *Wave Motion* **19**, 391–405, 1994.

King, M.E., Vakakis, A.F., Mode localization in a system of coupled flexible beams with geometric nonlinearities, *ZAMM* **75**, 127–139, 1995a.

King, M.E., Vakakis, A.F., A very complicated structure of resonances in a system with cyclic symmetry, *Nonl. Dynam.* **7**, 85–104, 1995b.

King, M.E., Vakakis, A.F., An energy-based approach to computing resonant nonlinear normal modes, *J. Appl. Mech.* **63**, 810–819, 1996.

Koon, W., Lo, M., Marsden, J., Ross, S., Resonance and capture of Jupiter comets. *Celest. Mech. Dyn. Astr.* **81**, 27–38, 2001.

Kosevitch, A.M., Kovalyov, A.S., *Introduction to Nonlinear Dynamics*, Naukova dumka, Kiev, 1989 [in Russian].

Lakrad, F., Belhaq, M., Periodic solutions of strongly nonlinear oscillators by the multiple scales method, *J. Sound Vib.* **258**(4), 677–700, 2002.

Le, T.P., Argoul, P., Continuous wavelet tranform for modal identification using free decay response, *J. Sound Vib.* **277**, 73–100, 2004.

Lee, C.-M., Goverdovskiy, V.N., Temnikov, A.I., Design of springs with 'negative' stiffness to improve vehicle driver vibration isolation, *J. Sound Vib.* **302**, 865–874, 2007.

Lichtenberg, A., Lieberman, M., *Regular and Stochastic Motions*, Springer-Verlag, Berlin/New York, 1983.

Lighthill, M.J., *Higher Approximations in Aerodynamic Theory*, Princeton University Press, Princeton, New Jersey, 1960.

Lochak, P., Meunier, C., *Multi-phase Averaging for Classical Systems*, Springer-Verlag, Berlin, 1988.

Lyapunov, A., *The General Problem of the Stability of Motion*, Princeton University Press, Princeton, New Jersey, 1947.

MacKay, R.S., Meiss, J.D. (Eds.), *Hamiltonian Dynamical Systems, A Reprint Selection*, Adam Hilger, Bristol and Philadelphia, 1987.

Manevitch, L.I., Complex representation of dynamics of coupled nonlinear oscillators, in *Mathematical Models of Non-Linear Excitations, Transfer Dynamics and Control in Condensed Systems and Other Media*, L. Uvarova, A. Arinstein and A. Latyshev (Eds.), Kluwer Academic/Plenum Publishers, 1999.

Manevitch, L.I., The description of localized normal modes in a chain of nonlinear coupled oscillators using complex variables, *Nonl. Dyn.* **25**, 95–109, 2001.

Manevitch, L.I., Mikhlin, Yu.V., On periodic solutions close to rectilinear normal vibration modes, *J. Appl. Math. Mech. (PMM)* **36**(6), 1051–1058, 1972.

Manevitch, L.I., Pinsky, M.A., On the use of symmetry when calculating nonlinear oscillations, *Izv. AN SSSR, MTT* **7**(2), 43–46, 1972a.

Manevitch, L.I., Pinsky, M.A., On nonlinear normal vibrations in systems with two degrees of freedom, *Prikl. Mech.* **8**(9), 83–90, 1972b.

Manevitch, L.I., Pervouchine, V.P., Transversal dynamics of one-dimensional chain on nonlinear asymmetric substrate, *Meccanica* **38**, 669–676, 2003.

Manevitch, L.I., Mikhlin, Yu.V., Pilipchuk, V.N., *The Method of Normal Oscillations for Essentially Nonlinear Systems*, Nauka, Moscow, 1989 [in Russian].

Meirovitch, L., *Elements of Vibration Analysis*, McGraw Hill, New York, 1980.

Mendonça, J.T., Bingham, R., Shukla, P.K., Resonant quasiparticles in plasma turbulence, *Phys. Rev. E.* **68**, 016406, 2003

Mikhlin, Yu.V., Resonance modes of near conservative nonlinear systems, *J. Appl. Math. Mech. (PMM)* **38**(3), 425–429, 1974.

Mikhlin, Yu.V., The joining of local expansions in the theory of nonlinear oscillations, *J. Appl. Math. Mech. (PMM)* **49**, 738–743, 1985.

Morozov, A. D., Shilnikov, L. P., On nonconservative periodic systems close to two-dimensional Hamiltonian, *J. Appl. Math. Mech. (PMM)* **47**(3), 327–334, 1984.

Moser, J.K., Periodic orbits near an equilibrium and a theorem, *Comm. Pure Appl. Math.* **29**, 727–747, 1976.

Moser, J.K., *Integrable Hamiltonian Systems and Spectral Theory*, Coronet Books, Philadelphia, PA, 2003.

Nayfeh, A.H., *The Method of Normal Forms*, Wiley Interscience, New York, 1993.

Nayfeh, A.H., *Nonlinear Interactions: Analytical, Computational and Experimental Methods*, Wiley Interscience, New York, 2000.

Nayfeh, A.H., Mook, D.T., *Nonlinear Oscillations*, Wiley Interscience, New York, 1995.

Nayfeh, A.H., Nayfeh, S.A., Nonlinear normal modes of a continuous system with quadratic nonlinearities, *J. Vib. Acoust.* **117**, 199–205, 1993.

Nayfeh, A.H., Chin, C., Nayfeh, A.H., On nonlinear normal modes of systems with internal resonance, *J. Vib. Acoust.* **118**, 340–345, 1996.

Neishtadt, A., Passage through a separatrix in a resonance problem with a slowly-varying parameter, *J. Appl. Math. Mech. (PMM)* **39**, 621–632, 1975.

Neishtadt, A., Scattering by resonances, *Cel. Mech. Dyn. Astr.* **65**, 1–20, 1997.

Neishtadt, A., On adiabatic invariance in two-frequency systems, in *Hamiltonian Systems with Three or More Degrees of Freedom*, NATO ASI Series C 533, Kluwer Academic Publishers, pp. 193–212, 1999.

Panagopoulos, P.N., Vakakis, A.F., Tsakirtzis, S., Transient resonant interactions of linear chains with essentially nonlinear end attachments leading to passive energy pumping, *Int. J. Solids Str.* **41**(22–23), 6505–6528, 2004.

Peeters, M., Viguié, R., Sérandour, G., Kerschen, G., Golinval, J.C., Nonlinear normal modes, Part II: Practical computation using numerical continuation techniques, *Mech. Syst. Signal Proc.*, 2008 (submitted).

Persival, I., Richards, D., *Introduction to Dynamics*, Cambridge University Press, Cambridge, UK, 1982.

Pesheck, E., *Reduced-Order Modeling of Nonlinear Structural Systems Using Nonlinear Normal Modes and Invariant Manifolds*, PhD Thesis, University of Michigan, Ann Arbor, MI.

Pesheck, E., Pierre, C., Shaw, S.W., A new Galerkin-based approach for accurate non-linear normal modes through invariant manifolds, *J. Sound Vib.* **249**(5), 971–993, 2002.

Pierre, C., Dowell, E.H., Localization of vibrations by structural irregularity, *J. Sound Vib.* **114**, 549–564, 1987.

Pierre, C., Jiang, D., Shaw, S.W., Nonlinear normal modes and their application in structural dynamics, *Math. Probl. Eng.* **10847**, 1–15, 2006.

Pilipchuk, V.N., Transient mode localization in coupled strongly nonlinear exactly solvable oscillators, *Nonl. Dyn.* **51**, 245–258, 2008.

Quinn, D., Resonance capture in a three degree-of-freedom mechanical system, *Nonl. Dyn.* **14**, 309–333, 1997a.

Quinn, D., Transition to escape in a system of coupled oscillators, *Int. J. Nonl. Mech.* **32**, 1193–1206, 1997b.

Rand, R.H., Nonlinear normal modes in two-degree-of-freedom systems, *J. Appl. Mech.* **38**, 561, 1971.

Rand, R.H., A direct method for nonlinear normal modes, *Int. J. Nonlinear Mech.* **9**, 363–368, 1974.

Rilling, G., Flandrin, P., Gonçalvès, P., On empirical mode decomposition and its algorithms, in *IEEE-Eurasip Workshop on Nonlinear Signal and Image Processing (NSIP-03)*, Grado, Italy, June 2003.

Rosenberg, R.M., On nonlinear vibrations of systems with many degrees of freedom, *Adv. Appl. Mech.* **9**, 155–242, 1966.

Sanders, J., Verhulst, F., *Averaging Methods in Nonlinear Dynamical Systems*, Springer-Verlag, New York, 1985.

Scott, A.S., Lomdahl, P.S., Eilbeck, J.C., Between the local-mode and normal-mode limits, *Chem. Phys. Lett.* **113**, 29–36, 1985.

Shaw, S.W., Pierre, C., Nonlinear normal modes and invariant manifolds, *J. Sound Vib.* **150**, 170–173, 1991.

Shaw, S.W., Pierre, C., Normal modes of vibration for nonlinear vibratory systems, *J. Sound Vib.* **164**, 85–124, 1993.

Touzé, C., Amabili, M., Non-linear normal modes for damped geometrically non-linear systems: Application to reduced-order modeling of harmonically forced structures, *J. Sound Vib.* **298**(4–5), 958–981, 2006.

Touzé, C., Thomas, O., Chaigne, A., Hardening / softening behaviour in nonlinear oscillations of structural systems using nonlinear normal modes, *J. Sound Vib.* **273**, 77–101, 2004.

Touzé, C., Amabili, M., Thomas, O., Camier, C., Reduction of geometrically nonlinear models of shell vibrations including in-plane inertia, in *Euromech Colloquium 483 on Geometrically Nonlinear Vibrations of Structures*, July 9–11, FEUP, Porto, Portugal, 2007a.

Touzé, C., Amabili, M., Thomas, O., Reduced-order models for large-amplitude vibrations of shells including in-plane inertia, Preprint, 2007b.

Tsakirtzis S., *Passive Targeted Energy Transfers From Elastic Continua to Essentially Nonlinear Attachments for Suppressing Dynamical Disturbances*, PhD Thesis, National Technical University of Athens, Athens, Greece, 2006.

Vakakis, A.F., Fundamental and subharmonic resonances in a system with a 1–1 internal resonance, *Nonl. Dyn.* **3**, 123–143, 1992.

Vakakis, A.F., Passive spatial confinement of impulsive responses in coupled nonlinear beams, *AIAA J.* **32**, 1902–1910, 1994.

Vakakis, A.F., Nonlinear mode localization in systems governed by partial differential equations, *Appl. Mech. Rev.* (Special Issue on 'Localization and the Effects of Irregularities in Structures', H. Benaroya, Ed.), **49**(2), 87–99, 1996.

Vakakis, A.F., Nonlinear normal modes and their applications in vibration theory: An overview, *Mech. Sys. Signal Proc.* **11**(1), 3–22, 1997.

Vakakis, A.F. (Ed.), *Normal Modes and Localization in Nonlinear Systems*, Kluwer Academic Publishers, 2002 [also, special issue o *Nonlinear Dynamics* **25**(1–3), 1–292, 2001].

Vakakis, A.F., Gendelman, O., Energy pumping in nonlinear mechanical oscillators: Part II – Resonance capture, *J. Appl. Mech.* **68**, 42–48, 2001.

Vakakis, A.F., King, M.E., Nonlinear wave transmission in a mono-coupled elastic periodic system, *J. Acoust. Soc. Am.* **98**, 1534–1546, 1995.

Vakakis, A.F., Rand, R.H., Normal modes and global dynamics of a two-degree-of-freedom nonlinear system – I. Low energies, *Int. J. Nonlinear Mech.* **27**(5), 861–874, 1992a.

Vakakis, A.F., Rand, R.H., Normal modes and global dynamics of a two-degree-of-freedom nonlinear system – II. High energies, *Int. J. Nonlinear Mech.* **27**(5), 875–888, 1992b.

Vakakis, A.F., Nayfeh, A.H., King, M.E., A multiple scales analysis of nonlinear, localized modes in a cyclic periodic system, *J. Appl. Mech.* **60**, 388–397, 1993.

Vakakis, A.F., King, M.E., Pearlstein, A.J., Forced localization in a periodic chain of nonlinear oscillators, *Int. J. Nonlinear Mech.* **29**(3), 429–447, 1994.

Vakakis, A.F., Manevitch, L.I., Mikhlin, Yu.V., Pilipchuk, V., Zevin, A.A., *Normal Modes and Localization in Nonlinear Systems*, J. Wiley & Sons, New York, 1996.

Vakakis, A.F., Manevitch, L.I., Gendelman, A., Bergman, L.A., Dynamics of linear discrete systems connected to local essentially nonlinear attachments, *J. Sound Vib.* **264**, 559–577, 2003.

Veerman, P., Holmes, P., The existence of arbitrarily many distinct periodic orbits in a two degree-of-freedom Hamiltonian system, *Physica D* **14**, 177–192, 1985.

Veerman, P., Holmes, P., Resonance bands in a two-degree-of-freedom Hamiltonian system, *Physica D* **20**, 413–422, 1986.

Verhulst, F., *Methods and Applications of Singular Perturbations*, Springer-Verlag, Berlin/New York, 2005.

Virgin, L.N., Santillan, S.T., Plaut, R.H., Vibration isolation using extreme geometric nonlinearity, in *Euromech Colloquium 483 on Geometrically Nonlinear Vibrations of Structures*, July 9–11, FEUP, Porto, Portugal, 2007.

Weinstein, A., Normal modes for nonlinear Hamiltonian systems, *Inv. Math.* **20**, 47–57, 1973.

Wiggins, S., *Introduction to Applied Nonlinear Dynamical Systems and Chaos*, Springer-Verlag, New York, 1990.

Yang, C.H., Zhu, S.M., Chen, S.H., A modified elliptic Lindstedt–Poincaré method for certain strongly nonlinear oscillators, *J. Sound Vib.* **273**, 921–932, 2004.

Yin, H.P., Duhamel, D., Argoul, P., Natural frequencies and damping estimation using wavelet transform of a frequency response function, *J. Sound Vib.* **271**(3–5), 999–1014, 2004.

Chapter 3
Nonlinear Targeted Energy Transfer in Discrete Linear Oscillators with Single-DOF Nonlinear Energy Sinks

In this chapter we initiate our study of passive *nonlinear targeted energy transfer* – *TET* (or, so-called nonlinear energy pumping) by considering discrete systems consisting of linear coupled oscillators (refered to from now on as 'primary systems') with single-DOF (SDOF) essentially nonlinear attachments. In later chapters we will extend this study to discrete and elastic continuous systems with SDOF or MDOF nonlinear attachments. We aim to show that under certain conditions, the nonlinear attachments are capable of passively absorbing and locally dissipating significant portions of vibration energy of the primary systems to which they are attached. Moreover, this passive targeted energy transfer will be shown to occur over broad frequency ranges, due to the capacity of the nonlinear attachments will be capable to engage in transient resonance (i.e., in transient resonance captures) with linear modes of the primary systems at arbitrary frequency ranges. Then, in essence, these essentially nonlinear attachments will act as *nonlinear energy sinks* (NESs).

By applying analytic methodologies especially developed for studying strongly nonlinear transient regimes (such as the CX-A method introduced in Section 2.4), performing numerical simulations, and post-processing the results by means of the signal analysis techniques discussed in Section 2.5, we will be able to study, model and understand the dynamical mechanisms governing passive nonlinear TET in the systems under consideration. Moreover, we will formulate appropriate measures for assessing the TET efficiency of different configurations of NESs, which, ultimately, will enable us to establish conditions for optimal TET in the systems considered. At the end of this chapter we will extend the study of TET to infinite-DOF chains with SDOF essentially nonlinear attachments and investigate TET generated by impeding elastic waves to boundary NESs.

3.1 Configurations of Single-DOF NESs

The realization of nonlinear targeted energy transfer – TET (or nonlinear energy pumping) was first observed by Gendelman (2001) who studied the transient dynamics of a two-DOF system consisting of a damped linear oscillator (LO) (designated as 'primary system') that was weakly coupled to an essentially (strongly) nonlinear, damped attachment, i.e., an oscillator with zero linearized stiffness. The need for *essentially nonlinearity* was emphasized, since linear or near-integrable nonlinear systems have essentially constant modal distributions of energy that preclude the possibility of energy transfers from one mode to another; moreover, such essentially nonlinear oscillators do not have preferential resonant frequencies of oscillation, which enables them to resonantly interact with modes of the primary system at arbitrary frequency ranges. Returning to the work by Gendelman (2001), he showed that, whereas input energy is imparted initially to the LO, a nonlinear normal mode (NNM) localized to the nonlinear attachment can be excited provided that the imparted energy is above a critical threshold. As a result, TET occurs and a significant portion of the imparted energy to the LO gets passively absorbed and locally dissipated by the essentially nonlinear attachment, which acts, in essence, as *nonlinear energy sink (NES)*.

This result was extended in other works. A slightly different nonlinear attachment was considered in Gendelman et al. (2001) and Vakakis and Gendelman (2001). In these papers (some results of which are reviewed in Section 2.3), the nonlinear oscillator (the NES) was connected to ground using an essential nonlinearity. This configuration (refered to as 'Configuration I' in Section 2.6) is depicted in Figure 3.1. TET was then defined as the one-way (irreversible on the average) channeling of vibrational energy from the directly excited linear primary structure to the attached NES. The underlying dynamical mechanism governing TET was found to be a transient resonance capture (TRC) (Arnold, 1988) of the dynamics of the nonlinear attachment on a 1:1 resonance manifold (see Section 2.3 for related definitions). An interesting feature of the dynamics discussed in these works is that a prerequisite for TET is damping dissipation; indeed, in the absence of damping, typically, the integrated system can only exhibit nonlinear beat phenomena (caused by internal resonances, see Section 2.3), whereby (the conserved) energy gets continuously exchanged between the linear primary system and the nonlinear attachment, but no TET can occur.

Nonlinear TET in two-DOF systems was further investigated in several recent studies. In Vakakis (2001), the onset of nonlinear energy pumping was related to the zero crossing of a frequency of envelope modulation, and a criterion (critical threshold) for inducing nonlinear energy pumping was formulated. The degenerate bifurcation structure of the NNMs, which reflects the high degeneracy of the underlying nonlinear Hamiltonian system composed of the undamped LO coupled to an undamped attachment with pure cubic stiffness nonlinearity, was explored by Gendelman et al. (2003). Vakakis and Rand (2004) discussed the resonant dynamics of the same undamped system under condition of 1:1 internal resonance and showed the existence of synchronous (NNMs) and asynchronous (elliptic orbits) periodic motions; the influence of damping on the resonant dynamics and TET phenomena in the damped system was studied in the same work. The structure and bifurcations of NNMs of the mentioned two-DOF system with pure cubic stiffness nonlinearity were analyzed in Mikhlin and Reshetnikova (2005).

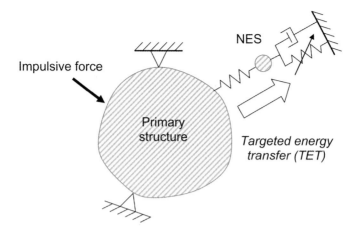

Fig. 3.1 Impulsively loaded primary structure weakly coupled to a grounded NES (referred to as 'Configuration I' in Section 2.6).

Kerschen et al. (2005) showed that the superposition of a frequency-energy plot (FEP) depicting the periodic orbits of the underlying Hamiltonian system, to the wavelet transform (WT) spectra of the corresponding weakly damped responses represents a suitable tool for analyzing energy exchanges and transfers taking place in the damped system. Goyal and Whalen (2005) considered a nonlinear energy sink design for mitigating vibrations of an air spring supported slab; the NES used in that work is similar to the grounded version of essentially nonlinear attachment (NES Configuration I) considered in this chapter. A procedure for designing passive nonlinear energy pumping devices was developed in Musienko et al. (2006), and the robustness of energy pumping in the presence of uncertain parameters was assessed in Gourdon and Lamarque (2006). Koz'min et al. (2007) performed studies of optimal transfer of energy from a linear oscillator to a weakly coupled grounded nonlinear attachment, employing global optimization techniques. Additional theoretical, numerical and experimental results on nonlinear TET were reported in recent works by Gourdon and Lamarque (2005) and Gourdon et al. (2007).

The first experimental evidence of nonlinear energy pumping was provided by McFarland et al. (2005a). TRCs leading to TET were further analyzed experimentally in Kerschen et al. (2007), whereas application of nonlinear energy pumping to problems in acoustics, was demonstrated experimentally by Cochelin et al. (2006).

In most of the above-mentioned studies, grounded and relatively heavy nonlinear attachments (NESs) were considered (i.e., Configuration I NESs – see Section 2.6), which clearly limits their applicability to practical applications. Gendelman et al. (2005) introduced a lightweight and ungrounded NES configuration (refered to as 'Configuration II' in Section 2.6) which led to efficient nonlinear energy pumping from the LO to which it was attached. This alternative configuration is depicted in Figure 3.2. Although there is no complete equivalence between the grounded and

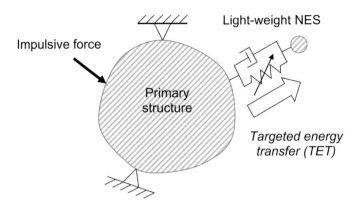

Fig. 3.2 Impulsively loaded primary structure connected to an ungrounded and lightweight NES (referred to as 'Configuration II' in Section 2.6).

ungrounded NES configurations depicted in Figures 3.1 and 3.2, it can be shown that, through a suitable change of variables the governing equations (and dynamics) of these two NES configurations may be related (Kerschen et al., 2005).

To show this, we consider the simplest possible system with an NES of Configuration II, namely a SDOF LO with a SDOF ungrounded nonlinear attachment,

$$\ddot{x} + x + C(x - v)^3 = 0$$
$$\varepsilon \ddot{v} + C(v - x)^3 = 0 \quad \text{(Config. II NES)}$$

and show that through a series of coordinate transformations it can be cast into a form that nearly resembles a primary system with an attached grounded NES of Configuration I. In the above system the lightweightness of the NES is ensured by requiring that $0 < \varepsilon \ll 1$; all other variables are treated as $O(1)$ quantities. Through the change of variables,

$$x = \varepsilon(z - w), \quad v = \varepsilon z + w$$

the above system is expressed as

$$\varepsilon(1 + \varepsilon)\ddot{z} + \varepsilon(z - w) = 0$$
$$(1 + \varepsilon)\ddot{w} + \varepsilon(w - z) + \frac{C(1 + \varepsilon)^4}{\varepsilon} w^3 = 0 \quad \text{(Config. I NES)}.$$

These equations correspond to a linear primary system (composed of a mass with no grounding stiffness) linearly coupled to an NES of Configuration I. Moreover a comparison between these two systems shows that an ungrounded NES (Config. II) with small mass ratio ε with respect to the mass of the primary system and coupled through essential nonlinearity to a LO (the primary system), is equivalent to a grounded NES (Config. I) with large mass ratio $(1 + \varepsilon)/\varepsilon$ and stiff grounding nonlinearity, that is weakly coupled to an ungrounded mass (the primary system) by means of a weak linear coupling stiffness. This result provides an initial hint on the fact that the mass of the NES affects differently the dynamics of the two considered configurations. Indeed, it will be shown in this work that NES Configuration II is most effective for relatively light mass (which makes it an excellent candidate as a practical vibration absorption device), whereas, on the contrary, NES Configuration I is effective for relatively heavier mass.

The dynamics of a two-DOF system composed of a linear primary oscillator coupled to an ungrounded and light-weight NES was analyzed in a series of recent papers. Lee et al. (2005) focused on the dynamics of the underlying Hamiltonian system. The different families of periodic orbits of the strongly nonlinear system were depicted in a frequency-energy plot (FEP) (see Section 2.1 for the appropriate definition), which was shown to possess: (i) a backbone curve with periodic orbits satisfying the condition of fundamental (1:1) internal resonance; and (ii) a countable infinity of subharmonic branches, with each branch corresponding to a different realization of an subharmonic resonance between the LO and the NES. In Kerschen et al. (2006a), the energy exchanges in the damped system were interpreted based on the topological structure and bifurcations of the periodic solutions of the underlying undamped system. It was observed that TET can be realized through two distinct mechanisms, namely fundamental and subharmonic TET. It was also noted that a third mechanism, which relies on the excitation of so-called impulsive periodic and quasi-periodic orbits, is necessary to initiate either one of the TET mechanisms through nonlinear beating phenomena. These impulsive orbits were studied using different analytic methods in Kerschen et al. (2008). These theoretical findings were validated experimentally in McFarland et al. (2005b).

Gendelman (2004) provided a different perspective of TET dynamics by computing the damped NNMs of a LO coupled to an NES using the invariant manifold approach. He showed that the rate of energy dissipation in this system is closely related to the bifurcations of the NNM invariant manifold. To complement this approach, Panagopoulos et al. (2007) analyzed how initial conditions determine the specific equilibrium point of the slow flow dynamics that is eventually reached by the trajectories of the system. Manevitch et al. (2007a, 2007b), Quinn et al. (2008) and Koz'min et al. (2008) discussed the conditions that should be satisfied by the system and forcing parameters for optimal TET to occur (i.e., so that the maximum portion of the vibration energy of the LO gets passively transferred and locally dissipated by the NES in the least possible time).

We conclude this bibliographical review on the dynamics of linear oscillators coupled to NESs by mentioning that alternative designs for SDOF NES have also been proposed. In (Georgiades et al., 2005) and (Karayannis et al., 2007), *TET at a fast time-scale* was achieved using NESs with non-smooth stiffness characteristics (clearances and impacts); NESs with non-smooth stiffness characteristics will be considered in detail in Chapter 7 of this work. In (Gendelman and Lamarque, 2005) and (Avramov and Mikhlin, 2006) an NES characterized by multiple states of equilibrium positions was considered (this was achieved through a snap-through stiffness element). Moreover, as reported in Laxalde et al. (2007), nonlinear energy which is the indication of the strongly nonlinear nature of its oscillation. Inspite of the fact that initially the energy is entirely stored in the LO, it quickly flows back and forth between the two oscillators. After $t = 15$ s, 87% of the instantaneous total energy is stored in the NES, but this number drops down to 3% immediately thereafter. Throughout this nonlinear beating phenomenon, a reversible energy transfer occurs, which, however, results in near optimal energy dissipation. At this intermediate energy regime, as much as 94.4% of the total input energy is dissipated by the damper of the NES. Another evidence of the nonlinear beating in this case is that the system performs fast oscillations with frequency close to 1 rad/s, modulated by a slowly-varying envelope (see Figure 3.6d for a close-up of the NES response); this is due to the fact that a 1:1 transient resonance capture (TRC) between the LO and the NES takes place. It is interesting to note that no *a priori* tuning of the NES parameters was necessary in order to achieve this result. It is the variation of the frequency of the NES due to damping dissipation that plays the role of 'tuning' (but also of 'detuning' at later times) for the realization of, and escape from TRC. This is markedly different from 'classical' nonlinear beat phenomena caused by internal resonances in Hamiltonian coupled oscillators with linearizable nonlinear stiffnesses, where the ratio of the linearized eigenfrequencies dictates the type of internal resonance between modes that is realized [see, for example, spring-pendulum systems in Nayfeh and Mook (1995)], and no escape from internal resonance is possible once it is initiated.

When the magnitude of the applied impulse is further increased ($X = 0.2$, Figure 3.7 and point C in Figure 3.4), the motion still localizes to the NES, but a different type of energy exchange is encountered. Indeed, during the initial stage of the motion (until approximately $t = 15$ s), a nonlinear beating phenomenon takes place as in the previous case. However, after this continuous energy exchange between the two oscillators takes place, an irreversible energy flow from the LO to the NES occurs, nonlinear energy pumping is triggered, and TET is realized. Figure 3.7d, which superposes both responses, illustrates that these are completely synchronized during this latter regime. In other words, they vibrate in an in-phase fashion with the same apparent frequency. The underlying dynamical phenomenon causing nonlinear TET can therefore be related to capture in the neighborhood of a 1:1 resonant manifold of the dynamics. The transient nature of the resonance capture is evident in Figure 3.7c, since energy is released back to the LO around $t = 300$ s. However, when this occurs, the total remaining energy level of the system is small compared to its initial value. Another manifestation of TET is that the envelopes of the displacements decrease monotonically (i.e., no modulation is observed in contrast to the beating regime), with the envelope of the LO decreasing faster than that of the NES. Overall, 87.6% of the total input energy is dissipated by the NES in this case.

At a higher-energy regime ($X = 0.5$, Figure 3.8 and point D in Figure 3.4), no further qualitative change appears in the dynamics. The nonlinear beating phenomenon dominates the early regime of the motion. A weaker but faster energy exchange is observed, since 32% of the total energy is transferred to the NES after $t = 4$ s. The triggering of TET still occurs, but the irreversible energy transfer from the LO to the NES is slower compared to the previous simulation. This is why energy dis-

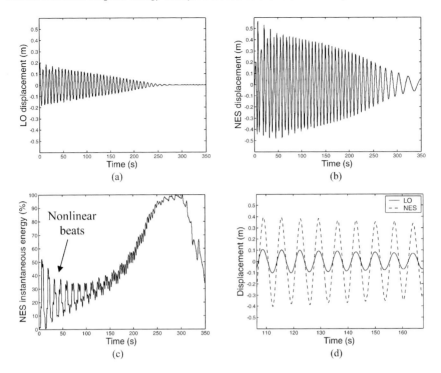

Fig. 3.7 Transient dynamics of the two-DOF system (moderate-energy level; $X = 0.2$): (a) LO displacement; (b) NES displacement; (c) percentage of instantaneous total energy in the NES and (d) superposition of both displacements during nonlinear TET.

sipation is less efficient than at the moderate-energy regime, and only 50% of the total input energy is dissipated by the NES.

Another interesting simulation is shown in Figure 3.9 for parameters $\varepsilon = 0.05$, $\omega_0^2 = C = 1$, $\lambda_1 = 0$, $\lambda_2 = 0.002$ and $X = 0.1039$. As in previous simulations, after an initial nonlinear beating (until approximately $t = 150$ s), a distinct regime of the transient dynamics is realized. As Figure 3.9d shows, the transient dynamics in the second regime is captured on a 1:3 resonant manifold of the dynamics, with the LO oscillating three times faster than the NES; it is noteworthy that the NES envelope grows during a few cycles, indicating that the NES is extracting energy from the LO. This simulation is further evidence that the NES has no preferential resonant frequency, and it can engage in (fundamental and subharmonic) resonance with the LO at multiple frequency ranges.

To highlight the fundamental difference between the SDOF NES and the classical linear TMD we compare their capacities for vibration absorption by performing an additional series of simulations. To this end, the two configurations depicted in Figure 3.10 are used in a parametric study where we vary, (i) the spring constant k_1 of the LO (and therefore the natural frequency of the LO, ω_0), and (ii) the magnitude X of the impulse applied to the LO. The three-dimensional plots in Figures 3.11 and

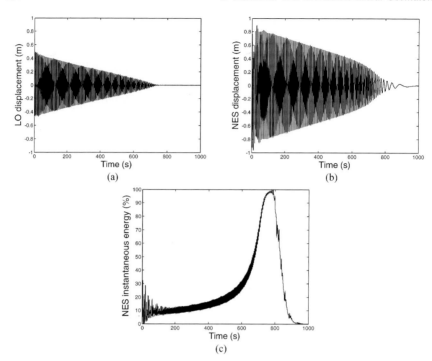

Fig. 3.8 Transient dynamics of the two-DOF system (high-energy level; $X = 0.5$): (a) LO displacement; (b) NES displacement and (c) percentage of instantaneous total energy in the NES.

3.12 display the energy dissipated by the TMD and the SDOF NES, respectively, as functions of parameters ω_0 and X. Due to the linear superposition principle, the normalized energy dissipated by the TMD does not depend on the impulse magnitude. There is a specific value of ω_0 for which the energy dissipation in the TMD is maximum (95.38% of the total input energy). Any deviation in the frequency content of the LO response from this regime decreases the TMD performance, signifying the well-known result that the TMD is only effective when it is tuned to the natural frequency of the LO.

Unlike the TMD, the NES performance depends critically on the impulse magnitude, which is an intrinsic limitation of this type of nonlinear absorbers. This is confirmed in Figure 3.12. This figure indicates that *the effectiveness of the NES is not significantly influenced by changes in the natural frequency of the LO*.

More precisely, for values of ω_0 beyond a critical threshold there exists impulse magnitudes for which the NES dissipates a significant amount of the total input energy. Moreover, there are alternative mechanisms by which the NES can induce TET in this system. For example, for $(\omega_0, X) = (2.3, 0.31)$ the response (depicted in Figure 3.13) reveals that the LO vibrates three times faster than the NES, i.e., similarly to what was observed in Figure 3.9. Hence, a 1:3 resonance between the LO and the NES seems to be the mechanism responsible for the sudden increase in

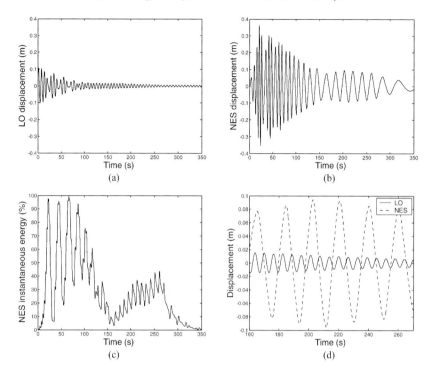

Fig. 3.9 Transient dynamics of the two-DOF system ($X = 0.1039$, $\lambda_1 = 0$): (a) LO displacement; (b) NES displacement; (c) percentage of instantaneous total energy in the NES and (d) superposition of both displacements during nonlinear energy pumping.

performance in this region. Such *subharmonic TET* is realized in the narrow zones of increased energy dissipation of the plot of Figure 3.12, which are quite distinct from the main region of increased energy dissipation where *fundamental TET* is realized through 1:1 TRCs. Therefore, it appears that *the NES can dissipate a substantial amount of the total input energy through fundamental* (1:1) *as well as subharmonic* (*m*:*n*) *resonances*.

In conclusion, even though the performance of the TMD is not affected by the level of total energy in the system, it is limited by its inherent sensitivity to uncertainties in the natural frequency of the primary system. In contrast, provided that the energy is above a critical threshold, the SDOF NES is capable of robustly absorbing transient disturbances over a broad range of frequencies. Hence, the NES may be regarded as an efficient passive absorber, possessing adaptivity to the frequency content of the vibrations of the primary system. This is due to its essential stiffness nonlinearity, which precludes the existence of any preferential resonance frequency.

It is also shown in this section, that a seemingly simple system comprising of a damped LO and an essentially nonlinear attachment may exhibit complicated dynamics and transitions, including fundamental and subharmonic resonances, nonlinear beating phenomena, multi-frequency responses and strong motion localization to

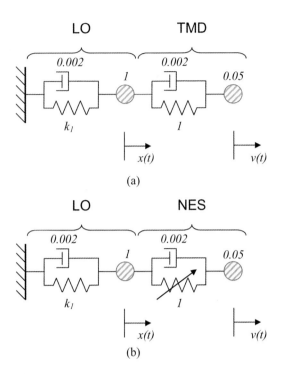

Fig. 3.10 Comparison of the linear and nonlinear energy absorbing devices: (a) TMD coupled to a LO; (b) NES coupled to a LO.

either oscillator. The most interesting feature of this system is arguably its capability to realize passive and irreversible energy transfer phenomena from the impulsively loaded LO to the NES, in spite of the relative lightness of the NES compared to the mass of the LO.

The complexity of the problem dictates a systematic study of the damped and undamped dynamics of the integrated discrete system composed of the linear oscillator with an essentially nonlinear attachment (the NES). In the following sections of this chapter we will employ a combination of numerical and analytical techniques, including direct numerical simulations; special analytical methodologies capable of modeling both qualitatively and quantitatively transient, strongly nonlinear damped transitions; and advanced signal processing techniques to analyze the resulting nonlinear and non-stationary signals. First, we will consider the system without damping. Even though damping is a prerequisite for TET (as the numerical results discussed in this section indicate), we will show that *for sufficiently weak damping the dynamics of* TET *is governed, in essence, by the underlying Hamiltonian dynamics*; the weak damping dissipation then controls the transient damped transitions along branches of NNMs of the Hamiltonian system. After studying the complex dynamics of the Hamiltonian system we will examine damped transient responses,

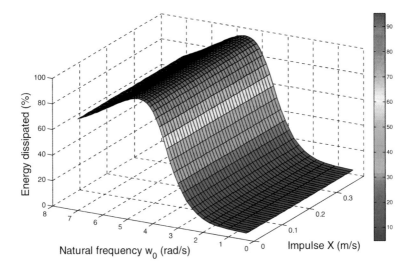

Fig. 3.11 TMD performance.

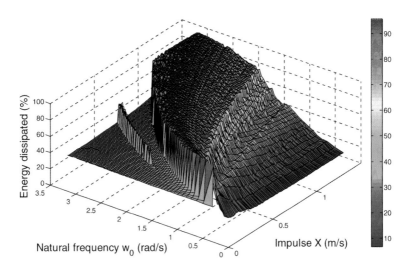

Fig. 3.12 NES performance.

modal energy exchanges and TET between the linear primary system and the attached NES. Experimental verification of the theoretical results will be presented later in this chapter.

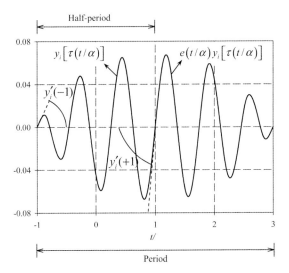

Fig. 3.15 Construction of the periodic solutions $v(t) = e(t/\alpha)y_1(\tau(t/\alpha))$ and $x(t) = e(t/\alpha)y_2(\tau(t/\alpha))$ over an entire normalized period $-1 \le t/\alpha \le 3$ from the solutions $y_i(\tau(t/\alpha))$, $i = 1, 2$ of the NLBVP (3.10), computed over the half normalized period $-1 \le t/\alpha \le 1$.

Referring to the general form of the periodic orbit depicted in Figure 3.15, we introduce the following classification:

(i) *Symmetric periodic orbits* (NNMs) $Snm\pm$ *correspond to orbits that satisfy the conditions,*

$$\dot{v}(-T/4) = \pm\dot{v}(+T/4) \Rightarrow y_1'(-1) = \pm y_1'(+1)$$

and

$$\dot{x}(-T/4) = \pm\dot{x}(+T)/4) \Rightarrow y_2'(-1) = \pm y_2'(+1)$$

with n being the number of half-waves in y_1 (and v), and m the number of half-waves in y_2 (and x) in the half-period interval $-T/4 \le t \le +T/4 \Leftrightarrow -1 \le \tau \le +1$. Hence, the periodic solutions on the branch of NNMs $Smn+$ ($Smn-$) pass through the origin of the configuration plane (x, v) with positive (negative) slope. The ratio $(m:n)$ indicates the order of the internal resonance realized during the given periodic motion. For instance, a 1:1 internal resonance is realized on both branches $S11\pm$, which means that both the LO and the NES vibrate with the same dominant frequency. Since the periodic orbits considered are nolinear, they will possess additional harmonics at multiples of this dominant frequency, but the amplitudes of these harmonics are expected to be small. Similarly, on branches $S13\pm$, a 1:3 internal resonance is realized between the two oscillators (with the LO oscillating 3 times faster than the NES), and there are two dominant harmonic components in the responses, one around

1 rad/s and one around 1/3 rad/s; as shown later, the amplitudes of these two harmonic components may vary along the branches $S13\pm$. Also, the (+) and (-) signs in the notations of these branches indicate whether the corresponding periodic solutions pass through the origin of the configuration plane with positive or negative slopes, respectively. For instance, an in-phase (out-of-phase) motion of the system is realized on $S11+$ ($S11-$).

(ii) *Unsymmetric periodic orbits* (NNMs) Upq are orbits that do not satisfy the conditions of the symmetric orbits. In particular, orbits $U(m+1)m$ bifurcate from the branch of symmetric NNMs $S11-$ at $T/4 \approx m\pi/2$, and exist approximately within the intervals $m\pi/2 < T/4 < (m+1)\pi/2$, $m = 1, 2 \ldots$. For example on branches $U21\pm$, a 2:1 internal resonance is realized between the two oscillators (with the NES oscillating two times faster than the LO), and there are two dominant harmonic components, one around 2 rad/s and one around 1 rad/s; we note that the magnitudes of these two harmonic components may vary along branches $U21\pm$.

As mentioned previously, the numerical solution of the two-point NLBVP (3.10) is constructed utilizing a shooting method, details of which can be found in Lee et al. (2006). In brief terms, the NLBVP is solved as follows. For a given nonlinear eigenvalue a (the quarter period of the NNM) the solutions of the NLBVP are computed at different energy levels; it is expected that at a given energy level there might co-exist multiple nonlinear periodic solutions sharing the same *minimal* period. Periodic orbits that correspond to synchronous motions of the two oscillators of the system, and pass through the origin of the configuration plane are termed nonlinear normal modes (NNMs) in Vakakis et al. (1996), but a more extended definition of NNMs is adopted in this work (see Section 2.1) to include all periodic motions (and not just synchronous ones).

The different families of computed periodic solutions are depicted in three types of plots. In the first two types of plots, we assume zero initial displacements $x(-T/4) = v(-T/4) = 0$, and depict the initial velocities $\dot{v}(-T/4) = y_1'(-1)/\alpha$ and $\dot{x}(-T/4) = y_2'(-1)/\alpha$ of the periodic orbits as functions of the quarter-period $\alpha = T/4$ of the (conserved) energy of that orbit:

$$h = (1/2)[\varepsilon\dot{v}^2(-T/4) + \dot{x}^2(-T/4)] = (1/2\alpha^2)[\varepsilon y_1'^2(-1) + y_2'^2(-1)].$$

In the third type of plots, we depict the frequencies of the periodic orbits as functions of their energies h. These plots clarify the bifurcations that connect, generate or eliminate the different branches (families) of periodic solutions (NNMs). As mentioned previously, the stability of the computed periodic orbits is determined by Floquet analysis and by performing direct numerical simulations of the equations of motion (3.6).

The numerical results correspond to system (3.6) with parameters and $\varepsilon = 0.05, \omega_0 = 1.0$ and $C = 1.0$, in the energy range $0 < h < 1$. In Figures 3.16 and 3.17 we depict the bifurcation diagrams of the initial velocities $\dot{v}(-T/4)$ and $\dot{x}(-T/4)$ of the computed periodic orbits for varying quarter-period α and energy h. Since the dynamical behavior of the system on the various branches of NNMs will be discussed in detail in the following sections, we make only some general and preliminary observations at this point. To illustrate the computational results, in Figure 3.18 we present time series of representative periodic motions on branches $S11+$, $S13+$ and $U21+$, together with the corresponding motions in the configuration plane of the system, (x, v). Figure 3.19 depicts the Fourier transforms of the time series to illustrate the frequency content of these periodic motions.

Considering the bifurcation diagrams of Figures 3.16 and 3.17 we make the following remarks. The NNM branches $Snn-$ exist in the quarter-period intervals $0 < \alpha < n\pi/2$, and their initial conditions satisfy the following limiting relationships (see Figure 3.16):

$$\lim_{\alpha \to 0}\{|\dot{v}(-\alpha)|, |\dot{x}(-\alpha)|\} = \infty \quad \text{and} \quad \lim_{\alpha \to n\pi/2}\{|\dot{v}(-\alpha)|, |\dot{x}(-\alpha)|\} = 0.$$

In the energy domain, these symmetric branches exist over the entire range $0 < h < 1$. We note that branches $Snn-$ are, in essence, identical to the branch $S11-$, since they are identical to it over the domain of their common minimal period (actually, the branches $Snn-$ are derived by branch $S11-$ by repeating it n times); similar remarks can be made regarding the branches $S(kn)km\pm$, k integer, which are identified with $Snm\pm$. Considering the neighborhoods of branches $S11\pm$ and referring to Figure 3.16, the branches $S11+$ and $U21$ bifurcate out at point $\alpha = \pi/2$ where $S11-$ disappears (similar behavior is exhibited by the branches S31, S21, ...). For $\pi/2 \leq \alpha \leq \pi$ a bifurcation from $S11+$ to $S13+$ takes place without change of phase; similar bifurcations take place at higher values of α for branches $S15+$, $S17+$, ... For $\alpha \approx 3\pi/2$ the branches $S13+$ and $S13-$ coalesce with branch $S11-$, with similar coalescences with branch $S11-$ taking place at higher values of α for the pairs of branches $S15\pm$, $S17\pm$, ...

The unsymmetric NNM branches $U(m+1)m$ bifurcate from the symmetric branches $S(m+1)(m+1)$ – at quarter-periods $\alpha = m\pi/2$. It turns out that certain periodic orbits on these branches, termed *impulsive orbits* – IOs, are of particular importance concerning TET in the damped system. The IOs satisfy the additional initial condition $y_1'(-1) \equiv \dot{v}(-\alpha) = 0$, and correspond to zero crossings of the branches $U(m+1)m$ in the bifurcation diagram of Figure 3.17a. Taking into account the formulation of the NLBVP (3.10), it follows that IOs satisfy initial conditions $v(-T/4) = \dot{v}(-T/4) = x(-T/4) = 0$, and $\dot{x}(-T/4) \neq 0$, which happen to correspond to the initial state of the undamped system (3.6) (being initially at rest) after application of an impulse of magnitude $\dot{x}(-T/4) = y_2'(-1)/\alpha$ to the linear oscillator.

This implies that if the LO of the system (being initially at rest) is forced impulsively and one of the stable IOs is excited, a portion of the imparted energy is transferred directly to the invariant manifold corresponding to that IO, and, as a result energy is passively transferred from the LO to the NES during the initial cycle of the motion; in subsequent cycles of the response energy gets continuously transferred back and forward between the NES and the LO, and a nonlinear beat phenomenon is formed. We will show that the *excitation of IOs provides one of the possible mechanisms for triggering* TET *in the damped system*. A detailed analy-

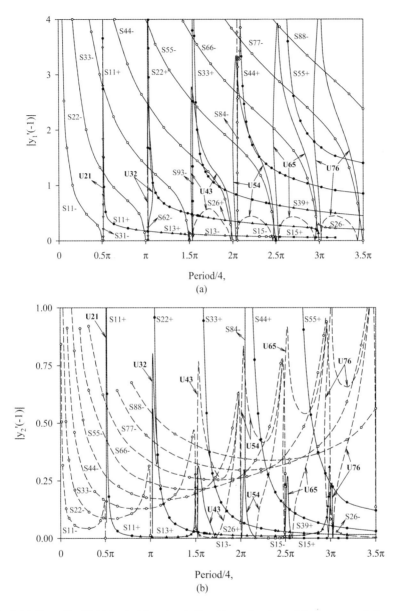

Fig. 3.16 Normalized initial velocities of periodic orbits as functions of the quarter-period α; solid (dashed) lines correspond to positive (negative) initial velocities: (a) $|y'_1(-1)|$ vs. α, (b) $|y'_2(-1)|$ vs. α ($S11$: ○, $S13$: △, $S15$: □, $S31$: ▽, $S21$: ◇ with in-phase as filled-in, and branches U without symbol).

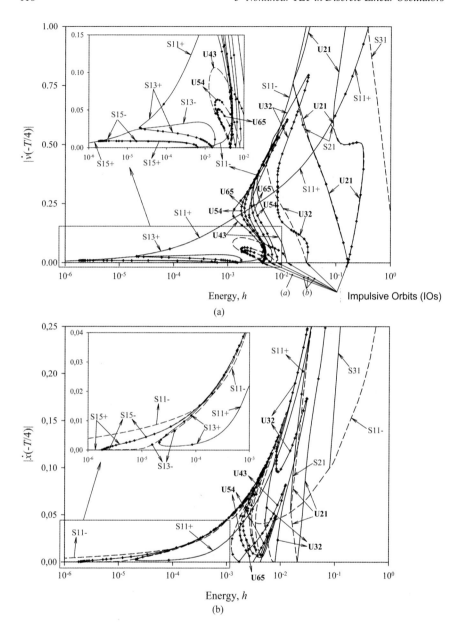

Fig. 3.17 Initial velocities of the periodic solutions as functions of energy; solid (dashed) lines correspond to positive (negative) initial velocities; unstable solutions are denoted by crosses: (a) $|\dot{v}(-T/4)|$ vs. α, (b) $|\dot{x}(-T/4)|$ vs. α.

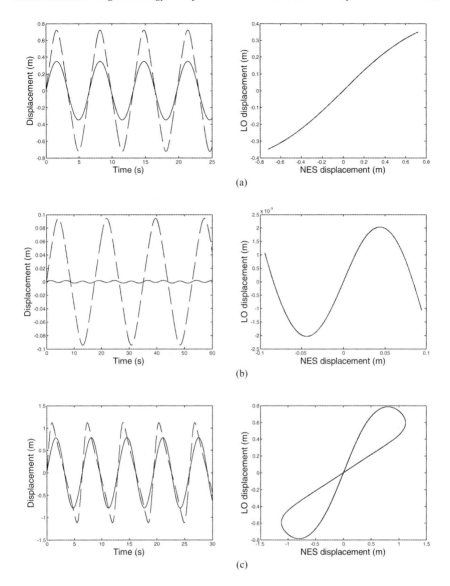

Fig. 3.18 Periodic motion on (a) $S11+$; (b) $S13+$ and (c) $U21+$ ($\varepsilon = 0.05$, $C = 1$); NES response - - -, LO response —.

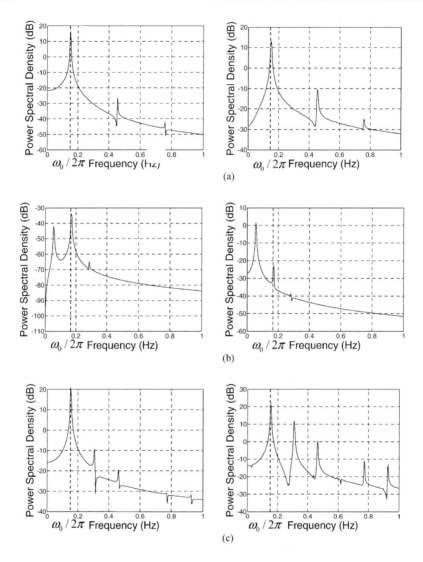

Fig. 3.19 Power spectral density of the periodic motion on (a) $S11+$; (b) $S13+$ and (c) $U21+$ ($\varepsilon = 0.05$, $C = 1$); left plots correspond to the LO response, and right plots to the NES response.

sis and discussion of the role of IOs on TET will be carried out in the following sections.

Similar classes of IOs can be realized also in a subclass of S- branches. In particular, this type of orbits can be realized on NNM branches $S(2k+1)(2p+1)\pm$, $k \neq p$, but not on periodic orbits that do not pass through the origin of the configuration plane (such as $S21$, $S12$, ...). NNM Branch $S11-$ is a particular case, where the

IO is realized only asymptotically, as the energy tends to zero, and the motion is localized completely in the linear oscillator.

3.3.1.2 Frequency-Energy Plots (FEPs)

A more suitable representation of the computed NNMs is to depict their frequency indices (FIs) as functions of their energies h in a frequency-energy plot (FEP). A first introduction of this type of plots was made in Section 2.1, where it was shown that they clearly depict and clarify the bifurcations that generate or eliminate the different branches of periodic solutions (NNMs) of a Hamiltonian system. To construct the FEP of the Hamiltonian system (3.6), the FI of a NNM on branches $Snm\pm$ and $Unm\pm$ is defined as the ratio of its two indices multiplied by the *driving* frequency ω_f of the system on the branch, i.e., $FI = n\omega_f/m$; the driving frequency is the frequency of the harmonic component closest to the natural frequency of the LO, $\omega_0 = 1$ rad/s, and slightly varies from one branch to another and even along the same branch. For instance, $S21\pm$ is characterized by the frequency index $FI = (2/1) \times 0.97 = 1.94$, as is $U21\pm$, and $S13\pm$ is characterized by $FI = (1/3) \times 1.05 = 0.35$. This rule holds for every branch, the only exception being the two branches $S11\pm$, which form the main backbones of the FEP. For these two backbone branches we utilize as FI the common dominant frequency of oscillation of the LO and the NES (as the condition of 1:1 resonance is satisfied pointwise on these branches).

In Figure 3.20 we depict the FEP of the Hamiltonian system (3.6) for parameters $\varepsilon = 0.05, \omega_0^2 = C = 1$. A periodic orbit (NNM) is represented by a point in the plot. A NNM branch, represented by a solid line, is a collection of periodic orbits possessing the same qualitative features. Bifurcation points are also indicated in that plot, with (+) and (o) used to indicate changes of stability. We note that, if the system was linear (i.e., if the essential cubic nonlinearity was replaced by a linear stiffness), the FEP would merely consist of two horizontal lines appearing passing through two natural frequencies of the corresponding two-DOF system. A consequence of the frequency convention (FI) adopted is that smooth transitions between certain branches translate to 'jumps' in the FEP (see for instance the dashed line between $S15\pm$ and $S11-$ in Figure 3.20). The complexity of the FEP is an indication of the complexity and richness of the nonlinear dynamics of this two-DOF Hamiltonian system. As discussed previously, this is a consequence of the high degeneracy of the dynamical system (3.6).

To understand the different types of periodic motions realized in this system, close-ups of several branches are provided in Figure 3.21. The corresponding periodic orbits are represented in the configuration plane (v, x) of the system. The aspect ratio is set so that increments on the horizontal and vertical axes are equal in size, enabling one to directly deduce whether the motion is localized to the LO (vertical line) or to the NES (horizontal line).

Although a systematic analytic study of the various types of periodic solutions of the FEP is postponed until in the next section, some preliminary remarks are due at this point. The backbones of the FEP are formed by NNM branches on which the

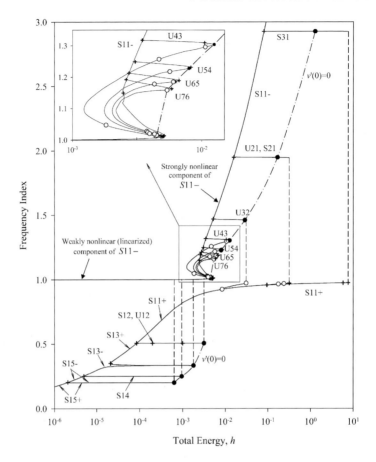

Fig. 3.20 Frequency-energy plot (FEP) depicting the periodic orbits of the Hamiltonian system (3.6); impulsive orbits (IOs) are denoted by bullets (•); bifurcation points are denoted by (+) when four Floquet multipliers are equal to +1, and (o) when two Floquet multipliers are equal to +1 and two to −1.

system response is nearly monochromatic (see, Figure 3.18a). Specifically, in- and out-of-phase synchronous vibrations of the two particles are realized on $S11+$ and $S11-$, respectively. These NNMs are strongly nonlinear analogs (continuations) of the in-phase and out-of-phase linear normal modes of the corresponding two-DOF linear system with all stifnesses being linear. However, unlike the classical linear normal modes, the shapes and frequencies of the NNMs are energy dependent.

The natural frequency of the LO ($\omega_0 = 1$ rad/s, identified by a frequency index equal to unity) divides naturally the NNMs into higher- and lower-frequency nonlinear modes. Figure 3.21a depicts the NNMs on the higher-frequency out-of-phase branch $S11-$. Due to their energy dependence, they become localized to the LO or to the NES as $\omega \to 1+$ or $\omega \gg 1$, respectively. Two saddle-node bifurcations can also be observed on this branch. In Figure 3.21b, the NNMs on the lower-frequency

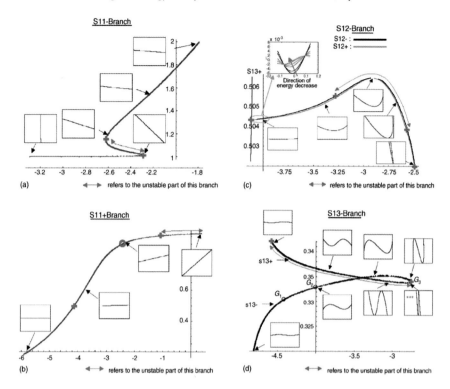

Fig. 3.21 Close-ups of specific branches of the FEP: (a) $S11-$; (b) $S11+$; (c) $S12\pm$; (d) $S13\pm$; (e) $S21\pm$; (f) $U21$; (g) $U43$; (h) $U65$; (i) $U12$; stability-instability boundaries are represented as in Figure 3.20 and IOs are indicated by triple asterisks; the plots for $U43$ and $U65$ consist of two nearly spaced branches, but only one of these is presented for clarity; since the motion is nearly identical on the two branches composing $S12$, $S21$, $U12$ and $U21$ (c, e, i, f), only the oscillations on one of the there branches are depicted in the configuration plane.

in-phase branch $S11+$ are depicted; these motions localize to the nonlinear attachment as the total energy in the system decreases. For further energy decrease, $S11+$ ceases to exist and is continued by $S13\pm$, $S15\pm$, etc., as shown in Figure 3.20.

There is a sequence of higher- and lower-frequency branches of subharmonic periodic motions $Snm\pm$ and $Unm\pm$ with $m \neq n$. These NNM branches are termed *subharmonic tongues*, and they bifurcate out from the backbone branches $S11\pm$. Unlike the NNMs on the backbones, the tongues consist of multi-frequency periodic solutions (see, i.e., Figures 3.19b, c). Specifically, each tongue occurs in the neighborhood of an internal resonance between the LO and the NES. Due to the essential nonlinearity of the system (3.6), there exists a countable infinity of tongues $Snm\pm$ and $Unm\pm$ in the FEP. This means that the NES is capable of engaging in every possible $n{:}m$ internal resonance with the LO, with n and m being relative prime integers; clearly, only a subset of these tongues can be represented in Figure 3.20. We mention at this point that the existence of a countable infinity of periodic orbits for

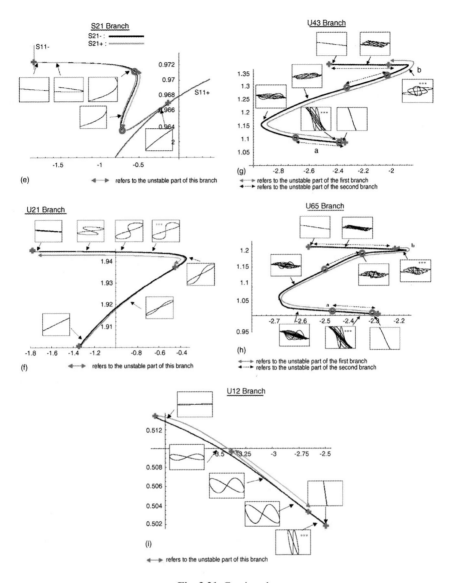

Fig. 3.21 Continued.

this system can be proved rigorously by applying subharmonic Melnikov analysis (Guckenheimer and Holmes, 1983; Wiggins, 1990). As discussed in Veerman and Holmes (1985, 1986) the generation of infinitely countable subharmonic orbits is related to the *non-integrability* (Lichtenberg and Lieberman, 1983) of the Hamiltonian system (3.6). Specifically, these countable infinities of subharmonic motions are generated from the breakdown of invariant KAM tori, and they give rise to low-scale chaotic layers close to the corresponding resonance bands of these motions. The generation and stability of subharmonic motions in non-integrable Hamiltonian systems can be studied through the use of averaging methodologies (Holmes and Marsden, 1982; Greenspan and Holmes, 1983; Veerman and Holmes, 1986; Wiggins, 1990).

A few symmetric tongues are now described in more detail. The periodic motions on these branches are also considered as NNMs according to our extended definition introduced in Section 2.1 (for a study of non-synchronous NNMs in systems with internal resonances in appropriately defined modal spaces, see Vakakis et al., 1996).

- The family $S1(2k+1)\pm$, $k = 1, 2$, etc., exists in neighborhoods of frequency indices $FI = 1/(2k+1)$. Each family is composed of two in-phase and out-of-phase branches. For fixed k each of the two branches $S1(2k+1)\pm$ is linked through a smooth transition with its neighboring branches $S1(2k-1)\pm$ or $S1(2k+3)\pm$, and exists only for a finite energy interval. The pair $S1(2k+1)\pm$ is eliminated through a saddle node bifurcation at a higher energy level, as illustrated in Figure 3.21d for branches $S13\pm$.
- The pairs of branches $S1(2k)\pm$, $k = 1, 2$, etc. bifurcate from the branches $S1(2k+1)\pm$. For instance, the coalescence of the pair of branches $S12\pm$ with the branch $S13+$ for decreasing energy is depicted in Figure 3.21c.
- The families $Sn1\pm$, $n = 2, 3$, etc. appear in neighborhoods of frequency indices $FI = n$, i.e., at progressively higher frequencies with increasing n. These tongues emanate from $S11-$ and coalesce with $S11+$ at higher energies. These coalescences seem to occur through jumps represented by dashed lines in the FEP of Figure 3.20, but as explained previously this is an artifact of the frequency convention (frequency indexing) adopted in the FEP.
- Focusing now on the unsymmetric tongues, the family $U(m+1)m$ bifurcates from branch $S11-$. At a higher energy level, the two branches composing the tongues are eliminated through saddle-node bifurcations. An additional family of unsymmetric solutions is $Um(m+1)$ and in Figure 3.20 this family is depicted only for frequency indices $FI < 1$. The shapes of these orbits in the configuration plane are similar to those of $U(m+1)m$, but rotated by $\pi/2$, as illustrated in Figures 3.21f, i. Periodic motions on the unsymmetric tongues are not NNMs because there exist non-trivial phases between the two oscillators, so they correspond to Lissajous curves in the configuration plane.

As mentioned previously, there exist special periodic orbits on the tongues that satisfy the conditions $\dot{x}(-T/4) \neq 0$ and $v(-T/4) = \dot{v}(-T/4) = x(-T/4) = 0$. These orbits, termed *impulsive orbits* (IOs), have important practical significance, since they correspond to impulsive forcing of the LO of system (3.6). These orbits are indicated by bullets in Figure 3.20 and triple asterisks in Figure 3.21. In principle, IOs can be realized on any subharmonic tongue, with the exception of tongues on which the periodic orbits do not pass through the origin of the configuration plane (for example, $S12\pm$). As far as the backbone curves are concerned, an IO on the out-of-phase branch $S11-$ is realized only asymptotically as the energy tends to zero, and the motion is completely localized to the LO; similarly, there is no finite-energy IO on branch $S11+$.

3.3.2 Analytic Study of Periodic Orbits (NNMs)

In an effort to better understand the dynamics and localization phenomena that occur in different frequency/energy ranges of system (3.6), we proceed to the analytical study of the periodic solutions shown in the FEP of Figure 3.20. Representative examples of this analysis will be given for periodic orbits on the backbone branches $S11\pm$ and on selected subharmonic tongues, namely $S13\pm$ and $U21\pm$.

Without loss of generality we assume that $\omega_0^2 = 1$ and express system (3.6) in the form:

$$\ddot{x} + x + C(x - v)^3 = 0$$
$$\varepsilon\ddot{v} + C(v - x)^3 = 0 \quad (3.11)$$

The analysis will be based on the complexification-averaging method (CX-A) first introduced by Manevitch (1999) and briefly outlined in Section 2.4. This technique will also be applied in later sections to analyze the strongly nonlinear transient dynamics of the damped version of system (3.6).

3.3.2.1 Backbone Branches $S11\pm$

The backbone branches $S11\pm$ correspond to motions where the two oscillators of the system possess identical dominant frequency components. The analytical study is performed by applying the CX-A methodology through a slow-fast partition of the dynamics. Following the method, we introduce the new complex variables:

$$\psi_1 = \dot{x} + j\omega x \quad \text{and} \quad \psi_2 = \dot{v} + j\omega v \quad (3.12)$$

where ω is the dominant (fast) frequency of oscillation and $j = (-1)^{1/2}$. Expressing the displacements and accelerations of the linear and nonlinear oscillators of the system in terms of the new complex variables, we obtain,

$$x = \frac{\psi_1 - \psi_1^*}{2j\omega}, \quad \ddot{x} = \dot{\psi}_1 - \frac{j\omega}{2}(\psi_1 + \psi_1^*)$$

$$v = \frac{\psi_2 - \psi_2^*}{2j\omega}, \quad \ddot{v} = \dot{\psi}_2 - \frac{j\omega}{2}(\psi_2 + \psi_2^*) \quad (3.13)$$

where the asterisk denotes complex conjugate.

Since nearly monochromatic (at fast frequency ω) periodic solutions of the equations of motion are sought, and since we make the assumption that the two oscillators vibrate with the same fast frequency, the previous complex variables are approximately expressed in terms of fast oscillations of frequency ω, $e^{j\omega t}$, modulated by slowly varying (complex) amplitudes $\phi_i(t)$, $i = 1, 2$:

$$\psi_1(t) = \phi_1(t)e^{j\omega t} \quad \text{and} \quad \psi_2(t) = \phi_2(t)e^{j\omega t} \tag{3.14}$$

This amounts to partitioning the dynamics into slow- and fast-varying components, with the modulations of the (approximately harmonic) fast oscillations $e^{j\omega t}$ providing the essential slow flow dynamics of the system.

Hence, the study of periodic orbits on NNM branches $S11\pm$ is reduced to studying the slow flow dynamics. This pattern of reducing the problem to the slow flow dynamics by means of CX-A analysis will be used throughout this work, as a means to separate the essential (slow flow) from the unessential (fast-flow) dynamics of the problem. Note that no *a priori* restriction was imposed on the frequency ω of the fast oscillation, which allows us to develop asymptotic approximations of the NNM branches over their entire domains of existence (since the fast frequency may vary within a branch of NNMs). At the same time, by the *ansatz* (3.14) we assume that the periodic motion is dominated by a *single* fast frequency harmonic (which holds for periodic motions on both backbone braches $S11\pm$). The analysis of more complex periodic motions (for example, NNMs on subharmonic tongues) dictates more complicated assumptions than (3.14); examples of such more involved analyses are provided in the following sections.

Substituting expressions (3.14) and (3.13) into (3.11) yields the following alternative expressions for the equations of motion, which are exact up to this point:

$$\dot{\phi}_1 e^{j\omega t} + \phi_1 j\omega e^{j\omega t} - \frac{j\omega}{2}(\phi_1 e^{j\omega t} + \phi_1^* e^{-j\omega t}) + \frac{\phi_1 e^{j\omega t} - \phi_1^* e^{-j\omega t}}{2j\omega}$$

$$+ C\left(\frac{\phi_1 e^{j\omega t} - \phi_1^* e^{-j\omega t} - \phi_2 e^{j\omega t} + \phi_2^* e^{-j\omega t}}{2j\omega}\right)^3 = 0$$

$$\varepsilon\left[\dot{\phi}_2 e^{j\omega t} + \phi_2 j\omega e^{j\omega t} - \frac{j\omega}{2}(\phi_2 e^{j\omega t} + \phi_2^* e^{-j\omega t})\right]$$

$$- C\left(\frac{\phi_1 e^{j\omega t} - \phi_1^* e^{-j\omega t} - \phi_2 e^{j\omega t} + \phi_2^* e^{-j\omega t}}{2j\omega}\right)^3 = 0 \tag{3.15}$$

The basic approximation related to the CX-A technique is that we perform averaging of equations (3.15) with respect to the fast frequency ω, after which only terms containing the fast frequency remain (to a first approximation). This leads to the following set of complex modulation equations, which constitute the approximate *slow flow reduction* of the dynamics:

$$\dot{\phi}_1 + (j\omega/2)\phi_1 - (j/2\omega)\phi_1 + (jC/8\omega^3)$$
$$\times (-3|\phi_1|^2\phi_1 + 3\phi_1^2\phi_2^* - 3\phi_2^2\phi_1^* + 3|\phi_2|^2\phi_2 + 6|\phi_1|^2\phi_2 - 6|\phi_2|^2\phi_1) = 0$$

$$\varepsilon[\dot{\phi}_2 + (j\omega/2)\phi_2] - (jC/8\omega^3)$$
$$\times (-3|\phi_1|^2\phi_1 + 3\phi_1^2\phi_2^* - 3\phi_2^2\phi_1^* + 3|\phi_2|^2\phi_2 + 6|\phi_1|^2\phi_2 - 6|\phi_2|^2\phi_1) = 0 \quad (3.16)$$

Introducing the polar representations $\phi_1 = Ae^{j\alpha}$ and $\phi_2 = Be^{j\beta}$ in equation (3.16), where A, B are real amplitudes and α, β real phases, and setting separately the real and imaginary parts of the resulting equations equal to zero, the following set of real modulation equations is obtained governing the slow evolution of the real amplitudes and phases of the modulations $\phi_i, i = 1, 2$:

$$\dot{A} + \frac{BC}{8\omega^3}[(3A^2 + 3B^2)\sin(\alpha - \beta) + 3AB\sin(2\beta - 2\alpha)] = 0$$

$$A\dot{\alpha} + \frac{\omega A}{2} - \frac{A}{2\omega} - \frac{3CA^3}{8\omega^3} - \frac{6AB^2C}{8\omega^3}$$
$$- \frac{BC}{8\omega^3}[(-9A^2 - 3B^2)\cos(\alpha - \beta) + 3AB\cos(2\beta - 2\alpha)] = 0$$

$$\varepsilon\dot{B} - \frac{AC}{8\omega^3}[(3B^2 + 3A^2)\sin(\alpha - \beta) + 3AB\sin(2\beta - 2\alpha)] = 0$$

$$\varepsilon B\dot{\beta} + \frac{\varepsilon\omega B}{2} - \frac{3B^3C}{8\omega^3} - \frac{6A^2BC}{8\omega^3}$$
$$- \frac{AC}{8\omega^3}[(-9B^2 - 3A^2)\cos(\alpha - \beta) + 3AB\cos(2\beta - 2\alpha)] = 0 \quad (3.17)$$

The first and third of equations (3.17) that describe the evolutions of the two real amplitude modulations, can be combined to yield

$$\dot{A} + \frac{\varepsilon B\dot{B}}{A} = 0 \Rightarrow A^2 + \varepsilon B^2 = N^2 \quad (3.18)$$

where N is a constant of integration. Clearly, (3.18) is a conservation-of-energy-like relation for the slow flow, as it is directly linked to conservation of total energy in the undamped system (3.11) during free oscillation. It follows that the modulation equations (3.17) can be reduced by one, with the addition of the algebraic relation (3.18).

The periodic solutions on the backbone branches $S11\pm$ are computed by setting the derivatives with respect to time in (3.17) equal to zero, i.e., by imposing stationarity conditions on the modulation equations. The resulting first and third equations are trivially satisfied if we assume identity of phases, $\alpha = \beta$, whereas the second and fourth equations become:

$$\frac{\omega A}{2} - \frac{A}{2\omega} - \frac{3CA^3}{8\omega^3} - \frac{6AB^2C}{8\omega^3} - \frac{BC}{8\omega^3}[-9A^2 - 3B^2 + 3AB]$$

$$= \frac{\omega A}{2} - \frac{A}{2\omega} - \frac{3C}{8\omega^3}(A-B)^3 = 0$$

$$\frac{\varepsilon\omega B}{2} - \frac{3B^3 C}{8\omega^3} - \frac{6A^2 BC}{8\omega^3} - \frac{AC}{8\omega^3}[-9B^2 - 3A^2 + 3AB]$$

$$= \frac{\varepsilon\omega B}{2} + \frac{3C}{8\omega^3}(A-B)^3 = 0 \qquad (3.19)$$

The amplitudes A and B can be estimated by combining these equations, which leads to the following analytic expressions for the periodic motions (NNMs) on the backbone branches $S11\pm$:

$$x(t) \approx X \cos \omega t = \frac{\psi_1 - \psi_1^*}{2j\omega} = (A/\omega)\cos\omega t$$

$$= \left(\frac{-\varepsilon\omega^2}{\omega^2-1}\right)\left\{\frac{4\omega^2\varepsilon(\omega^2-1)^3}{3C[(1+\varepsilon)\omega^2-1]^3}\right\}^{1/2}\cos\omega t$$

$$v(t) \approx V \cos \omega t = \frac{\psi_2 - \psi_2^*}{2j\omega} = (B/\omega)\cos\omega t$$

$$= \left\{\frac{4\omega^2\varepsilon(\omega^2-1)^3}{3C[(1+\varepsilon)\omega^2-1]^3}\right\}^{1/2}\cos\omega t \qquad (3.20)$$

Since a single fast frequency was assumed in the slow-fast partitions (3.14), and only terms containing this fast frequency were retained after averaging the complex equations (3.15), the analytical expressions (3.20) are only approximations of the original dynamics of (3.11).

It is interesting to note that the ratios of the amplitudes of the linear and nonlinear oscillators on both branches $S11\pm$ are given approximately by the following simple form:

$$\frac{X}{V} = \frac{-\varepsilon\omega^2}{\omega^2 - 1} \qquad (3.21)$$

This relation shows that if the mass ε of the NES is small (as assumed in this study), and the frequency is not close to unity, the motion is always localized to the NES. Indeed, as one would expect intuitively, the oscillation localizes to the LO sufficiently close to its resonant frequency $\omega = \omega_0 = 1$. This result is compatible to the fact that NNM *branches of the* FEP *with large curvatures represent strongly nonlinear oscillations, as they correspond to strong dependence of frequency on energy.*

There is a region in the frequency domain, $\sqrt{1/(1+\varepsilon)} < \omega < 1$, where the coefficients X and V are imaginary, indicating that no in-phase or out-of-phase NNMs can occur there. Indeed, the in-phase backbone branch $S11+$ exists only for $\omega \leq \sqrt{1/(1+\varepsilon)}$, whereas the out-of-phase branch $S11-$ for $\omega \geq 1$. Moreover, the analytical approximations of branches in the FEP are computed by noting that the conserved energy of the system is given by

A better estimate can be obtained if an additional third harmonic component is included in the *ansatz* (3.31–3.33). From (3.38) we conclude that the frequency of the IO depends essentially on the mass ratio ε; moreover, it can be proven that it does not depend on the coefficient of the nonlinearity of the attachment. As $\varepsilon \to 0$ it can be shown that $\omega_{IO}^{(U21)} \to 1$, and the frequency of the IO tends to the natural frequency of the linear oscillator.

The degree of localization of the IO can be estimated by considering the ratio V/Y of the maximum amplitudes attained by the nonlinear attachment and the linear oscillator during one period of the motion. It can be shown that this ratio is independent of the coefficient of nonlinearity of the attachment, and the stiffness of the linear oscillator, but depends only on the mass ratio ε. Moreover, stronger localization to the nonlinear attachment occurs for small mass ratios, with $V/Y \to 1.65$ as $\varepsilon \to 0$. It is interesting to note that this localization limit appears to be independent of the actual parameters of the system, and depends only on its configuration. These results show that *best localization results for the IO are realized for light attachments, and that the degree of localization obtained in the limit of small mass ratios reaches a parameter-independent limit.* As mentioned in the previous section, if a stable, localized impulsive orbit is excited by external forcing or by the initial conditions of the system, then during the first cycle of the motion energy is rapidly transferred from the directly excited LO to the nonlinear attachment, and from there on a continuous exchange of energy between the two oscillators occurs in the form of a nonlinear beat phenomenon; as shown in the next section, the excitation of such nonlinear beats provides conditions for the realization of efficient TET in the damped, impulsively forced system. This issue will be studied in detail in the following exposition.

Similar analysis can be performed to model the dynamics of nonlinear beat phenomena on the other unsymmetric branches $Um(m + 1)$ and $U(m + 1)m$. We note that due to the essential nonlinearity of the system considered, the nonlinear beat phenomena on the U-branches do not require any *a priori* 'tuning' of the nonlinear attachment, since at specific frequency-energy ranges the nonlinear attachment passively 'tunes itself' in an internal resonance with the linear oscillator. This represents a significant departure from the 'classical' nonlinear beat phenomena observed in coupled oscillators with linearizable nonlinear stiffnesses [for example, in a spring-pendulum system (Nayfeh and Mook, 1985)], where the ratio of the linearized natural frequencies of the components dictates the possible types of internal resonances that can be realized. This observation further highlights the enhanced versatility of the NES as vibration absorber due to its essential stiffness nonlinearity.

A systematic analytical study of IOs of the Hamiltonian system (3.11) is postponed until Section 3.3.4, whereas a numerical study of these special orbits is performed in the next section.

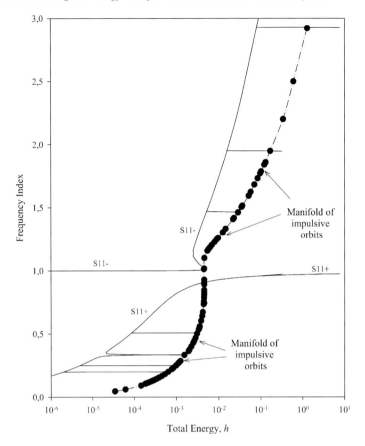

Fig. 3.25 Manifold of impulsive orbits (IOs) represented in the FEP; periodic impulsive orbits are denoted by bullets (•).

3.3.3 Numerical Study of Periodic Impulsive Orbits (IOs)

In Section 3.3.1.2 we discussed the existence of periodic and quasi-periodic IOs, which correspond to non-zero initial velocity of the LO with all other initial conditions zero. Since these are the exact orbits that are directly excited after the application of an impulsive excitation to the LO, they have an important significance in practical applications of TET.

An extensive series of computations of IOs was carried out employing the numerical algorithm described in Section 3.3.1.1 with the additional restriction that $\dot{v}(0) = 0$. The results are presented in Figure 3.25. Because the NES is capable of engaging in a countable infinity of $n : m$ internal resonances with the LO, with n and m being relative prime integers, there exists a countable infinity of periodic IOs, which are aligned along a smooth curve in the FEP. In addition, one can reason-

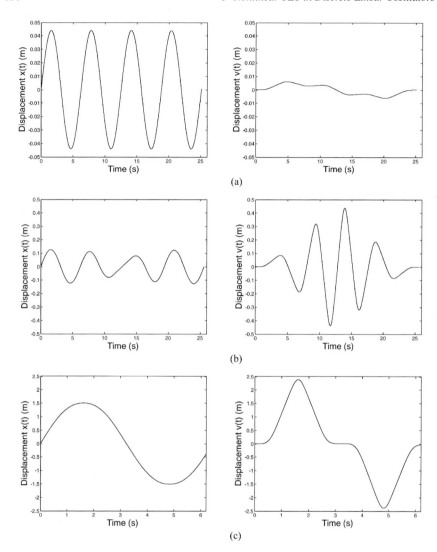

Fig. 3.26 Time series of representative periodic IOs for $\varepsilon = 0.05$, $C = 1$: (a) low-energy IO $U14$; (b) moderate-energy IO $U54$; (c) high-energy IO $S31$.

ably assume that there exists an uncountable infinity of quasi-periodic IOs, which correspond to irrational ratios of frequencies of oscillation of the LO and the NES. The periodic and quasi-periodic IOs form a smooth manifold of solutions in the FEP, which is of significant practical importance. This is due to the fact that this manifold provides the impulse magnitude needed to excite an IO for a specified frequency.

As shown below a subset of periodic IOs represents stable oscillations of the system that strongly localize to the NES. It follows that if such a stable periodic IO is excited by an external shock, strong energy transfer from the directly excited LO to the NES takes place over a period of the oscillation. In the Hamiltonian system such an IO is repeated as time progresses, and energy gets continuously exchanged between the LO and the NES; however, in the weakly damped system, an initial excitation of a stable, periodic IO that is strongly localized to the NES leads to strong TET, and, in fact, as shown in the following sections this represents one of three possible mechanisms for generating TET in the damped system.

The time series of three representative periodic IOs are depicted in Figure 3.26. Comparing the relative magnitudes attained by the linear and nonlinear oscillators in each of the IOs depicted in that Figure, we note the following. The low-energy periodic IO $U14$, which corresponds to a 1:4 internal resonance between the two oscillators, is localized to the LO (see Figure 3.26a). If this orbit is excited by an external shock, a very small fraction of the input energy is transferred to the NES during the nonlinear beating phenomenon. The moderate-energy IO $U54$ (see Figure 3.26b) is strongly localized to the NES. The excitation of this orbit channels a major portion of the induced energy from the directly excited LO to the nonlinear attachment during a period of the oscillation. Regarding the high-energy periodic IO $S31$ (see Figure 3.26c), the NES still undergoes a motion with a larger amplitude than that of the LO, but localization to the NES is less pronounced compared to the IO $U54$.

In Figure 3.27 representative periodic IOs are depicted in the configuration plane (v, x) of the Hamiltonian system. By construction, these IOs have a common feature: each orbit passes with vertical slope through the origin of the configuration plane. These plots indicate that low-energy periodic IOs with $X \leq 0.078$ are localized to the LO (where $\dot{x}(0) = X$ is the only non-zero initial condition of the IO). In contrast, moderate-energy periodic IOs in the range $X \in [0.104, 0.158]$ are localized to the NES. As far as the high-energy periodic IOs with $X \geq 0.58$ are concerned, energy is shared between the two oscillators.

Due to the significance of IOs as a basic underlying mechanism for realizing TET in the damped system, in the following section we provide an extensive analytical study of the manifold of IOs in the FEP of system (3.11). Due to the complexity of the problem, it turns out that we need to perform three separate analytical studies of IOs, in the high-, moderate- and low-energy regimes, respectively.

3.3.4 Analytic Study of Periodic and Quasi-Periodic IOs

Motivated by the numerical results of the previous section, low-energy (i.e., $S1m$ and $U1m, m > 1$), moderate-energy (i.e., $U(k+1)k, k > 1$) and high-energy (i.e., $Sn1$ and $Un1, n > 1$) impulsive orbits will be analyzed separately. To this end, we reconsider the undamped Hamiltonian system,

$$\ddot{x} + x + C(x - v)^3 = 0$$

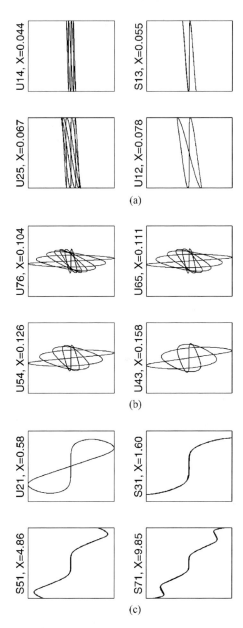

Fig. 3.27 Representative periodic IOs in the configuration plane (for $\varepsilon = 0.05$, $C = 1$): (a) low-; (b) moderate- and (c) high-energy orbits; the horizontal and vertical axes represent the NES and LO displacements, respectively, and their aspect ratio is set so that increments on the horizontal and vertical axes are equal in size, enabling one to directly deduce whether the motion is localized to the LO (near vertical) or to the NES (near horizontal).

$$\varepsilon \ddot{v} + C(v-x)^3 = 0 \tag{3.39}$$

where, as previously, we assume that $0 < \varepsilon \ll 1$, indicating a lightweight nonlinear attachment. We recall that an IO of the dynamical system (3.39) is defined as the orbit corresponding to initial conditions $v(0) = \dot{v}(0) = x(0) = 0$ and $\dot{x}(0) \neq 0$. The singularity in the second of equations (3.39) (since as $\varepsilon to 0$ the highest derivative is eliminated) can be removed by introducing the following rescalings:

$$x \to (8\varepsilon/C)^{1/2} x, \quad v \to (8\varepsilon/C)^{1/2} v \tag{3.40}$$

so that (3.39) can be transformed into the form

$$\ddot{x} + x + 8\varepsilon(x-v)^3 = 0$$
$$\ddot{v} + 8(v-x)^3 = 0 \tag{3.41}$$

subject to initial conditions $v(0) = \dot{v}(0) = x(0) = 0$ and $\dot{x}(0) = X$. The additional coordinate transformation,

$$y_1 = x + \varepsilon v, \quad y_2 = x - v \tag{3.42}$$

renders the dynamical system into the following final form,

$$\ddot{y}_1 + \frac{y_1 + \varepsilon y_2}{1+\varepsilon} = 0$$
$$\ddot{y}_2 + \frac{y_1 + \varepsilon y_2}{1+\varepsilon} + 8(1+\varepsilon)^3 y_2^3 = 0 \tag{3.43}$$

subject to initial conditions:

$$\dot{y}_1 = \dot{y}_2 = Y \neq 0, \quad y_1 = y_2 = 0 \tag{3.44}$$

Note that for notational consistency we have replaced in (3.44) the initial condition X by Y. In physical terms, the new coordinate y_1 denotes the motion of the center of mass of the two oscillators, whereas coordinate y_2 their relative response. These new coordinates are natural for describing and studying TET phenomena in the corresponding weakly damped system, since the capacity of the nonlinear attachment to passively absorb and locally dissipate energy from the LO depends on the relative displacement y_2 and its derivative, rather on the absolute response v.

The dynamical system (3.43–3.44) is equivalent to systems (3.39) and (3.41), and has the advantage that the small parameter does not multiply any of the time derivatives of the dependent variables. Hence, system (3.43–3.44) is considered in the following analytical study of IOs. Examining (3.44) we deduce that, correct to first order, the center of mass of the system undergoes a linear oscillation of unit frequency, whereas the relative motion between the LO and the nonlinear attachment is governed by a strongly nonlinear ordinary differential equation with cubic nonlinearity. This $O(1)$ partition of the linear and nonlinear dynamics is one addi-

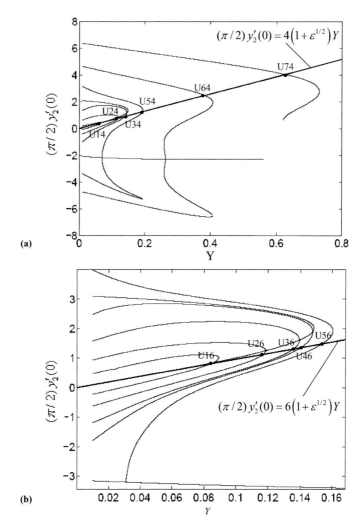

Fig. 3.28 Graphic computation of periodic IOs (•) from the bifurcation diagram of periodic orbits: (a) $k = 4$, $\varepsilon = 0.05$; (b) $k = 6$, $\varepsilon = 0.05$.

tional advantage for considering the transformed dynamical system (3.43–3.44) in the following analysis.

Introducing the rescaled time $\tau = \omega t$, where ω is a characteristic frequency of the motion, solving the first of equations (3.43) and substituting into the second, the dynamical system is reduced to the following form:

$$y_1(\tau) = (1+\varepsilon)^{1/2} Y \sin k\tau + O(\varepsilon)$$

$$y_2''(\tau) + 8(1+\varepsilon)^2 k^2 y_2^3(\tau) = -k^2 (1+\varepsilon)^{1/2} Y \sin k\tau + O(\varepsilon)$$

$$y_2(0) = 0, \quad y_2'(0) = Y/\omega, \quad k = \omega^{-1}(1+\varepsilon)^{-1/2} \qquad (3.45)$$

where primes denote differentiation with respect to τ. We note that in terms of the normalized time the LO performs approximate harmonic oscillations with normalized frequency k, and the problem of computing the IOs of system (3.39) is reduced to solving the second of equations (3.45). We note at this point that the reduced system approximates well the original system (3.39) only at moderate or large energies of the motion, i.e., in response regimes where the $O(1)$ approximations dominate over the (omitted) $O(\varepsilon)$ corrections. At low energies, however, $O(\varepsilon)$ terms are expected to play a dominant role in the response, so the reduced system (3.45) may not be used to approximate the response in these regimes.

Approximations to the periodic IOs are computed by imposing on (3.45) the periodicity condition $y_2(\tau) = y_2(\tau + 2\pi)$, $\forall \tau \in R^+$, and the additional initial conditions $y_2(0) = 0$ and $y_2'(0) = Y/\omega$. Note that by imposing the 2π-periodicity condition on $y_2(\tau)$ we impose the additional restriction of integer values for $k \in N^+$. In Figure 3.28 we present the graphic computation of periodic IOs. The bifurcation diagrams in these plots depict the 2π-periodic solutions of $y_2(\tau)$ that satisfy only the initial condition $y_2(0) = 0$, $\varepsilon = 0.05$ and $k = 4, 6$; the corresponding normalized initial conditions $(\pi/2) y_2'(0)$ are depicted *versus* the initial condition Y. The periodic IOs are then computed as intersections of the plots in these bifurcation diagrams with the lines $(\pi/2) y_2'(0) = Y/\omega = k(1+\varepsilon^{1/2})Y$ since at these intersections the second initial condition in (3.45) is satisfied as well. The classification of the impulsive periodic orbits follows the notation introduced in Section 3.3.1.2 for symmetric and unsymmetric periodic orbits ($S-$ or $U-$ orbits, respectively).

In Figure 3.29 we depict some representative periodic IOs reconstructed from the approximations $y_1(\tau)$ and $y_2(\tau)$ of the reduced system (3.45), and compare them to the exact IOs computed numerically from the original equations (3.39). It can be observed that the reduced system approximates well the original system at moderate and high energies, but not at small ones. Of particular interest is the impulsive orbit $U54$ depicted in Figure 3.29b, which is in the form of a modulated signal or beat (that is, a 'fast' oscillation modulated by a 'slow' envelope). This orbit occurs at a moderate energy level, and its fast frequency is close to the eigenfrequency of the linear oscillator, so that near 1:1 internal resonance between the linear oscillator and the nonlinear attachment occurs. In the next section it will be shown that such IOs possess two close, rationally related frequency components, which when superimposed produce the observed beating behavior. It follows that for such moderate-energy impulsive orbits one can approximately partition the dynamics into slow and fast components and employ averaging arguments. No such slow-fast partition, however, of the dynamics is possible for the other types of impulsive periodic orbits depicted in Figures 3.29a, c.

When integrated by quadratures subject to the initial condition (3.51c) equation (3.51b) can be recast into the following form:

$$\int_{(Y/\omega)^2}^{N^2} \frac{du}{\left[u - (Y/\omega)^2\right]^{1/2} [I_3 + I_2 u + I_1 u^2 - u^3]^{1/2}} = \pm \int_0^\tau \frac{\beta d\xi}{2} \tag{3.52}$$

where the coefficients of the denominator of the integrand of the left-hand side are defined as follows:

$$I_1 = -(Y/\omega)^2 + (4\alpha/\beta), \quad I_2 = -(4\alpha^2/\beta^2) + (Y/\beta\omega)[8\rho + \beta(Y/\omega)^3]$$

$$I_3 = [(4\rho/\beta) - (2\alpha Y/\beta\omega) + (Y/\omega)^3]^2 \tag{3.53}$$

The definite integral (3.52) can be expressed in terms of elliptic functions (Gradshteyn and Ryzhik, 1980):

$$\int_b^y \frac{du}{(a-u)^{1/2}(u-b)^{1/2}[(u-c)(u-c^*)]^{1/2}} = g \, \text{cn}^{-1}(\cos\phi, m) = g \, F(\phi, m) \tag{3.54}$$

with

$$b < y \leq a, \quad c = b_1 + ja_1, \quad c^* = b_1 - ja_1$$

$$g = (AB)^{-1/2}, \quad m = [(a-b)^2 - (A-B)^2]/4AB$$

$$A^2 = (a - b_1)^2 + a_1^2, \quad B^2 = (b - b_1)^2 + a_1^2$$

$$\phi = \cos^{-1}\left[\frac{(a-y)B - (y-b)A}{(a-y)(B+(y-b)A)}\right] \tag{3.55}$$

In the above expressions $\text{cn}^{-1}(\bullet, \bullet)$ is the inverse Jacobi elliptic cosine, $F(\bullet, \bullet)$ the incomplete elliptic function of the second kind, and m the modulus. Expression (3.54) can be applied to solve (3.52) by assigning the parameter value $b = (Y/\omega)^2$, and computing a and (c, c^*) as the (single) real and complex pair of roots of the equation $I_3 + I_2 u + I_1 u^2 - u^3 = 0$, respectively. As a result, the solution of (3.52) is given by:

$$\cos\phi = \text{cn}\left(\frac{\beta\tau}{2g}, m\right) \Rightarrow \frac{(a-N^2)B - (N^2-b)A}{(a-N^2)B + (N^2-b)A} = \text{cn}\left(\frac{\beta\tau}{2g}, m\right)$$

$$\Rightarrow N^2(\tau) = \frac{(aB+bA) - (aB-bA)\text{cn}\left(\frac{\beta\tau}{2g}, m\right)}{(B+A) + (A-B)\text{cn}\left(\frac{\beta\tau}{2g}, m\right)} \tag{3.56}$$

This expression computes the amplitude squared of the slow modulation, N^2, as a function of the normalized time τ for the moderate-energy IO. It can be easily verified that the above expression satisfies the initial condition (3.51c), i.e.,

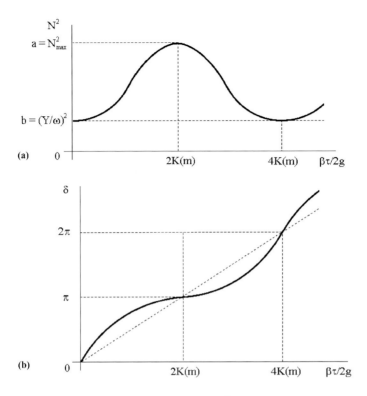

Fig. 3.30 Slow evolutions of (a) the amplitude squared $N^2(\tau)$, and (b) phase $\delta(\tau)$ of the modulation (envelope) of $y_2(\tau)$ for a moderate-energy IO under conditions of 1:1 internal resonance.

$N^2(0) = b = (Y/\omega)^2$. Once $N^2(\tau)$ is approximated through (3.56), the phase $\delta(\tau)$ of the modulation is computed through (3.51a). Schematics of the evolutions of $N^2(\tau)$ and $\delta(\tau)$ over one cycle of the IO are depicted in Figure 3.30. The analytic approximation of the response $y_2(\tau)$ is then computed by combining the expressions (3.46), (3.47) and (3.50),

$$y_2(\tau) \approx -\frac{jN(\tau)}{2k} e^{j[k\tau+\delta(\tau)]} + cc = \underbrace{-(j/2k)N(\tau)e^{j\delta(\tau)}}_{\text{Slow component}} \underbrace{e^{jk\tau}}_{\text{Fast component}} + cc$$

$$= \frac{N(\tau)}{k} \sin[k\tau + \delta(\tau)] \qquad (3.57)$$

with cc denoting the complex conjugate, and the normalized time defined according to $\tau = [k(1+\varepsilon)^{1/2}]^{-1}t$. The solution (3.57) has normalized frequency $\tilde{\Omega}(k) \approx k + \delta'(\tau)$; since $\delta'(\tau)$ is a slowly varying quantity, the normalized frequency can be approximated further as $\tilde{\Omega}(k) \approx k + \langle \delta'(\tau) \rangle_\tau$, where $\langle \bullet \rangle_\tau$ denotes average with respect to normalized time τ.

The analytic expression (3.57) approximates moderate-energy IOs of system (3.45) under condition of 1:1 internal resonance. It is interesting to note that this expression is valid for periodic as well as quasi-periodic IOs, since no periodicity condition has yet been imposed on the solution (as the initial condition Y is yet undetermined and k is assumed to be real but not necessarily integer). To compute *periodic* IOs in the region of 1:1 internal resonance, we must impose additional 2π-periodicity conditions on $y_1(\tau)$ and $y_2(\tau)$. Considering the first of expressions (3.45), 2π-periodicity of $y_1(\tau)$ implies that k must be a positive integer; this, however, does not imply necessarily that 2π is the *minimal* period of $y_1(\tau)$. Considering the approximation (3.57) for $y_2(\tau)$, a *minimal* 2π-normalized period is imposed on the amplitude $N^2(\tau)$ and phase $\delta(\tau)$; from (3.51a) and (3.56) this implies that $\text{cn}(\beta\tau/2g, m)$ must be 2π-periodic. Combining all previous arguments, we conclude that periodic moderate-energy periodic IOs are obtained provided that the following conditions are enforced:

$$k \in N^+ \quad \text{and} \quad 4K(m)\frac{2g}{\beta} = 2\pi \quad \text{(Periodic IOs)} \qquad (3.58)$$

where $K(m)$ is the complete elliptic integral of the first kind (see Figure 3.30b).

We note that by the conditions (3.58) $y_2(\tau)$ has a *minimal* normalized period equal to $2\pi/\tilde{\Omega}(k) \approx 2\pi/(k + \langle\delta'(\tau)\rangle_\tau) = 2\pi/(k+1)$, and $y_1(\tau)$ a minimal normalized period equal to $2\pi/k$. Hence, a $(k+1):k$ internal resonance occurs between the LO and the NES for the computed moderate-energy IO, which, for large values of k, satisfies the initial assumption of near 1:1 internal resonance. It follows that for sufficiently large integers k, the two oscillators of system (3.41) [and of the original dynamical system (3.39) with appropriate rescalings] execute oscillations, $x(\tau) = [y_1(\tau) + \varepsilon y_2(\tau)]/(1+\varepsilon)$ and $v(\tau) = [y_1(\tau) - y_2(\tau)]/(1+\varepsilon)$; these are indeed in the form of beats, since they represent the superposition of two signals with near identical normalized frequencies equal to k and $k+1$. Moreover, by the above construction of the IOs, the higher the positive integer k is, the closer the IO satisfies the condition of 1:1 internal resonance, and the more valid the beat assumption (and the slow-fast partition) for the IO becomes.

Summarizing, the procedure for computing an analytic approximation for a moderate-energy periodic IO is outlined below:

- Select the order of the internal resonance k
- Determine the coefficients α, β and ρ in (3.53)
- Consider a specific initial condition Y
- Compute the denominator of the integrand (3.52) using expressions (3.53)
- Compute the roots a, b, c and c^* of the denominator of the integrand of (3.54) by solving the algebraic equation $I_3 + I_2 u + I_1 u^2 - u^3 = 0$
- Compute the coefficients g and m, hence compute the coefficient $4K(m)(2g/\beta)$
- If $4K(m)(2g/\beta)$ is equal to 2π, the periodicity condition for $y_2(\tau)$ is satisfied, and the periodic IO $U(k+1)k$ is realized. If not, modify Y and return to step 4

Table 3.1 Initial conditions for moderate-energy periodic IOs.

Periodic IO		$\dot{x}(0)$ (exact)	$\dot{x}(0)$ (analytic)
$U21$	($k = 1$)	0.5794	0.2697
$U32$	($k = 2$)	0.2398	0.1675
$U43$	($k = 3$)	0.1581	0.1288
$U54$	($k = 4$)	0.1263	0.1099
$U65$	($k = 5$)	0.1115	0.0999
$U76$	($k = 6$)	0.1039	0.0944
$U87$	($k = 7$)	0.1000	0.0914
$U98$	($k = 8$)	0.0977	0.0898
$U10\text{--}9$	($k = 9$)	0.0965	0.0889

- Using (3.51a) and (3.56), compute the amplitude $N(\tau)$ and the phase $\delta(\tau)$ of the envelope of the IO
- From (3.45) and (3.57), compute $y_1(\tau)$ and $y_2(\tau)$, and transform them back to the original variables $x(t)$ and $v(t)$, taking into account the rescalings (3.40).

In Table 3.1 we present a comparison between the exact and analytically predicted initial conditions for certain moderate-energy unsymmetric periodic IOs, for the system (3.39) with $\varepsilon = 0.05$, $C = 1$. Apart from $U21$, satisfactory agreement between theory and numerics is obtained, which confirms that the previous analysis is valid near the region of 1:1 internal resonance; indeed, as predicted, the accuracy of the analytical predictions for the family of IOs $U(k+1)k$ is expected to improve with increasing k. In Figure 3.31 we present comparisons between analytical and numerical time series of the responses $x(t)$ and $v(t)$ for the periodic IOs $U43$ and $U65$. The analytical approximations were computed based on the previous analysis, whereas the numerical simulations by directly integrating the governing equations (3.39). The analytical periodic IOs can also be represented in the FEP of the Hamiltonian system, when noting that the (conserved) energy of each IO is given by $Y^2/2$, and the corresponding frequency index by

$$\omega = \frac{\tilde{\Omega}(k)}{k(1+\varepsilon)^{1/2}} \approx \frac{1}{(1+\varepsilon)^{1/2}} + \frac{1}{k(1+\varepsilon)^{1/2}} \quad (3.59)$$

The analytically predicted IOs are shown in Region I of the plot of Figure 3.32a, which when compared to the exact result of Figure 3.32b, validates the previous analytical methodology.

We end this section with some remarks. First, the periodicity conditions (3.58) can be generalized by substituting the second of these relations with the more general relation $4K(m)(2g/\beta)p = 2\pi$, $p \in N^+$, which amounts to p waveforms for $y_2(\tau)$ in the normalized interval $\tau \in [0, 2\pi]$; however, in order to ensure that the modulations $N^2(\tau)$ and $\delta(\tau)$ are still slow compared to the fast oscillation $\exp(jk\tau)$, we must require that $k \gg p$. The second remark concerns the fact that IOs not satisfying the periodicity conditions (3.58) are quasi-periodic beats that can still be partitioned in terms of slow-fast components. Indeed, by varying k one obtains a

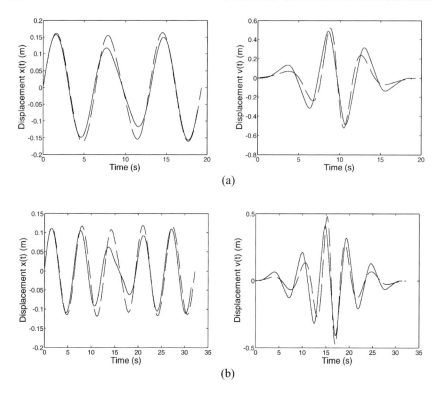

Fig. 3.31 Comparisons between analytical approximations (dashed lines) and direct numerical simulations (solid lines) of moderate-energy periodic IOs: (a) $U43$; (b) $U65$.

one-dimensional manifold possessing an uncountable infinity of quasi-periodic impulsive orbits, and a countable infinity of periodic impulsive orbits imbedded onto it. In this case, the quantity $4K(m)(2g/\beta)$ with k non-integer defines the (slow) frequency of the envelope modulation of the quasi-periodic response $y_2(\tau)$, which is a function of the initial condition Y.

As a final remark we note that relations (3.56) may be used to estimate the maximum amplitude attained by the slow envelope, $N_{\max} = a^{1/2}$, where a was defined previously as one of the real roots of the integrand in (3.54). This measure (which is valid for periodic as well as quasi-periodic IOs) is directly related to the energy passively transferred from the LO to the nonlinear attachment during a cycle of the nonlinear beat. Moreover, although during the nonlinear beat (i.e., the moderate-energy IO) energy is continuously exchanged between the LO and the nonlinear attachment, when damping is added to the Hamiltonian system (3.39) this energy exchange is replaced by targeted energy transfer (TET) to the attachment (Kerschen et al., 2005). Hence, the *maximum amplitude $N_{\max}$ of the slow envelope directly affects the effectiveness of* TET *in the system under consideration*. It can be shown that $N_{\max}$ increases with increasing k, as the 1:1 resonance region is approached

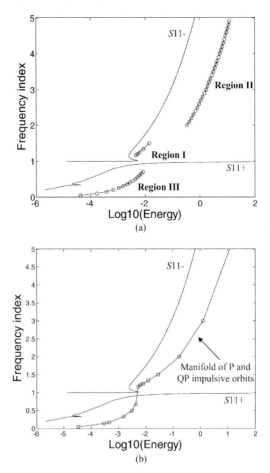

Fig. 3.32 IOs represented in the FEP of the system: (a) analytic predictions; (b) exact results; regions I, II and III correspond to moderate-, high- and low-energies, respectively.

from higher frequencies, though this increase reaches a definite limit (Lee et al., 2005). The relation between moderate-energy IOs of the Hamiltonian system and TET in the weakly damped one will be discussed in detail in later sections.

3.3.4.2 IOs at High-Energy Levels

We now proceed to analyze high-energy IOs of the general form $Sn1$ and $Un1$. Judging from the results depicted in Figure 3.26, high-energy IOs have distinctly different waveforms than moderate-energy ones, since they do not appear in the form of beats. Hence, the analytical methodology of the previous section cannot be applied for analyzing this class of IOs, and a separate analysis must be developed.

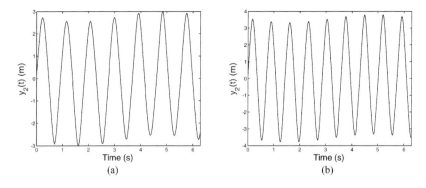

Fig. 3.33 Response $y_2(t)$ for the high-energy IOs, (a) $S71$; (b) $S91$.

To this end, the approximate dynamical system (3.45) is expressed in terms of the original time variable t, yielding:

$$y_1(t) = (1+\varepsilon)^{1/2} Y \sin[(1+\varepsilon)^{-1/2} t] + O(\varepsilon)$$
$$\ddot{y}_2(t) + 8(1+\varepsilon)y_2^3(t) + (1+\varepsilon)^{-1} y_1(t) = 0 + O(\varepsilon),$$
$$y_2(0) = 0, \quad \dot{y}_2(0) = Y \qquad (3.60)$$

At sufficiently high energy levels, the essentially nonlinear coupling stiffness behaves almost as a rigid connection. It is therefore reasonable to assume that $x(t) \approx v(t) \Rightarrow |y_1(t)| \gg |y_2(t)|$ in this regime. Then, the relative displacement is expressed as a superposition of slow and fast components

$$y_2(t) \approx \underbrace{s(t)}_{\text{Slow component}} + \underbrace{f(t)}_{\text{Fast component}} \qquad (3.61)$$

where for high-energy IOs it is natural to assume that $|f(t)| \gg |s(t)|$. This is illustrated in Figure 3.33 for the high-energy IOs S71 and S91.

Substituting (3.61) into the second of equations (3.60) and the accompanying initial conditions, yields the following differential equation possessing slow and fast varying parts:

$$\ddot{f} + \ddot{s} + 8(1+\varepsilon)(f^3 + 3f^2 s + 3fs^2 + s^3)$$
$$= -\frac{Y}{(1+\varepsilon)^{1/2}} \sin\left[\frac{t}{(1+\varepsilon)^{1/2}}\right]$$
$$s(0) + f(0) = 0 \Rightarrow s(0) = f(0) = 0,$$
$$\dot{s}(0) + \dot{f}(0) = Y \Rightarrow \dot{s}(0) = 0, \quad \dot{f}(0) = Y \qquad (3.62)$$

Setting separately equal to zero the fast and slow components of (3.62), and taking into account that, $|f(t)| \gg |s(t)|$ we find that the fast dynamics is governed by an unforced oscillator of Duffing-type,

$$\ddot{f} + 8(1+\varepsilon)f^3 = 0, \quad f(0) = 0, \quad \dot{f}(0) = Y \tag{3.63}$$

the solution of which is readily obtained in closed form,

$$f(t) = -A \operatorname{cn}\left[\eta\left(t + \frac{K(1/2)}{\eta}\right), \frac{1}{2}\right] \tag{3.64}$$

where $A = 2^{-1}(1+\varepsilon)^{1/2}Y$ and $\eta = A[8(1+\varepsilon)]^{1/2}$. The expressions $\operatorname{cn}(\bullet,\bullet)$ and $K(1/2)$ in (3.64) denote the Jacobi elliptic cosine function and the complete elliptic integral of the first kind, respectively.

Substituting (3.64) into (3.62) and averaging out the fast dynamics we obtain the following approximate dynamical system governing the slow dynamics

$$\ddot{s} + 8(1+\varepsilon)[3\langle f^2\rangle_T s + s^3] = -(1+\varepsilon)^{1/2} Y \sin[(1+\varepsilon)^{-1/2} t], \quad s(0) = \dot{s}(0) = 0 \tag{3.65}$$

where the average of the fast oscillation $\langle f^2\rangle_T$ can be explicitly computed according to (Gradshteyn and Ryzhik, 1980):

$$\langle f^2\rangle_T = \frac{1}{T}\int_0^T A^2 \operatorname{cn}^2(\eta t, 1/2) dt = \frac{A^2}{K(1/2)}[E(\pi, 1/2) - 2K(1/2)] \tag{3.66}$$

where $T = 4K(1/2)/2\eta$, and $E(\bullet,\bullet)$ is the incomplete elliptic function of the second kind.

Because $|s(t)| \ll 1$, to a first approximation the cubic term can be neglected in (3.65), so the slow flow dynamical system can be reduced approximately to the following linear system that can be solved explicitly:

$$s(t) \approx \frac{-Y(1+\varepsilon)^{1/2}}{24(1+\varepsilon)^2\langle f^2\rangle_T - 1} \sin[(1+\varepsilon)^{-1/2} t]$$

$$+ \frac{Y \sin[24(1+\varepsilon)\langle f^2\rangle_T t]}{[24(1+\varepsilon)^2\langle f^2\rangle_T - 1][24(1+\varepsilon)\langle f^2\rangle_T]} \tag{3.67}$$

Combining the solutions (3.64) and (3.67), the relative displacement $y_2(t)$ for the high-energy IO can be approximated by the analytical expression:

$$y_2(t) \approx \underbrace{-A \operatorname{cn}\left[\eta\left(t + \frac{K(1/2)}{\eta}\right), \frac{1}{2}\right]}_{\text{Fast component}} \tag{3.68}$$

Table 3.2 Initial conditions for high-energy periodic IOs.

Periodic IO	$\dot{x}(0)$ (exact)	$\dot{x}(0)$ (analytic)
$U21\ (n=2)$	0.58	0.82
$S31\ (n=3)$	1.59	1.84
$S51\ (n=3)$	4.85	5.12
$S71\ (n=7)$	9.75	10.03
$U81\ (n=8)$	12.80	13.10
$S91\ (n=9)$	16.30	16.58

$$+ \underbrace{\frac{-Y(1+\varepsilon)^{1/2}}{24(1+\varepsilon)^2 \langle f^2 \rangle_T - 1} \sin[(1+\varepsilon)^{-1/2} t] + \frac{Y \sin[24(1+\varepsilon)\langle f^2 \rangle_T\, t]}{[24(1+\varepsilon)^2 \langle f^2 \rangle_T - 1][24(1+\varepsilon)\langle f^2 \rangle_T]}}_{\text{Slow component}}$$

Then, the IO in terms of the original variables can be evaluated by combining the first of expressions (3.60) and (3.68), and inversing the coordinate transformations $y_1 = x + \varepsilon v$, $y_2 = x - v$.

To compute the initial condition Y corresponding to a specific high-energy periodic IO, a periodicity condition similar to that for the moderate-energy case should be imposed. This periodicity condition is formulated as follows:

$$n \frac{4K(1/2)}{\eta} = 2\pi(1+\varepsilon)^{1/2}, \quad n \in N^+ \quad \text{(Periodic IOs)} \tag{3.69}$$

and amounts to a $n:1$ internal resonance between the LO and the NES. This condition requires that the period of the slow component $s(t)$ is n times the period of the fast component $f(t)$, with the overall (not necessarily) minimal period of $y_2(t)$ being equal to $2\pi(1+\varepsilon)^{1/2}$ [i.e., equal to the period of $y_1(t)$]. From (3.69) the corresponding initial condition for Y is computed:

$$Y(n) = \frac{K^2(1/2)n^2}{\pi^2(1+\varepsilon)^{3/2}} \left(\frac{8\varepsilon}{C}\right)^{1/2} \tag{3.70}$$

where the rescaling (3.40) is taken into account.

Table 3.2 presents the comparison between the predicted and exact initial conditions for a few symmetric and unsymmetric high-energy IOs for a system with $\varepsilon = 0.05$ and $C = 1$. Good agreement between theory and numerics is noted. In Figure 3.34 we depict the analytical time series for the IOs S71 and S91 and compare them to the corresponding exact solutions derived by direct integrations of the equations of motion (3.39). Overall, satisfactory agreement is obtained, particularly when the order n of the internal resonance is increased. The total energy of the IO is computed as $E = Y^2/2$, whereas the frequency index of an orbit is given by $\omega \approx n$. Employing (3.70), an analytic expression for the locus of high-energy IOs in the FEP can be derived as

$$E = \frac{4\varepsilon K^4(1/2)\omega^4}{C\pi^4(1+\varepsilon)^3} \tag{3.71}$$

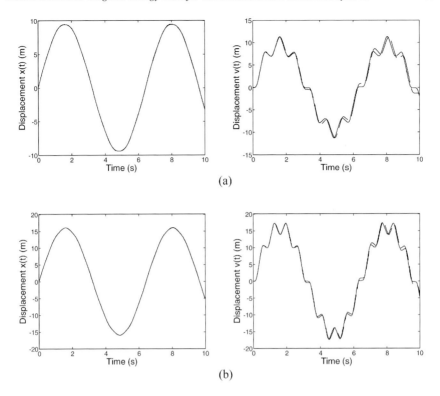

Fig. 3.34 Comparisons between analytical approximations (dashed lines) and direct numerical simulations (solid lines) of high-energy periodic IOs: (a)$S71$; (b) $S91$.

This approximation is presented in Region II of Figure 3.32a and compares well with the exact high-energy IO manifold of Figure 3.32b.

3.3.4.3 IOs at Low-Energy Levels

The low-energy periodic IOs $S1m$ and $U1m$ are finally analyzed. As mentioned previously, at low energies, $O(\varepsilon)$ terms are expected to play a dominant role in the response, so the reduced system (3.45) may not be used to approximate the IOs in this case. Instead the rescaled dynamical system (3.41) is reconsidered,

$$\ddot{x} + x + 8\varepsilon(x - v)^3 = 0$$
$$\ddot{v} + 8(v - x)^3 = 0$$
$$\dot{x}(0) = Y, \quad x(0) = v(0) = \dot{v}(0) = 0, \quad 0 < \varepsilon \ll 1 \quad (3.72)$$

where for coherence with the previous two sections, the initial condition is denoted by Y. Figure 3.29a illustrates that low-energy IOs are characterized by (i) motions of the two oscillators with very small amplitudes, and (ii) a much larger amplitude of oscillation of the LO; motivated by these numerical results we assume that in low-energy IOs it holds that $|v(t)| \ll |x(t)| \ll 1$.

Taking into account this assumption it appears that an appropriate ansatz for the low-energy IOs is

$$x(t) = Y \sin t + \cdots, \quad v(t) = \underbrace{B \sin t}_{\text{Fast component}} + \underbrace{s(t)}_{\text{Slow component}} + \cdots \quad (3.73)$$

with $|B| \ll |Y| \ll 1$ and $|s(t)| \ll |Y| \ll 1$. In contrast to the analysis of the previous section, the component of the NES response with frequency close to unity is regarded as the fast component, whereas the second component $s(t)$ is regarded as the slow component of the solution. Substituting (3.73) into the second of equations (3.72) yields the following differential equation:

$$-B \sin t + \ddot{s}(t)$$
$$+ 8[(B-Y)^3 \sin^3 t + 3(B-Y)^2 s(t) \sin^2 t$$
$$+ 3(B-Y) s^2(t) \sin t + s^3(t)] = 0 \quad (3.74)$$

Setting separately equal to zero the slow and fast components of (3.74), we partition the dynamics into the following slow and fast components:

$$-B \sin t + 8(B-Y)^3 \sin^3 t + 24(B-Y) s^2(t) \sin t = 0$$
$$\ddot{s}(t) + 24(B-Y)^2 s(t) \sin^2 t + 8 s^3(t) = 0 \quad (3.75)$$

The method of harmonic balance is applied to the first of equations (3.75), i.e., to the fast component of the dynamics, leading to the relation:

$$-B + 6(B-Y)^3 + 24(B-Y) f^2(t) = 0$$
$$\Rightarrow -B + 6(-Y)^3 \approx 0 \Rightarrow B \approx -6Y^3 \quad (3.76)$$

Focusing now in the slow component of the dynamics [the second of equations (3.75)], the fast term $\sin^2 t$ is averaged out to yield the following averaged slow flow dynamical system:

$$\ddot{s}(t) + 12(B-Y)^2 s(t) + 8 s^3(t) \approx 0 \quad (3.77)$$

since $\langle \sin^2 t \rangle_T = 1/\pi \int_0^\pi \sin^2 t \, dt = 1/2$. In view of the fact that $|B| \ll |Y|$ and $|s(t)| \ll |Y| \ll 1$, expression (3.77) may be approximated by the simplified linear equation

$$\ddot{s}(t) + 12 Y^2 s(t) \approx 0 \quad (3.78)$$

which is readily solved, by imposing the initial conditions for the impulsive orbit:

Table 3.3 Initial conditions for low-energy periodic IOs.

Periodic IO		$\dot{x}(0)$ (exact)	$\dot{x}(0)$ (analytic)
U1–22	($m = 22$)	0.0083	0.0083
S19	($m = 9$)	0.0201	0.0203
U2–15	($m = 15/2$)	0.0241	0.0243
U16	($m = 6$)	0.0299	0.0304
S13	($m = 3$)	0.0555	0.0609
U12	($m = 2$)	0.0781	0.0913
U34	($m = 4/3$)	0.0942	0.1369

$$s(t) \approx \frac{6Y^2}{\sqrt{12}} \sin \sqrt{12}\, Yt \qquad (3.79)$$

Combining the previous results, the low-energy IOs of system (3.72) are analytically approximated as follows:

$$x(t) \approx Y \sin t, \quad v(t) \approx -6Y^3 \sin t + \frac{6Y^2}{\sqrt{12}} \sin \sqrt{12}\, Yt \qquad (3.80)$$

Depending on the non-zero initial condition Y, relations (3.80) describe either periodic or quasi-periodic low-energy IOs. As in the analytical derivations of the previous two sections, the periodicity of the solution (3.80) is ensured by applying a periodicity condition, i.e., by imposing a $1 : m$ internal resonance between the LO and the nonlinear attachment:

$$\sqrt{12}\, Y = \frac{1}{m}, \quad m \in N^+ \quad \text{(Periodic IO)} \qquad (3.81)$$

Because of the slow-fast partition in the *ansatz* (3.73), the analytic approximation (3.80) is expected to be in better agreement with the exact solution for large integers m, that is, for sufficiently small energies. Taking the rescaling (3.40) into account an approximation of the low-energy periodic IO of the original dynamical system (3.39) is obtained in the following form:

$$x(t) \approx \frac{\sqrt{8\varepsilon}}{2\sqrt{3C}\, m} \sin t, \quad v(t) \approx \frac{\sqrt{8\varepsilon}}{4\sqrt{3C m^3}} [m \sin(t/m) - \sin t] \qquad (3.82)$$

Table 3.3 presents a comparison between predicted and exact low-energy periodic IOs for the system with $\varepsilon = 0.05$ and $C = 1$. Again, good agreement between the analytical and exact values is observed. Figure 3.32 depicts the analytical and exact time series for the IOs $U1$–22 and $S13$, from which good agreement is noted. The total energy of a low-energy IO is equal to $E = Y^2/2$, whereas its frequency index is $\omega \approx 1/m$. Employing the resonance condition (3.81), a surprisingly simple but accurate analytic approximation of the locus of low-energy IOs in the FEP is obtained:

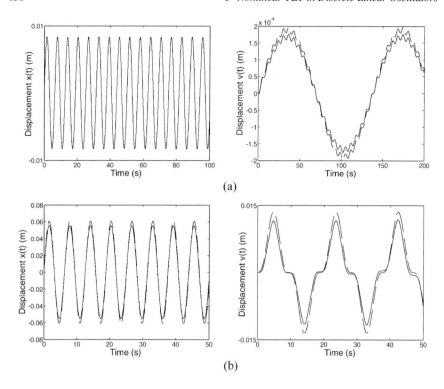

Fig. 3.35 Comparisons between analytical approximations (dashed lines) and direct numerical simulations (solid lines) of low-energy periodic IOs: (a) $U1$–22; (b) $S13$.

$$E = \frac{\varepsilon \omega^2}{3C} \tag{3.83}$$

The locus of IOs is depicted in Region III of Figure 3.32a. Overall, good agreement is obtained between the predictions and the exact results, which demonstrates the accuracy of the analysis.

In summary, we studied the periodic and quasi-periodic IOs of the strongly nonlinear Hamiltonian system (3.39). These are responses of the system initially at rest and excited by an impulsive force applied to the linear oscillator. As shown in later sections IOs directly affect the TET capacity of the damped system, i.e., the capacity of the nonlinear attachment to passively absorb broadband energy from the linear oscillator in a one-way, irreversible fashion. The manifold of quasi-periodic and periodic IOs in the FEP was analytically studied by considering separately the high-, moderate- and low-energy regimes. Different analytical methods were applied to analyze the IOs in these regimes. Of particular interest are moderate-energy IOs in the neighborhood of 1:1 internal resonance of the system which are in the forms of nonlinear beats, with the motion localized mainly to the nonlinear oscillator. As shown in a later section the excitation of an IO in the 1:1 internal resonance regime represents a very effective dynamical mechanism for strong passive TET from the linear oscillator to the nonlinear attachment.

3.3.5 Topological Features of the Hamiltonian Dynamics

In this section we focus in the intermediate-energy region, and provide some remarks on the topological features of the dynamics in phase space under conditions of 1:1 internal resonance. Our aim is to relate solutions, such as NNMs on branches $S11\pm$ and IOs, to certain global topological features of the Hamiltonian dynamics of system (3.6). Through a suitable change of variables we will reduce the isoenergetic dynamics to a three-dimensional sphere, and discuss how the critical energy threshold required for TET in the damped system (discussed in Section 3.2) can be directly related to a similar critical energy threshold in the Hamiltonian system, above which the IOs are in the form of nonlinear beats with strong energy exchanges between the LO and the nonlinear attachment. Finally, we will discuss how the topology of the phase space close or away from a homoclinic connection of the slow flow dynamics affects the qualitative features of the IOs discussed in Sections 3.3.3 and 3.3.4. The following exposition follows closely (Quinn et al., 2008).

Considering again the two-DOF Hamiltonian system (3.6) and setting (without loss of generality) $\omega_0 = 1$,

$$\ddot{x} + x + C(x - v)^3 = 0$$
$$\varepsilon \ddot{v} + C(v - x)^3 = 0 \quad (3.84)$$

we recall from Section 3.3.2.1, that solutions in the neighborhoods of the two backbone branches $S11\pm$ of the FEP can be analytically modeled using the CX-A technique. Indeed, assuming the following *ansatz* for these solutions:

$$x(t) \approx \frac{A(t)}{\omega} \cos[\omega t + \alpha(t)], \quad v(t) \approx \frac{B(t)}{\omega} \cos[\omega t + \beta(t)] \quad (3.85)$$

we obtain the set of four modulation equations (3.17) that govern the slow evolution of the amplitudes $A(t), B(t)$ and phases $\alpha(t), \beta(t)$ of the two oscillators. Note that the ansatz (3.85) indicates that conditions of 1:1 internal resonance are realized in the dynamics, so that the harmonic components of frequency ω in the response of the two oscillators. dominate over all other higher harmonics (this would not occur, for example, in neighborhoods of, or on subharmonic and superarmonic tongues, see Sections 3.3.1 and 3.3.2).

Introducing the phase difference $\phi = \alpha - \beta$, the slow flow equations (3.17) can be reduced to the following three-dimensional autonomous dynamical system on the cylinder $(R^+ \times R^+ \times S^1)$,

$$\dot{a}_1 = \frac{-3a_2 C}{8} \sin\phi[(a_1^2 + a_2^2) - 2a_1 a_2 \cos\phi]$$

$$\dot{a}_2 = \frac{3a_1 C}{8\varepsilon} \sin\phi [(a_1^2 + a_2^2) - 2a_1 a_2 \cos\phi]$$

$$\dot{\phi} = \frac{1}{2} - \frac{3C}{8}[(a_1^2 + a_2^2) - 2a_1 a_2 \cos\phi]$$

$$\times \left[\left(\frac{1}{\varepsilon}\right)\left[1 - \frac{a_1}{a_2}\cos\phi\right] - \left[1 - \frac{a_2}{a_1}\cos\phi\right]\right] \quad (3.86)$$

where the notation $a_1 = A$, $a_2 = B$ was utilized.

In Section 3.3.2.1, the analytic modeling of periodic orbits that satisfy the *exact* 1:1 internal resonance condition was considered; moreover, since we were interested on steady state solutions, we imposed stationarity conditions to the derived modulation equations (i.e., the terms containing derivatives with respect to time were set equal to zero). In this section, a more general analysis is carried out in the sense that fast oscillations with frequencies $\omega \approx 1$ and modulated by slowly-varying envelopes are sought. In other words, we are primarily interested in the dynamics near the region of 1:1 internal resonance, which corresponds to the intermediate-energy regime of the FEP in the notation of the previous sections.

It turns out that the autonomous dynamical system (3.86) is fully integrable, as it possesses the following two independent first integrals of motion:

$$a_1^2 + (\sqrt{\varepsilon}a_2)^2 \equiv r^2$$

$$\frac{a_1^2}{2} + \varepsilon \frac{a_2^2}{4} + \frac{3C}{32}(a_1^2 + a_2^2 - 2a_1 a_2 \cos\phi)^2 \equiv h \quad (3.87)$$

The first equation is a consequence of energy conservation in (3.84), and enables us to introduce a second angle ψ into the problem, defined by

$$\tan\left(\frac{\psi}{2} + \frac{\pi}{4}\right) = \frac{a_1}{\sqrt{\varepsilon}a_2}, \quad \psi \in \left[-\frac{\pi}{2}, \frac{\pi}{2}\right] \quad (3.88)$$

Taking into account the first integrals of (3.87) and introducing the new angle into the problem, the slow flow dynamical system (3.86) can be further reduced to system on a three-dimensional sphere,

$$\dot{r} = 0$$

$$\dot{\psi} = \frac{-3Cr^2}{8\varepsilon^{3/2}}[(1+\varepsilon) - (1-\varepsilon)\sin\psi - 2\sqrt{\varepsilon}\cos\psi\cos\phi]\sin\phi$$

$$\dot{\phi} = \frac{1}{2} - \frac{3Cr^2}{16\varepsilon^2}[(1+\varepsilon) - (1-\varepsilon)\sin\psi - 2\sqrt{\varepsilon}\cos\psi\cos\phi]$$

$$\times \left[(1-\varepsilon) - 2\varepsilon^{1/2}\frac{\sin\psi\cos\phi}{\cos\psi}\right] \quad (3.89)$$

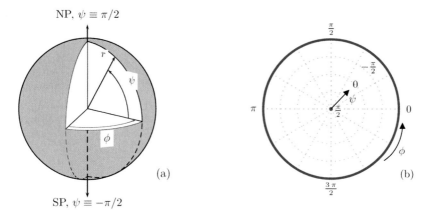

Fig. 3.36 Topology of the reduced phase space: (a) three-dimensional sphere $(r, \phi, \psi) \in (R^+ \times S^1 \times S^1)$, (b) projection of the reduced dynamics onto the unit disk.

where $(r, \phi, \psi) \in (R^+ \times S^1 \times S^1)$ (see Figure 3.36). Then, the second of the first integrals of motion (3.87) can be expressed in the form

$$\frac{r^2}{8}\left\{3 + \sin\psi + \frac{3Cr^2}{16\varepsilon^2}[(1+\varepsilon) - (1-\varepsilon)\sin\psi - 2\varepsilon^{1/2}\cos\psi\cos\phi]^2\right\} = h$$
(3.90)

Considering the isoenergetic dynamical flow corresponding to $r = $ const, the orbits of the system lie an a topological two-sphere, and follow the level sets of the first integral of motion (3.90).

Projections of the isoenergetic reduced dynamics onto the unit disk at different energy levels are depicted in Figure 3.37. The north pole (NP) at $\psi = \pi/2$ lies at the center of the disk, while the south pole (SP) $\psi = -\pi/2$ is mapped onto the entire unit circle. In this projection, trajectories that pass through the SP approach the unit circle at $\phi = \pi/2$ and are continued at $\phi = -\pi/2$. If the response is localized to the LO, so that $a_2 \ll a_1$, the phase variable ψ lies close to $+\pi/2$. In contrast, a localized response in the nonlinear attachment (i.e., $a_1 \ll a_2$) implies that $\psi \approx -\pi/2$.

Before we examine the dynamics near the region of 1:1 internal resonance, we reconsider the periodic motions on branches $S11\pm$, corresponding to the equilibrium points of the slow flow (3.89). These equilibrium points are explicitly evaluated by the following expressions:

$$\dot{\psi} = 0 \Rightarrow \sin\phi_{eq} = 0 \Rightarrow \phi_{eq} = 0, \pi$$

$$\dot{\phi} = 0 \Rightarrow \cos\psi_{eq} - \frac{3Cr^2}{8\varepsilon}(1+\varepsilon)^2[1 - \sin(\psi_{eq} + \gamma_{eq})]\cos(\psi_{eq} + \gamma_{eq}) = 0$$

(3.91)

with

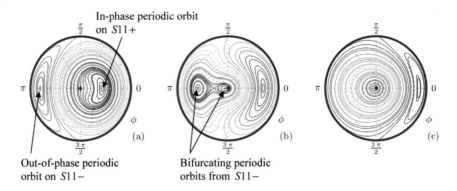

Fig. 3.37 Projection of the dynamics of the isoenergetic manifold onto the unit disk at different energy levels ($\varepsilon = 0.1$, $C = 2/15$); (a) $r = 1.00$, (b) $r = 0.375$, (c) $r = 0.25$.

$$\tan \gamma_{eq} = \frac{2\sqrt{\varepsilon} \cos \phi_{eq}}{1 - \varepsilon} \quad (3.92)$$

Equilibrium points satisfying $\phi_{eq} = 0$ correspond to in-phase periodic motions and generate the backbone branch $S11+$ for varying frequency and energy; those corresponding to $\phi_{eq} = \pi$, represent out-of-phase periodic motions and generate the other backbone $S11-$. In the projections of the phase space shown in Figure 3.37, periodic motions (NNMs) on $S11+$ appear as equilibrium points that lie on the horizontal axis to the right of the origin, whereas periodic motions on $S11-$ as equilibrium points that lie on the horizontal axis to the left of the origin.

With increasing energy, i.e., as $r \to \infty$, both equilibrium points approach the value

$$\lim_{r \to \infty} \psi_{eq} = \arctan\left(\frac{1 - \varepsilon}{2\sqrt{\varepsilon} \cos \phi_{eq}}\right) \quad (3.93)$$

so that, for $0 < \varepsilon \ll 1$ and in the limit of high energies we have that $\psi_{eq, S11+} > 0$ and $\psi_{eq, S11-} < 0$. With increasing energy the in-phase NNMs on $S11+$ localize to the LO, while the out-of-phase NNMs on $S11-$ localize to the nonlinear attachment (the NES). The degree of localization is controlled only by the mass ratio ε, and for small but finite values of this ratio the high-energy localization is *incomplete*, as the limiting values of $\psi_{eq, S11+}$ and $\psi_{eq, S11-}$ do not attain $\pi/2$ in magnitude.

Considering now the low-energy limit, it is easily shown that for sufficiently small values of r the equilibrium equation for ψ_{eq} degenerates to the simple limiting relation $\cos \psi_{eq} \to 0$. Therefore, we conclude that as $r \to 0+$, the following values are attained by the equilibrium value for ψ:

$$\lim_{r \to 0+} \psi_{eq, S11+} = -\pi/2 \quad \text{and} \quad \lim_{r \to 0+} \psi_{eq, S11-} = +\pi/2 \quad (3.94)$$

It follows that in the limit of small energies the in-phase NNM on $S11+$ localizes to the nonlinear oscillator, while the out-of-phase NNM on $S11-$ to the LO. However,

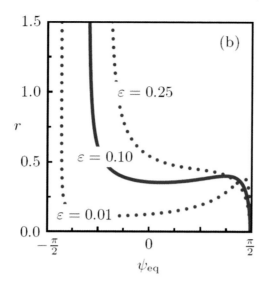

Fig. 3.38 Topology of the branch $S11-$ for varying ε and $C = 2/15$.

unlike the high-energy limits (3.93), as $r \to 0$ localization is *complete* to either the LO or the nonlinear attachment.

In the transition from high to low energies, the branch of out-of-phase NNMs $S11-$ undergoes two saddle-node bifurcations. In the first bifurcation, a new pair of stable-unstable equilibrium points is generated near $\psi = +\pi/2$. As energy decreases a second (inverse) saddle-node bifurcation occurs that anhiliates the unstable equilibrium generated by the first bifurcation, together with the stable branch of $S11-$ that existed for higher energy values. It should be noted, however, that these bifurcations occur only below a certain critical mass ratio ε, i.e., only for sufficiently light attachments. This is demonstrated in Figure 3.38, which depicts the variation of the out-of-phase branch $S11-$ in the (ψ_{eq}, r) plane for three values of the mass ratio ε; note that no bifurcations occur for the higher value of for ε. Figures 3.37a, b, c depict the above-mentioned bifurcations in projections of the phase space of the isoenergetic dynamics. Projections of the topological structure of the phase space of the system before the first (higher energy) bifurcation, in between the two bifurcations, and below the second (lower energy) bifurcation are depicted in Figures 3.37a, b and c, respectively. An alternative representation of these bifurcations in the FEP was depicted in Figures 3.20 and 3.21a for branch $S11-$.

We now focus on the topology of the impulsive orbits (IOs) in the neighborhood of $\omega = 1$, under conditions of 1:1 internal resonance. From the discussion of Sections 3.3.3 and 3.3.4, it is clear that an IO corresponds to the initial condition $a_2(0) = 0 \Rightarrow \psi(0) = \pi/2$. In terms of the spherical topology of the isoenergetic flow, an IO is therefore coincident with a trajectory passing through the NP, which renders this graphical representation particularly attractive. The IO computed from

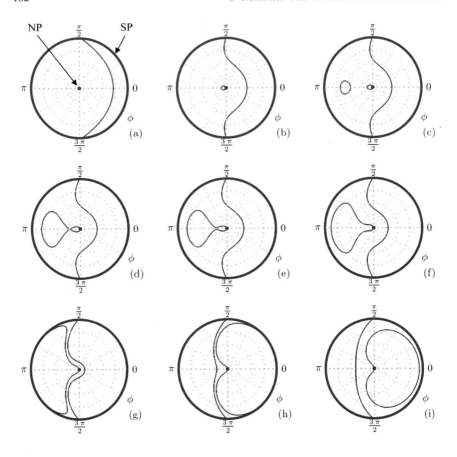

Fig. 3.39 IOs passing through the NP (the origin of the projection), and orbits passing through the SP for $\varepsilon = 0.1$ and $C = 2/15$: (a) $r = 0.25$, (b) $r = 0.36$, (c) $r = 0.37$, (d) $r = 0.386$, (e) $r = 0.387$, (f) $r = 0.40$, (g) $r = 0.44$, (h) $r = 0.46$, (i) $r = 0.50$; the shift of the IO from the left to the right between (g) and (h) is an artifact of the projection.

the slow flow (3.89), together with the trajectory passing through the SP (corresponding to the orbit having as only non-zero initial condition the velocity of the LO) are shown in Figure 3.39 for varying values of the energy-like parameter r (on different isoenergetic manifolds). We note that the depicted IOs may be either periodic or quasi-periodic. In Figures 3.39c, d a third isolated trajectory is seen which lies on the same energy level as the trajectory passing through the NP.

Starting from the low-energy isoenergetic manifold of Figure 3.39a, we note that the IO makes a small excursion in the spherical phase space, and remains localized close to $\psi = +\pi/2$; it follows that in this case, the energy exchange between the LO to the nonlinear attachment is insignificant, and the oscillation remains confined predominantly to the LO. The same qualitative behavior is preserved until the critical energy $r = r_{cr} = 0.3865$ (occurring between Figures 3.39d and 3.39e), for which the IO coincides with two homoclinic loops in phase space; these turn out to be the homoclinic loops of the unstable hyperbolic equilibrium (NNM) on $S11-$ which exists between the two saddle-node bifurcations discussed previously. For $r > r_{cr}$ the topology of the IO changes drastically, as it makes much larger excursions in phase space; this means continuous, strong energy exchange between the LO and the nonlinear attachment in the form of nonlinear beats. At an even higher value of energy, $r \approx 0.4495$, the IO passes through both the NP and SP (this occurs between Figures 3.39g, h), and 100% of the energy is transferred back and forth between the LO and the nonlinear attachment during the occurring nonlinear beats.

We conclude that for fixed mass ratio ε and nonlinear coefficient C, the geometries of the IOs undergo significant changes for varying energy: for low energies, the IOs are localized to the LO, whereas above a critical energy threshold the IOs appear as nonlinear beats, whereby energy gets continuously exchanged between the LO and the nonlinear attachment. Moreover, at specific energy levels almost the entire (conserved) energy of the motion gets transferred back and forth between the linear and nonlinear oscillators.

It turns out that the critical value of the energy-like variable, r_{cr}, can be directly related to the energy threshold required for TET in the weakly damped system. Indeed, as we recall from the numerical results of Section 3.2, strong TET phenomena in the damped system (3.2) occur only when the external impulsive excitation applied to the LO (i.e., the initial energy of the system) exceeds a certain critical value. The threshold for TET in the damped system can be directly related to the existence of a critical energy level (signified by r_{cr}) in the underlying Hamiltonian system, above which the IO makes large excursions in phase space and nonlinear beats corresponding to strong energy exchanges between the LO and the nonlinear attachment are initiated. Moreover, conditions for optimal TET in the damped system can be formulated by studying the topology of the IOs in the neighborhood of the homoclinic loops in the slow flow of the Hamiltonian system. These remarks provide a first indication of the intricate relation between IOs and TET, and of the importance of understanding the Hamiltonian dynamics in order to correctly interpret strongly nonlinear transitions and TET in the weakly dissipative system. A systematic study of the dynamics of the damped system will carried out starting from the next section.

Figure 3.40 depicts the maximum excursion attained by an IO from the NP (i.e., the measure $||\psi_{NP}|| = |\pi/2 - \psi_{NP}|$), as function of r and different values of the mass ratio ε; as discussed above this measure provides a good picture of the energy exchange that occurs between the linear and nonlinear oscillators. Considering the results of Figure 3.40 there are two interesting findings. First, below a critical mass ratio there occurs a discontinuity in this energy exchange. For instance, for $\varepsilon = 0.25$, the variation of $||\psi_{NP}||$ is continuous with r (Figure 3.40d); the reason is that the branch $S11-$ does not undergo any saddle-node bifurcations for this mass ratio (see Figure 3.38), so no homoclinic loops exist (and, hence, no significant topological change in the shape of the IOs occurs) as r varies. On the contrary, for smaller mass ratios, the IOs undergo significant topological changes as r varies (see Figure 3.39), which leads to the discontinuities in energy exchanges noted in Figures 3.40a–c.

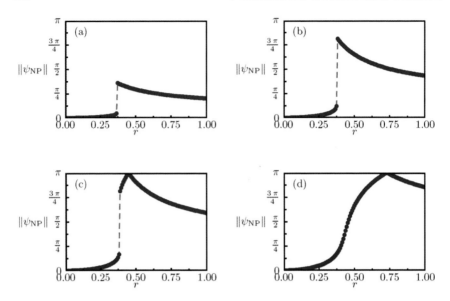

Fig. 3.40 Amplitude of the IO as function of r for $C = 2/15$ and: (a) $\varepsilon = 0.01$, (b) $\varepsilon = 0.05$, (c) $\varepsilon = 0.10$, (d) $\varepsilon = 0.25$.

The second interesting finding is that the mass ratio has a critical influence on the capacity of the nonlinear attachment to passively absorb energy from the LO during a cycle of the motion. Specifically, we note that for $\varepsilon = 0.01$, only a small amount of energy is transferred from the LO to the nonlinear attachment, as evidenced by the small value of $||\psi_{NP}||$ in Figure 3.40a. However, for $\varepsilon = 0.1$ and $\varepsilon = 0.25$, complete energy exchange between the two oscillators takes place (i.e., the upper bound $||\psi_{NP}|| = \pi$ is reached for a specific value of r) during a cycle of the motion (see Figures 3.40c, d). The energy level $r = r_{complete}$ for which complete energy exchange occurs between the LO and the nonlinear attachment during the beating phenomenon is related to the energy of the impulsive orbit,

$$h_{NP} = \frac{r^2}{2} + \frac{3Cr^4}{32} \tag{3.95}$$

and to the energy of the trajectory passing through the SP:

$$h_{SP} = \frac{r^2}{4} + \frac{3Cr^4}{32\varepsilon^2} \tag{3.96}$$

Equating these two energies, we ensure that an orbit initiated from the NP (i.e., an IO) passes also from the SP, signifying that there occurs complete energy transfer from the LO to the nonlinear attachment during a cycle of the ensuing nonlinear beat. This provides the sought after critical value for $r_{complete}$ as follows:

$$h_{\text{NP}} = h_{\text{SP}} \Rightarrow r_{\text{complete}} = \left[\frac{8\varepsilon^2}{3C(1-\varepsilon^2)}\right]^{1/2} \quad (3.97)$$

According to this expression, for $\varepsilon = 0.1$ and $C = 2/15$, there is complete energy exchange between the two oscillators when $r = r_{\text{complete}} = 0.4495$, which is in agreement with the results depicted in Figure 3.40. Because no complete energy exchange can be achieved for small mass ratios, expression (3.97) only holds for sufficiently large values of ε.

These results conclude our numerical and analytical study of the dynamics of the Hamiltonian system (3.6). In the next section we start our systematic study of the dynamics of the weakly dissipative system, which will include a detailed discussion of damped transitions and of targeted energy transfer (TET) phenomena. We will show that for sufficiently weak damping (which is a reasonable and practical assumption for typical mechanical systems and structural components) the underlying Hamiltonian dynamics govern, in essence the damped responses, with damping playing a rather parasitic role, in the sense that it does not 'produce' to any new dynamics; this observation, however, is not intended to diminish the important role that damping plays on TET phenomena, as discussed below. Viewed in this context, we will then argue that the excitation of stable IOs giving rise to strong energy exchanges between the LO and the nonlinear attachment, provides an important mechanism for strong TET in the weakly damped system. Moreover, conditions for optimal TET will be closely related to the topology of orbits of the underlying Hamiltonian system, and especially to the topology of the manifold of IOs. Hence, the response of the Hamiltonian system and the analysis presented in the previous sections provide the necessary framework for understanding and analyzing the responses of the weakly damped system, for interpreting complex nonlinear modal interactions and transitions, and, more importantly, for designing NESs with optimal TET capacities.

3.4 SDOF Linear Oscillators with SDOF NESs: Transient Dynamics of the Damped Systems

Based on our knowledge of the Hamiltonian dynamics, we initiate our study of the transient dynamics of the weakly damped system (3.2), which is reproduced here for convenience:

$$\ddot{x} + \lambda_1 \dot{x} + \lambda_2(\dot{x} - \dot{v}) + \omega_0^2 x + C(x - v)^3 = 0$$
$$\varepsilon \ddot{v} + \lambda_2(\dot{v} - \dot{x}) + C(v - x)^3 = 0 \quad (3.98)$$

Again we will assume that the nonlinear attachment is lightweight, $0 < \varepsilon \ll 1$. In an initial series of numerical simulations we demonstrate the intricate relation between the weakly dissipative and Hamiltonian dynamics.

3.4.1 Nonlinear Damped Transitions Represented in the FEP

The aim of this section is to show that the previously studied structure of the underlying Hamiltonian dynamics of (3.98) greatly influences the transient dynamics of the weakly damped system. When viewed from this perspective, one can systematically interpret complex multi-frequency transitions between different nonlinear normal modes (NNMs) in the damped dynamics, by relating them to transitions between different branches of NNMs in the FEP of Figure 3.20. Unless otherwise noted, in the following simulations of this section we consider system (3.98) with parameters $\varepsilon = 0.05$, $\omega_0 = 1.0$, $C = 1.0$, and weak damping, $\lambda_1 = 0$, $\lambda_2 = 0.0015$.

In the first numerical simulation (see Figure 3.41) we initiate the motion on the high-energy unstable IO on branch $U21$ corresponding to initial conditions $v(-T/4) = \dot{v}(-T/4) = x(-T/4) = 0$ and $\dot{x}(-T/4) = X = -0.579$. Even though the excited IO is unstable, there is strong targeted energy transfer (TET) from the (directly excited) LO to the NES, as evidenced by the rapid and strong build-up of the oscillation amplitude of the NES (note that the NES is initially at rest). Moreover, due to the instability of the excited IO the motion escapes immediately from branch $U21$ to land on $S11+$ through a frequency transition (jump). As energy further decreases due to viscous dissipation the motion follows a multi-mode transition visiting the branches $S13+$, $S13-$, $S15-$, $S15+$, ..., i.e., it follows the basic backbone curve of the frequency-energy plot (FEP) of Figure 3.20a. This is shown in Figure 3.41c where the wavelet transform (WT) spectrum of the relative displacement $(v - x)$ is superimposed to the FEP of the underlying Hamiltonian system. Although this plot provides a purely phenomenological interpretation of the damped transitions in terms of the undamped Hamiltonian dynamics, it validates our previous assertion regarding the parasitic role of weak damping in the transient dynamics. Indeed, damping does not generate any new dynamics, but merely influences the damped transitions (jumps) between different branches of NNMs of the Hamiltonian system. Clearly, by depicting the damped dynamics on the FEP, we are able to interpret complex multi-frequency transitions such as the ones shown in Figures 3.41a, b, involving the participation of multiple nonlinear modes in the transient response. A more detailed consideration of this nonlinear damped transition can be found in Lee et al. (2006).

In the second simulation we initiate the motion on the moderate-energy stable IO on branch $U76$ (corresponding to the non-zero initial condition $X = -0.1039$). In Figures 3.42a, b we depict the transient responses of the LO and the NES, indicating that there occurs stronger TET to the NES in this case. Moreover, since the initially excited special orbit on $U76$ is stable, there occurs a prolonged initial oscillation of the system on that branch at the early stage of the motion (see Figure 3.42c). As energy decreases due to damping dissipation there occurs a transition (jump) to the stable branch $S13-$, where the NES engages into a transient 1:3 internal resonance with the LO; this is referred to as a 1:3 *transient resonance capture* (TRC) (Arnold, 1988; Quinn, 1997 – see also Section 2.3). As energy decreases even further due to viscous dissipation there occurs escape from 1:3 TRC, and the motion evolves along branches S15, S17, ... as in the previous simulation.

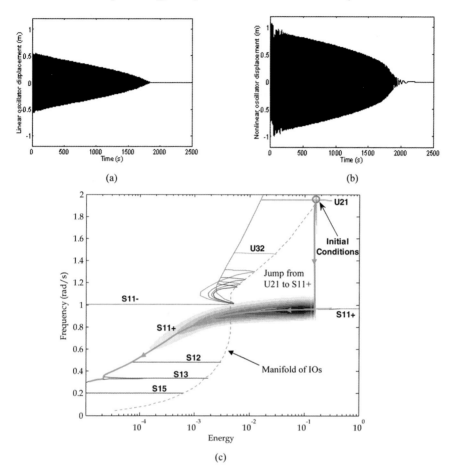

Fig. 3.41 Damped transition initiated on the unstable IO on branch $U21$: transient responses of (a) the LO and (b) the nonlinear oscillator (NES); (c) WT spectrum of $(v-x)$ superimposed to the Hamiltonian FEP.

This second simulation provides the first numerical evidence that the excitation of a stable IO close to the 1:1 resonance manifold represents one of the mechanisms for strong TET in system (3.98). Lee et al. (2006) showed that the strongly nonlinear damped transitions depicted in Figures 3.41 and 3.42, are sensitive to damping, since for small damping variation a qualitatively different series of multi-modal transitions may result. An additional observation drawn from these numerical simulations is that the excitation of a stable IO prolongs the initial phase of nonlinear beats between the LO and the NES, resulting in strong TET to the NES. Indeed, by comparing the time series of Figures 3.41a, b and 3.42a, b we conclude that when an unstable IO is initially excited (so that no significant initial beating occurs), TET from the LO to the NES is weaker.

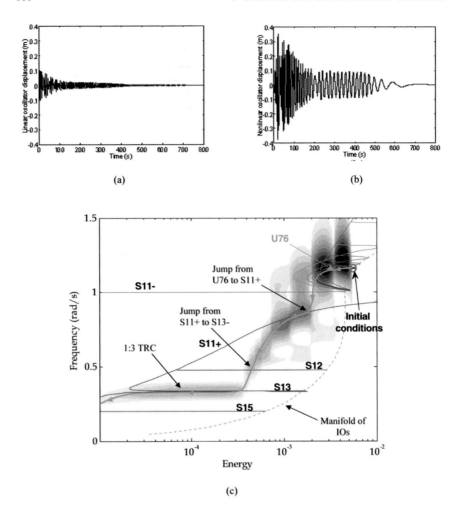

Fig. 3.42 Damped transition initiated on the stable IO on branch $U76$: transient responses of (a) the LO and (b) the nonlinear oscillator (NES); (c) WT spectrum of $(v-x)$ superimposed to the Hamiltonian FEP.

In the third series of damped transitions depicted in Figure 3.43 we study damped transitions initiated by exciting low-, moderate- and high-energy IOs of the system with $\lambda_1 = \lambda_2 = 0.005$. The qualitative differences between these transitions are evident, indicating the sensitivity of the dynamics of system (3.98) on the initial conditions (or, equivalently, on the initial energy of the motion). For initial condition $X = 0.05$ (corresponding to a low-energy IO, Figure 3.43a) the response possesses a frequency component around $\omega = 0.2$ rad/s during the initial stage of the motion, which indicates excitation of the low-energy IO. As discussed in Sections 3.3.3 and 3.3.4, such an IO is localized to the LO, and this is why a transition to $S11-$ is observed after a short multi-frequency initial transient. Eventually, only

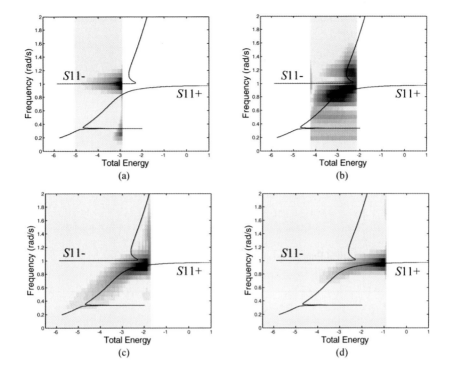

Fig. 3.43 WT spectra of the transient damped response $(v - x)$ of the two-DOF system (3.98) interpreted in the FEP for excitation of: (a) a low-energy IO, $X = 0.05$; (b) a moderate-energy IO, $X = 0.12$; (c) a moderate-energy IO, $X = 0.2$, and (d) a high-energy IO, $X = 0.5$.

a small portion of vibration energy is transferred to, and dissipated by the NES in this case, a result which is compatible with the fact that passive TET is 'triggered' only above a critical energy threshold (Section 3.2). Figure 3.43a also illustrates that the dynamics is *weakly nonlinear* at this low-energy level, since after the initial transients the dominant frequency component of the damped motion is near the linearized frequency $\omega_0 = 1$, and the response is narrowband.

Qualitatively different transient dynamics is encountered for initial condition $X = 0.12$ and excitation of a moderate-energy IO (see Figure 3.43b). Strong and sustained harmonic components appear in this case, and the damped motion never fully enters into the domain of attraction of the 1:1 resonant manifold; instead, the damped response is in the form of a prolonged nonlinear beat, which results in strong TET from the LO to the NES. This regime of motion is *strongly nonlinear*, as revealed by the appearance of multiple strong sustained harmonics over a relatively broadband frequency range.

Increasing further the initial condition to $X = 0.2$ (and exciting still a moderate-energy IO, see Figure 3.43c), gives rise to a different damped transition scenario. Specifically, there occurs a rapid transition of the damped dynamics from the IO to

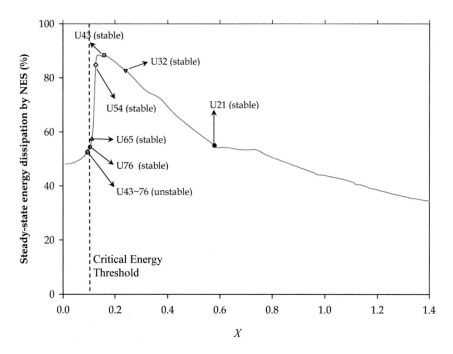

Fig. 3.44 EDM when an IO is excited, as function of the non-zero initial condition of that IO.

branch $S11+$, where sustained 1:1 TRC is initiated. This transition is similar to that encountered in the numerical simulation of Figure 3.41, and results in moderate TET from the LO to the NES. A similar transition is noted for the initial condition $X = 0.5$ corresponding to excitation of a high-energy IO, and shown in Figure 3.43d.

In summary, different transition scenarios are realized in the damped dynamics depending on the energy of the IOs that are initially excited. These different transitions may result in enhanced (or weaker) TET from the LO to the NES, depending on the excitation (or lack of) of nonlinear beat pheneomena leading to strong localization of the motion to the NES. To further emphasize this point, in Figure 3.44 we depict the energy dissipation measure (EDM) (i.e., the percentage of input energy dissipated by the NES) when an IO is excited, as function of the non-zero initial condition X of that IO; the system parameters for these simulations are selected as $\varepsilon = 0.5, \omega_0^2 = 1, C = 1, \lambda_1 = \lambda_2 = 0.01$. The positions of some representative (stable and unstable) IOs are indicated in that plot as well. Low-energy impulsive orbits are located below the critical energy threshold, and their excitation results in weak TET. Optimal TET is associated with the excitation of moderate-energy IOs, located just above the energy threshold and satisfying conditions of near 1:1 internal resonance between the LO and the NES (i.e., $U54, U43, \dots$). By further increasing the initial condition of the IO we get deterioration of TET, as we leave the regime of 1:1 internal resonance so that less pronounced nonlinear beats are realized when an IO is excited (see Figure 3.26 and the analysis of Section 3.3.4).

The results of this section show the clear relation between TET and the strongly nonlinear multi-mode (and multi-frequency) transitions that take place in the FEP. This naturally leads to a detailed discussion of the alternative mechanisms for the realization of TET in system (3.98), a task addressed in the next section.

3.4.2 Dynamics of TET in the Damped System

We now study the capacity for targeted energy transfer (TET) of the lightweight ungrounded NES considered in the previous sections; that is, its capacity to passively absorb and locally dissipate vibration energy from the SDOF linear oscillator (LO), without 'spreading back' the absorbed energy. We will show that key to understanding TET in the weakly damped system is our knowledge of the topological structure of the orbits of the underlying Hamiltonian system, as it is the undamped dynamics that influences in a essential way the weakly damped transitions and the resulting strongly nonlinear modal interactions.

The first mechanism for TET, *fundamental TET or fundamental energy pumping*, is due to 1:1 transient resonance capture (TRC) of the dynamics, and is realized when the damped motion traces approximately the in-phase backbone curve $S11+$ of the FEP of Figure 3.20, at relatively low frequencies $\omega < \omega_0$. The second mechanism, *subharmonic* TET, resembles the first, but is realized when the motion takes place along a lower frequency subharmonic tongue Snm, $n < m$ of the FEP; it is due to $n : m$ TRC, and is less efficient than fundamental TET. The third mechanism, TET *through nonlinear beats*, is the most powerful mechanism for TET, as it involves the initial excitation of an IO at a higher frequency tongue, at frequencies $\omega > \omega_0$. In the following sections we will discuss each TET mechanism separately through numerical simulations and analysis.

3.4.2.1 TET through Fundamental Transient Resonance Capture (TRC)

The first mechanism for TET involves excitation of the branch of in-phase NNMs $S11+$, where the LO and the NES oscillate with identical frequencies in the neighborhood of the fundamental frequency ω_0. In Figure 3.21b we depict a detailed plot of branch $S11+$ of the Hamiltonian system i.e., the set (3.98) with $\lambda_1 = \lambda_2 = 0$], and note that at higher energies the in-phase synchronous periodic oscillations (NNMs) are spatially extended (involving finite-amplitude oscillations of both the LO and the nonlinear attachment). However, since the *nonlinear mode shapes of NNMs on $S11+$ strongly depend on the level of energy, and as energy decreases they become localized to the nonlinear attachment.*

This low-energy localization is a basic characteristic of the two-dimensional NNM invariant manifold corresponding to $S11+$; moreover, this localization property is preserved in the weakly damped system, where the motion takes place on a two-dimensional *damped* NNM *invariant manifold* (Shaw and Pierre, 1991, 1993).

This means that when the initial conditions of the damped system place the motion on the damped NNM invariant manifold corresponding to $S11+$, for decreasing energy the mode shape of the resulting oscillation makes a transition from being initially spatially extended to being localized to the NES. This, in turn, leads to passive transfer of energy from the LO to the NES.

As shown below, the underlying dynamical phenomenon governing fundamental TET is TRC on a 1:1 *resonance manifold* of the damped system. As discussed in Section 2.3, TRC is a form of transient nonlinear resonance between two modes of a system, followed by escape from the capture regime. TRCs and sustained resonance captures (SRCs) have been studied extensively in weakly varying Hamiltonian systems and in non-conservative oscillators [(Kevorkian, 1971, 1974; Gautesen, 1974; Neishtadt, 1975, 1986, 1987, 1997, 1999; Haberman, 1983; Kath, 1983; Arnold, 1988; Bosley and Kevorkian, 1992; Quinn et al., 1995; Bosley, 1996; Quinn, 1997; Vakakis and Gendelman, 2001; Vainchtein et al., 2004); see also the discussion in Wiggins (1990) on the interaction of resonance bands in weakly damped oscillators using geometrical methods]. Regarding the study of energy exchanges and nonlinear dynamical interactions caused by TRCs, we mention the work by Neishtadt (1975) on the transition of a Hamiltonian system across a separatrix (separatrix crossing) caused by periodic parametric excitation due to a slowly varying frequency; the work by Friedland (1997) on trapping into resonance in adiabatically varying systems driven by externally launched pump waves; on continuous resonant growth of induced nonlinear waves (Aranson et al., 1992); on the excitation of an oscillatory nonlinear system to high energy by weak chirped frequency forcing (Marcus et al., 2004); and on a method based on resonance capture to control transitions between different regimes of Hamiltonian systems (Vainchtein and Mezic, 2004). However, with the exception of the paper by Quinn et al. (1995) these works deal with systems without damping; on the other hand, in contrast to the results reported in this work Quinn et al. (1995) did not consider strong inertial asymmetry, which as shown below is a necessary condition (along with weak dissipation) for realizing TET through TRC.

We note that in the absence of damping, no TET, i.e., irreversible energy transfer, can occur on motions initiated on branch $S11+$. The reason is that in the absence of energy dissipation the distribution of energy between the linear and nonlinear components is 'locked' (due to the invariance of the NNM manifold $S11+$), so no localization can occur to either one of these system components. In addition, unlike the phenomenon of internal resonance encountered in conservative oscillators, during TRC the frequency of oscillation of the NES varies with time, depending on the amount of energy transferred from the LO; therefore, it is indeed possible to escape from the fundamental resonance capture regime if the frequency of the NES departs away from the neighborhood of the natural frequency of the LO, ω_0. Finally, we note that although the NES has no preferential resonant frequency (as it possesses nonlinearizable stiffness nonlinearity), it may synchronize with the LO along $S11+$ due to the invariance properties of the damped NNM manifold, and this occurs passively, without the need to 'tune' the NES parameters. This demonstrates the enhanced versatility of the systems with essential nonlinearities considered in this work.

Numerical evidence of fundamental TET in the damped system (3.98) is presented in Figure 3.45 for $\omega_0 = 1, C = 1, \varepsilon = 0.05, \lambda_1 = \lambda_2 = 0.002$. Weak damping is considered in order to better highlight the TET phenomenon, and the motion is initiated on a NNM on $S11+$ corresponding to initial conditions $x(0) = v(0) = 0, \dot{x}(0) = 0.175, \dot{v}(0) = 0.386$. Considering the transient responses depicted in Figures 3.45a, b, we note that the envelope of the response of the LO decays more rapidly than that of the NES. The detail of the response presented in Figure 3.45c indicates that motion along $S11+$ corresponds to in-phase vibration of the two masses with identical fast frequency, confirming that the transient dynamics is locked into 1:1 transient resonance capture (TRC). The percentage of instantaneous energy stored in the NES is presented in Figure 3.45d, confirming that as the damped motion follows branch $S11+$ with decaying energy, an irreversible and complete energy transfer takes place from the LO to the NES, at least until escape from resonance capture occurs around $t \approx 300$ s. We commend that the reversal in instantaneous energy suffered by the NES for $t > 300$ s occurs at the very late stage of the response where the energy of the system has almost completely been dissipated by damping. Finally, in Figure 3.45e, the Morlet WT spectrum of the relative response between $v(t)-x(t)$ is superposed to the backbone of the Hamiltonian FEP, confirming that the in-phase branch $S11+$ is approximately traced by the damped transient response. This validates our previous conjecture that the TET dynamics in the damped system is mainly governed by the topological structure and bifurcations of the periodic (and quasi-periodic) motions of the underlying Hamiltonian system.

We now proceed to analytically study the fundamental TET mechanism by analyzing system (3.98) through the complexification-averaging (CX-A) technique discussed in Sections 2.4 and 3.3.2. Even though (3.98) is a strongly nonlinear system of coupled oscillators, analytical modeling of its transient dynamics leading to TET can still be performed. Indeed, motivated by the time series of the transient responses of Figures 3.45a, b we will partition the transient dynamics into slow and fast components, and then reduce our study to the investigation the corresponding slow flow dynamics of the system. The slow flow governs the essential (important) dynamics of the weakly damped system, as well as the nonlinear modal interactions that occur between the LO and the NES and lead to fundamental TET.

As discussed in Sections 2.4 and 3.3.2 the CX-A technique is especially suited for studying TET, as it can be applied to the analysis of transient, strongly nonlinear responses that possess multiple distinct fast frequencies, yielding the reduced slow flow dynamics that govern the slow modulations of these fast components (namely, their amplitudes and phases). Clearly, the CX-A approach provides a good approximation of the exact dynamics only as long as the corresponding assumptions of the analysis are satisfied, and within the time domain of validity of the associated averaging operations [see (Sanders and Verhulst, 1985) and the discussion in Section 2.4].

There are important motivations for reducing the dynamics of (3.98) to the slow flow. First, as mentioned above, the slow flow-dynamics can be regarded as the im-

$$\dot{a}_2 + (\lambda/2)a_2 - (\lambda/2)a_1\cos\phi - (3C/8\varepsilon)(a_1^2 + a_2^2 - 2a_1a_2\cos\phi)a_1\sin\phi = 0$$

$$\dot{\phi} + (\lambda/2)[(\varepsilon a_2/a_1) + (a_1/a_2)]\sin\phi - 1/2 + (3C/8)(a_1^2 + a_2^2 - 2a_1a_2\cos\phi)$$
$$\times \{(1/\varepsilon)[1 - (a_2/a_1)\cos\phi] - [1 - (a_1/a_2)\cos\phi]\} = 0 \qquad (3.103)$$

The variables a_1 and a_2 represent the (real) amplitudes of the slowly-varying envelopes of the linear and nonlinear responses, respectively, whereas $\phi(t)$ the phase difference of the evolutions of these envelopes.

The reduced dynamical system (3.103) governs the slow flow dynamics of the fundamental TET. In particular, 1:1 TRC, the *underlying dynamical mechanism of TET*, *is associated with non-time-like evolution of the phase angle ϕ* or, equivalently, failure of the averaging theorem with respect to that angle (Sanders and Verhulst, 1985; Verhulst, 2005). Indeed, in case that ϕ would exhibit time-like behavior, we could regard it as a fast angle and apply the averaging theorem over ϕ to prove that the amplitudes a_1 and a_2 decay exponentially with time, nearly independently from each other (see also the discussion in Section 2.4). Then, no significant energy exchanges between the linear and nonlinear oscillators would take place, and no TET would be possible.

Figure 3.46a depicts the dynamics of 1:1 TRC in the slow flow phase plane $(\dot{\phi}, \phi)$ for system (3.103) with $\varepsilon = 0.05$, $\lambda = 0.01$, $C = 1$, $\omega_0 = 1$ and initial conditions $a_1(0) = 0.24$, $a_2(0) = 0.01$, $\phi(0) = 0$. The oscillatory behavior of the phase variable in the neighborhood of the in-phase limit $\phi = 0+$ confirms the occurrence of 1:1 TRC in the neighborhood of the in-phase NNM branch $S11+$. As evidenced by the build-up of amplitude a_2 of the envelope of the NES depicted in Figures 3.46b, d, this leads to fundamental TET from the LO to the NES. Escape from the 1:1 TRC is associated with time-like behavior of ϕ and rapid decrease of the amplitudes a_1 and a_2, as predicted by applying averaging in (3.103). A comparison of the analytical approximations (3.101–3.103) with direct numerical simulation of (3.98) subject to the previous initial conditions is presented in Figure 3.4c confirming the accuracy of the analysis. The discrepancy between analysis and numerical simulation noted for $T > 50S$ is attributed to the escape of the dynamics from the regime of 1:1 TRC, where the assumptions of the analysis are not valid any more. Moreover, due to the averaging operations associated with the CX-A technique, the resulting analytical approximation is not expected to be valid for relatively large times (see the discussion on the relation between averaged and exact dynamics in Section 2.4).

3.4.2.2 TET through Subharmonic TRC

Subharmonic TET involves excitation of a low-frequency subharmonic S-tongue of NNMs for frequencies $\omega < \omega_0$. As mentioned in Section 3.3.1.2, by low-frequency tongues we mean families of NNMs of the underlying Hamiltonian system with the nonlinear attachment engaging in m:n internal resonance with the LO (where m, n are integers with $m < n$). Another feature of a low-frequency tongue Smn, $m < n$ is that it is represented by a nearly horizontal line in the FEP, since on the tongue

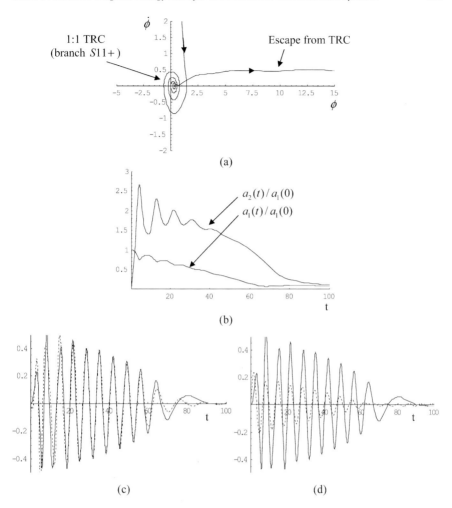

Fig. 3.46 Dynamics of fundamental TET: (a) 1:1 TRC in the slow flow; (b) normalized amplitude modulations; (c) comparison between analytical approximation (dashed line) and direct numerical simulation (solid line) of NES response (v_t); (d) system responses, [dashed line $x(t)$, solid line $v(t)$].

the strongly nonlinear response resembles that of a linear system with the NES oscillating slower than the LO and the ratio of their frequencies being approximately equal to $m/n < 1$ (see the discussion about oscillations on tongues $S13\pm$ in Section 3.3.2.2). Moreover, to each rational number $m/n, m > n$ there corresponds a pair of closely spaced tongues, composed of in-phase ($Smn+$) and an out-of-phase ($Smn-$) periodic motions, respectively; finally, these tongues exist over finite energy ranges. Hence, a countable infinity of low-frequency subharmonic tongues exists over finite energy ranges of the Hamiltonian system corresponding to

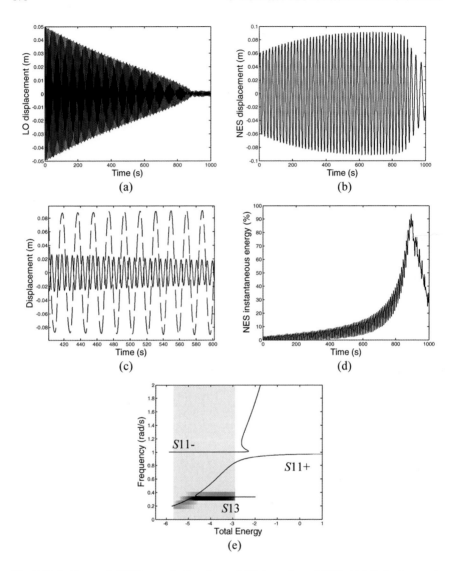

Fig. 3.47 Subharmonic TET ($\omega_0 = 1$, $C = 1$, $\varepsilon = 0.05$, $\lambda_1 = \lambda_2 = 0.001$): (a) LO displacement; (b) NES displacement; (c) superposition of system displacements (solid line: LO; dashed line: NES); (d) percentage of instantaneous total energy in the NES; and (e) WT spectrum of the relative response $(v - x)$ superposed to the backbone of FEP of the underlying Hamiltonian system.

$\lambda_1 = \lambda_2 = 0$ in (3.98). As mentioned in Section 3.3.1.2 this is a direct sequence of the non-integrability of this strongly nonlinear Hamiltonian system under examination.

To explain subharmonic TET in the damped system (3.98), we focus in the particular pair of lower tongues $S13\pm$, and refer to Figure 3.21d. As discussed in Section 3.3.2.2, at the extremity of this tongue (i.e., at the maximum energy of the tongue), the oscillation is localized to the LO. However, as in the case of fundamental TET, the *reduction of energy by damping dissipation leads to gradual delocalization of the motion from the* LO *and localization to the* NES; as a result, passive energy transfer from the LO to the NES, i.e., subharmonic TET, takes place. It follows that, as in the case of fundamental TET, it is the change of shape of NNMs on $S13\pm$ that eventually leads to subharmonic TET in the damped system. Again, one can invoke arguments of invariance and persistence of the damped NNM manifold resulting from the perturbation due to weak damping of the corresponding NNM invariant manifolds $S13\pm$ of the underlying Hamiltonian system. In this case, the underlying dynamics causing TET is an *m:n* TRC that occurs in the neighborhood of an *m:n* resonance manifold of the dynamics, as discussed later in this section.

The transient dynamics for motion initiated on the stable branch $S13-$ (with initial conditions $x(0) = v(0) = 0$, $\dot{x}(0) = -0.0497$, $\dot{v}(0) = 0.0296$) is displayed in Figure 3.47 for $\omega_0 = 1$, $C = 1$, $\varepsilon = 0.05$, and $\lambda_1 = \lambda_2 = 0.001$. Despite the presence of viscous dissipation, the NES response grows continuously as it passively absorbs and locally dissipates vibration energy from the LO whose amplitude rapidly decreases. Figure 3.47d shows that subharmonic TET takes place until approximately $t = 900$ s, during which almost complete energy transfer from the LO to the NES is realized. The WT spectrum of Figure 3.47e demonstrates clearly that the damped response traces approximately the subharmonic tongue $S13-$ until it reaches the backbone curve of the FEP, after which it traces that branch. This provides further evidence of the close relation of the weakly damped and Hamiltonian dynamics, and highlights the mechanism governing TET in this case. It is interesting to note that for the specific 1:3 subharmonic TET shown in Figure 3.47, the LO oscillates with a frequency approximately three times that of the NES. Moreover, due to the stability properties of the tongues $S13\pm$, subharmonic TET can only take place for out-of-phase relative motions between the LO and the NES (i.e., for excitation of the stable out-of-phase NNMs on tongue $S13-$), and not for in-phase ones, since the in-phase tongue $S13+$ is unstable (see Figure 3.21d).

To demonstrate the analysis of the dynamics governing subharmonic TET, we focus on 1:3 TRC in the neighborhood of tongue $S13-$. However, similar analysis can be applied to other cases of subharmonic resonance captures leading to TET. Due to the fact that motion in the neighborhood of $S13-$ possesses two main harmonic components with frequencies ω and $\omega/3$, the transient damped responses of system (3.98) are expressed as

$$x(t) = x_1(t) + x_{1/3}(t), \quad v(t) = v_1(t) + v_{1/3}(t) \tag{3.104}$$

where the indices 1 and 1/3 indicate that the respective terms possess dominant frequencies equal to ω and $\omega/3$, respectively. As in the case of fundamental TET, we introduce the following new complex variables:

$$\psi_1(t) = \dot{x}_1(t) + j\omega x_1(t) \equiv \varphi_1(t)e^{j\omega t},$$
$$\psi_3(t) = \dot{x}_{1/3}(t) + j(\omega/3)x_{1/3}(t) \equiv \varphi_3(t)e^{j(\omega/3)t}$$

$$\psi_2(t) = \dot{v}_1(t) + j\omega v_1(t) \equiv \varphi_2(t)e^{j\omega t},$$

$$\psi_4(t) = \dot{v}_{1/3}(t) + j(\omega/3)v_{1/3}(t) \equiv \varphi_4(t)e^{j(\omega/3)t} \qquad (3.105)$$

Again slow-fast partitions of the dynamics are introduced, but this is performed in a different way than in the case the fundamental TET case, to reflect the existence of two fast frequencies ω and $\omega/3$ in the responses during 1:3 TRC. Although $\omega \approx 1$ during 1:3 TRC in the neighborhood of tongue $S13-$, we opt to keep ω as a yet undetermined frequency parameter for the time being. In (3.105) the variables $\varphi_i(t), i = 1, \ldots, 4$ represent slowly varying complex modulations of the fast oscillations with frequencies ω and $\omega/3$. Expressing the responses x and v and their time derivatives in terms of the new complex variables, i.e.,

$$x = \frac{\psi_1 - \psi_1^*}{2j\omega} + \frac{3(\psi_3 - \psi_3^*)}{2j\omega}, \quad v = \frac{\psi_2 - \psi_2^*}{2j\omega} + \frac{3(\psi_4 - \psi_4^*)}{2j\omega} \qquad (3.106)$$

and substituting the resulting expressions into (3.98), we perform averaging over each of the two fast frequencies ω and $\omega/3$, and derive the following set of complex coupled differential equations governing the slow evolutions of the four complex modulations,

$$\dot{\varphi}_1 + (j\omega/2 - j/2\omega)\varphi_1 + (\varepsilon\lambda/2)(2\varphi_1 - \varphi_2)$$
$$+ (jC/8\omega^3)\{3[9\varphi_3^3 - 27\varphi_3^2\varphi_4 - 9\varphi_4^3 - (\varphi_1 - \varphi_2)|\varphi_1 - \varphi_2|^2 + 27\varphi_3\varphi_4^2$$
$$- 18(\varphi_1 - \varphi_2)|\varphi_3 - \varphi_4|^2]\} = 0$$

$$\dot{\varphi}_3 + (j\omega/6 - 3j/2\omega)\varphi_3 + (\varepsilon\lambda/2)(2\varphi_3 - \varphi_4)$$
$$+ (jC/8\omega^3)\{-9[\varphi_1(2(\varphi_3 - \varphi_4)(\varphi_1^* - \varphi_2) - 3(\varphi_3^* - \varphi_4^*)^2)$$
$$+ \varphi_2(2(\varphi_4 - \varphi_3)(\varphi_1^* - \varphi_2) + 3(\varphi_3^* - \varphi_4^*)^2) + 9(\varphi_3 - \varphi_4)|\varphi_3 - \varphi_4|^2]\} = 0$$

$$\dot{\varphi}_2 + (j\omega/2)\varphi_2 + (\lambda/2)(\varphi_2 - \varphi_1) - (jC/\varepsilon 8\omega^3)\{3[9\varphi_3^3 - 27\varphi_3^2\varphi_4 - 9\varphi_4^3$$
$$- (\varphi_1 - \varphi_2)|\varphi_1 - \varphi_2|^2 + 27\varphi_3\varphi_4^2 - 18(\varphi_1 - \varphi_2)|\varphi_3 - \varphi_4|^2]\} = 0$$

$$\dot{\varphi}_4 + (j\omega/6)\varphi_4 + (\lambda/2)(\varphi_4 - \varphi_3)$$
$$- (jC/\varepsilon 8\omega^3)\{-9[\varphi_1(2(\varphi_3 - \varphi_4)(\varphi_1^* - \varphi_2) - 3(\varphi_3^* - \varphi_4^*)^2)$$
$$+ \varphi_2(2(\varphi_4 - \varphi_3)(\varphi_1^* - \varphi_2) + 3(\varphi_3^* - \varphi_4^*)^2)$$
$$+ 9(\varphi_3 - \varphi_4)|\varphi_3 - \varphi_4|^2]\} = 0 \qquad (3.107)$$

where it is assumed that $\lambda_1 = \lambda_2 = \lambda$.

The complex amplitudes are expressed in polar form, $\varphi_i(t) = a_i(t)e^{j\beta_i(t)}$, $i = 1, \ldots, 4$, which when substituted into (3.107) and upon separation of real and imaginary parts lead to an autonomous set of seven slow flow real modulation equations in terms of the amplitudes $a_i = |\varphi_i|, i = 1, \ldots, 4$, and three phase differences defined as $\phi_{12} = \beta_1 - \beta_2$, $\phi_{13} = \beta_1 - 3\beta_3$, and $\phi_{14} = \beta_1 - 3\beta_4$. Due to its complexity, the autonomous system that governs the slow flow of 1:3 TRC is not reproduced in its entirety here, but is only expressed in the following compact form:

$$\dot{a}_1 + (\varepsilon\lambda/2)(2a_1 - a_2) + g_1(\underline{a}, \underline{\phi}) = 0$$

$$\dot{a}_3 + (\varepsilon\lambda/2)(2a_3 - a_4) + g_3(\underline{a}, \underline{\phi}) = 0$$

$$\dot{a}_2 + (\lambda/2)(a_2 - a_1) + g_2(\underline{a}, \underline{\phi})/\varepsilon = 0$$

$$\dot{a}_4 + (\lambda/2)(a_4 - a_3) + g_4(\underline{a}, \underline{\phi})/\varepsilon = 0$$

$$\dot{\phi}_{12} + f_{12}(\underline{a}) + g_{12}(\underline{a}, \underline{\phi}; \varepsilon) = 0$$

$$\dot{\phi}_{13} + f_{13}(\underline{a}) + g_{13}(\underline{a}, \underline{\phi}) = 0$$

$$\dot{\phi}_{14} + f_{14}(\underline{a}) + g_{14}(\underline{a}, \underline{\phi}; \varepsilon) = 0 \quad (3.108)$$

In the system above, g_i and g_{ij} are 2π-periodic functions in terms of the phase angles $\underline{\phi} = (\phi_{12} \; \phi_{13} \; \phi_{14})^T$, and $\underline{a}$ is the (4×1) vector of amplitudes,

$$\underline{a} = [a_1 \; a_2 \; a_3 \; a_4]^T.$$

As in the case of fundamental TET, strong energy exchanges between the LO and the NES can occur only if a subset of phase angles ϕ_{ij} does not exhibit time-like behavior, that is, when some phase angles possess non-monotonic behavior with respect to time. This can be deduced from the structure of the slow flow (3.108), where it is clear that if all phase angles exhibit time-like behavior and functions g_i are small, averaging over these phase angles (which could then be regarded as fast angles) would lead to decaying amplitudes. In that case no significant energy exchanges between the LO and the NES could take place. As a result, 1:3 *subharmonic TET is associated with non-time-like behavior of (at least) a subset of the slow phase angles* ϕ_{ij} *in (3.108)*.

Figure 3.48 depicts the results of the numerical simulation of the slow flow (3.107) for $\varepsilon = 0.05$, $\lambda = 0.03$, $C = 1$ and $\omega_0 = 1$. The motion is initiated on branch S13− with initial conditions $v(0) = x(0) = 0$, $\dot{v}(0) = 0.01499$, and $\dot{x}(0) = -0.059443$. The issue of computing the corresponding initial conditions for the slow flow (3.107) is non-trivial and indeterminate, as this system possesses more dimensions than the exact problem. The discussion of this issue is postponed until Section 9.2.2.2 in Chapter 9, and here it suffices to state that the initial conditions for the complex amplitudes and the value of the frequency of the slow flow model (3.107) are computed by minimizing the difference between the analytical and numerical responses of the system in the interval $t \in [0, 100]$:

$$\varphi_1(0) = -0.0577, \quad \varphi_2(0) = 0.0016, \quad \varphi_3(0) = -0.0017$$

$$\varphi_4(0) = 0.0134, \quad \omega = 1.0073 \quad (3.109)$$

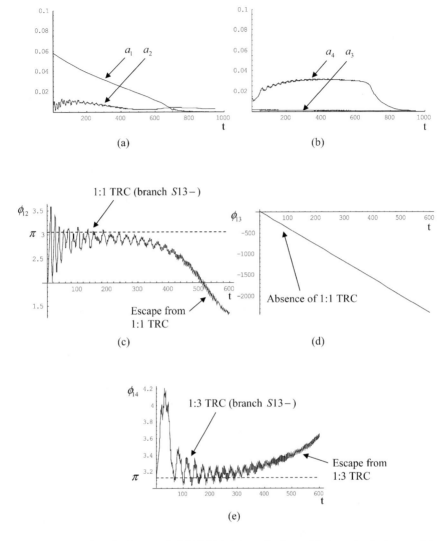

Fig. 3.48 Dynamics of subharmonic 1:3 TET: (a, b) amplitude modulations; (c–e) phase modulations.

This result proves that indeed frequency ω is close to unity, in accordance to our previous discussion.

Before proceeding with discussing the numerical results, we mention that the initial conditions required for the solution of the set modulations (3.107) exceeds in number the available initial conditions of the original problem (3.98); the reason, of course, is that, due to decompositions (3.104, 3.105) we are in need to define initial conditions *separately* for each of the harmonic components at frequencies

ω and $\omega/3$. The method of defining the initial conditions adopted above, although not conceptually elegant and non-unique, nevertheless provides satisfactory initial conditions for the slow flow as judged by the following numerical results.

The initial conditions (3.109) indicate that the energy at $t = 0$ is almost entirely stored in the fundamental frequency component of the LO. Figures 3.48a, b depict the slow evolutions of the amplitudes a_i. As judged from the build-up of amplitude a_4 and the corresponding decay of a_1, it becomes evident that 1:3 *subharmonic* TET *involves primarily energy transfer from the fundamental component of the LO to the 1/3 subharmonic component of the NES*. Considering the evolution of the amplitude a_2, we conclude that a smaller amount of energy is transferred from the fundamental component of the LO to the fundamental component of the NES.

These conclusions are supported by the plots of Figures 3.48c–e, where the temporal evolutions of the phase differences $\phi_{12} = \beta_1 - \beta_2$, $\phi_{13} = \beta_1 - 3\beta_3$, and $\phi_{14} = \beta_1 - 3\beta_4$ are presented. Absence of strong energy exchanges between the fundamental and 1/3 subharmonic components of the LO response is associated with the time-like behavior of the corresponding phase difference ϕ_{13}, whereas strong energy transfer from the fundamental component of the LO response to both fundamental and 1/3 subharmonic components of the NES response, is associated with early-time oscillatory (i.e., non-time-like) behavior of the corresponding phase differences ϕ_{12} and ϕ_{14}. Oscillatory behaviors of ϕ_{12} and ϕ_{14} signify 1:1 and 1:3 TRCs, respectively, between the fundamental component of the LO response and the fundamental and 1/3 subharmonic components of the NES response. With progressing time, the phase variables become eventually time-like, signifying escapes from the corresponding TRCs. We note that the oscillations of ϕ_{12} and ϕ_{14} take place in the neighborhood of π, which confirms that, in this particular example, 1:3 subharmonic TET involves out-of-phase relative motions between the LO and the NES (since they take place in the neighborhood of tongue $S13-$).

The predictive capacity of the analytical slow flow model (3.107, 3.108) in the regime of 1:3 subharmonic TET is demonstrated by the result depicted in Figure 3.49. It can be observed that the analytically predicted NES response is in satisfactory agreement with the exact response obtained by direct simulation of equations (3.98); this, in spite of the fact that transient and strongly nonlinear dynamics is considered. However, the analytic model fails to accurately model the response in the later regime, where escape from 1:3 TRC occurs. This occurs because during this regime the damped response leaves the neighborhood of tongue $S13-$ and approximately evolves along the backbone curve of the FEP. Eventually, the next tongue $S15$ is reached, and at that point the motion cannot be described by the *ansatz* (3.104, 3.105) anymore, since the 1/3 subharmonic component gradually diminishes becoming unimportant and a new 1/5 subharmonic component enters into the dynamics. As a result, the considered analytical model looses validity.

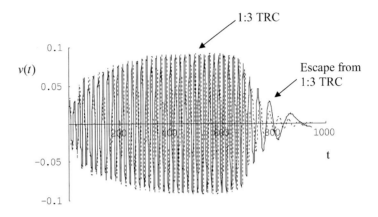

Fig. 3.49 Transient damped response of the NES during 1:3 subharmonic TET: comparison between analytical slow-flow approximation (dashed line) and direct numerical simulation (solid line).

3.4.2.3 TET through Nonlinear Beats

The previous two TET mechanisms cannot be 'triggered' with the NES being initially at rest, since both require non-zero initial velocity for the NES, i.e., $\dot{v}(0) \neq 0$. This means that neither fundamental nor subharmonic TET can occur immediately after the application of an impulsive excitation to the LO. An alternative TET mechanism, however, TET *through nonlinear beats*, not only surpasses this limitation, but proves to be the most powerful TET mechanism since it is capable of initiating stronger energy transfers from the LO to the NES compared to the above-mentioned two TET mechanisms. This TET mechanism is based on the initial excitation of IOs (especially, moderate-energy ones, close to the 1:1 resonance manifold) which have been discussed in detail in Sections 3.3.3 and 3.3.4.

As mentioned previously, the excitation of stable localized IOs in the regime of 1:1 internal resonance of the Hamiltonian system (with the system being initially at rest subject to impulsive excitations of the LO – equivalently, with initial conditions $\dot{x}(0) \neq 0$ and $v(0) = \dot{v}(0) = x(0) = 0$), leads to rapid transfer of energy from the LO to the NES during a cycle of the motion. This transfer is realized through nonlinear beats. We will show that in the weakly damped system, such IOs play the role of *transient bridging orbits* that direct the damped motion into the domain of attraction of a resonant manifold, which eventually leads to (triggers) either fundamental or subharmonic TET.

Recalling the analysis of Section 3.3.4.1, the class of moderate-energy IOs occurs only above a critical energy threshold. It follows, that the corresponding triggering mechanism for TET is effective only for input energies above this critical threshold. Indeed, as shown in Section 3.3.3, low-energy (or equivalently low-frequency) IOs transfer a small fraction of the input energy from the LO to the NES, so they cannot induce TET. It should also be noted that, due to the essential (nonlinearizable) nonlinearity of the NES the considered nonlinear beating phenomena do not require any *a priori* tuning of the nonlinear attachment: at a specific frequency-energy range corresponding to n:m resonance capture, the essential nonlinearity of the NES passively adjusts the amplitude to fulfill the required resonance conditions. This represents a significant departure from classical nonlinear beat phenomena observed in coupled oscillators with linearizable nonlinear stiffnesses where the ratio of the linearized natural frequencies of the components dictates the type of internal resonance that can be realized.

To validate our conjecture, we perform a numerical simulation where system (3.98) is initiated at the IO on $U21$ (corresponding to initial conditions $x(0) = v(0) = \dot{v}(0) = 0$, and $\dot{x}(0) = 0.5794$, for system parameters $\omega_0 = 1, C = 1, \varepsilon = 0.05, \lambda_1 = \lambda_2 = 0.005$). As evidenced in the instantaneous energy plot of Figure 3.50c, a nonlinear beating phenomenon takes place in the initial stage of the motion until approximately $T = 50$ s; this corresponds to the initial excitation of the damped analogue of the IO on $U21$. During the nonlinear beat phenomenon, the relative displacement $v(t) - y(t)$ possesses two main frequency components (around 1 and 2 rad/s), but the higher harmonic is barely visible in the WT spectrum plot of Figure 3.50d. After this initial nonlinear energy exchange between the two oscillators, the dynamics makes a transition to the damped in-phase NNM manifold $S11+$, and the dynamics is captured into the domain of attraction of the 1:1 resonant manifold. Eventually, fundamental TET takes place. We note that TET through nonlinear beats also occurs in the numerical simulation depicted in Figure 3.42; in that case, however, the initial beats due to excitation of the IO on $U76$ lead, first to a transition to small duration fundamental TET, and then to a second transition to a more prolonged 1:3 subharmonic TET. This underlines the fact that although damping cannot generate new dynamics in the system, it critically influences the damped transitions between branches of solutions of the underlying Hamiltonian system.

Finally, we note that TET through nonlinear beats proves to be the most efficient TET mechanism. Further discussion of this TET mechanism is postponed until Section 3.4.2.4 where conditions for optimal TET are discussed. In the next section we discuss TET from the alternative view of damped NNM manifolds, which highlights more clearly the role of damping on TET.

3.4.2.4 Damped NNM Manifolds and Fundamental TET

In this section we wish to further demonstrate the important role of damping on fundamental TET. Although the analysis will be carried out under the assumption of 1:1 resonance capture leading to fundamental TET, it can be extended to the more complicated case of m:n subharmonic TET, with appropriate modifications. Reconsidering equations (3.98), which describe the two-DOF damped dynamics of an essentially nonlinear system, it is clear that they cannot be solved exactly (i.e., in explicit analytic form). However, as shown in Section 3.4.2.1 fundamental TET can be approximately analyzed by performing averaging in the vicinity of the 1:1 reso-

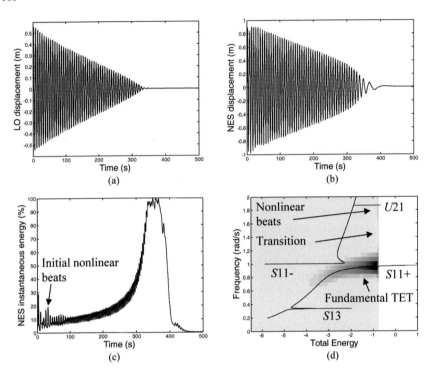

Fig. 3.50 TET through nonlinear beats, excitation of IO $U21$ ($\omega_0 = 1$, $C = 1$, $\varepsilon = 0.05$, $\lambda_1 = \lambda_2 = 0.005$): (a) LO displacement; (b) NES displacement; (c) percentage of instantaneous total energy in the NES; and (d) WT spectrum of the relative response $(v - x)$ superposed to the backbone of FEP of the underlying Hamiltonian system.

nance manifold (or for the more complicated case of subharmonic TET, by multiphase averaging in the neighborhoods of the corresponding resonance manifolds – see Section 3.4.2.2). We note that even the resulting reduced averaged system (3.102–3.103) is still too complicated to be solved analytically, although its state space may be reduced to three dimensions, unlike the exact system (3.98).

Approximate solutions of the averaged system governing fundamental TET may be computed based on two different approximations, each of which is now discussed. In the following analysis we will relax the condition $\lambda_1 = \lambda_2 = \lambda$ enforced in (3.102), and instead adopt independent values for both damping constants. The first option to analyze the averaged system in the regime of 1:1 resonance capture is to suppose that the damping coefficient λ is small; it follows that the zeroth-order approximation to solving (3.102) is the undamped system which is completely integrable as discussed previously. The effect of non-zero damping may then be described by application of appropriate asymptotic procedures. Such an approach, however, does not seem meaningful for studying TET, since as shown below TET strongly depends on the value of damping, so that the mentioned low-order perturbation scheme cannot be expected to describe the details of this strong dependence.

The second perturbation approach for analyzing the averaged dynamics, is based on the assumption of strong mass asymmetry between the LO and the NES, as described by the small parameter ε in (3.98); this means that we will focus on linear oscillators with lightweight NESs. This approach does not necessarily assume small damping, and instead relies on perturbation analysis considering the NES mass ε as the small parameter. This approach is considered in this section, for a system with parameters, $\lambda_1 = 0, \lambda_2 = \varepsilon\lambda, C = 4\varepsilon/3$ and $\omega_0 = 1$. The two latter conventions do not affect the generality of the analysis, since they may be satisfied by appropriate rescalings of the dependent and independent variables of the averaged system.

We start our analysis of fundamental TET by considering the system of averaged (complex modulation) equations (3.102). Introducing the following change of complex variables,

$$\chi_1 = \frac{\varphi_1 + \varepsilon\varphi_2}{1 + \varepsilon}$$

$$\chi_2 = \varphi_1 - \varphi_2$$

the modulation equations (3.102) take the form:

$$\dot{\chi}_1 + \frac{j\varepsilon}{2(1+\varepsilon)}(\chi_1 - \chi_2) = 0$$

$$\dot{\chi}_2 + \frac{j}{2(1+\varepsilon)}(\chi_2 - \chi_1) + \frac{\lambda(1+\varepsilon)}{2}\chi_2 - \frac{j(1+\varepsilon)}{2}|\chi_2|^2 \chi_2 = 0 \quad (3.110)$$

We recall that the slow flow system (3.102), and, hence (3.110) was derived under the assumption of 1:1 resonance between the LO and the NES, and so this model is valid only in the neighborhood of the 1:1 resonance manifold of the underlying Hamiltonian system. As in (3.102) the complex coordinates χ_1 and χ_2 describe the oscillations of the center of mass of, and the relative displacement between the LO and the NES, respectively. By successive differentiation and simple algebra, the above averaged system may be reduced to the following single modulation equation governing the slow flow of 1:1 resonance capture in the damped dynamics:

$$\frac{d^2\chi_2}{dt^2} + \frac{d}{dt}\left[\frac{j}{2}\chi_2 + \frac{\lambda(1+\varepsilon)}{2}\chi_2 - \frac{j(1+\varepsilon)}{2}|\chi_2|^2 \chi_2\right]$$

$$+ \frac{j\varepsilon}{4}(\lambda\chi_2 - j|\chi_2|^2 \chi_2) = 0 \quad (3.111)$$

This equation is integrable for $\lambda = 0$, but here we are interested in the damped case $\lambda > 0$. More precisely, we assume that $\lambda \gg \varepsilon$, so we treat λ as an $O(1)$ quantity.

Equation (3.111) may be analyzed by the multiple scales approach (Nayfeh and Mook, 1995). To this end, we introduce the new time scales, $\tau_i = \varepsilon^i t, i = 0, 1 \ldots$, which are treated as distinct independent variables in the following analysis. Expressing the time derivatives in (3.111) as

$$\frac{d}{dt} = \frac{\partial}{\partial \tau_0} + \varepsilon \frac{\partial}{\partial \tau_1} + O(\varepsilon^2), \quad \frac{d^2}{dt^2} = \frac{\partial^2}{\partial \tau_0^2} + 2\varepsilon \frac{\partial^2}{\partial \tau_0 \partial \tau_1} + O(\varepsilon^2) \quad (3.112)$$

substituting (3.112) into (3.111), and retaining only $O(1)$ terms we derive the following first-order modulation equation,

$$\frac{\partial^2 \chi_2}{\partial \tau_0^2} + \frac{\partial}{\partial \tau_0}\left[\frac{j}{2}\chi_2 + \frac{\lambda}{2}\chi_2 - \frac{j}{2}|\chi_2|^2 \chi_2\right] = 0 \quad (3.113)$$

which possesses the following exact first integral of motion:

$$\frac{\partial \chi_2}{\partial \tau_0} + \left[\frac{j}{2}\chi_2 + \frac{\lambda}{2}\chi_2 - \frac{j}{2}|\chi_2|^2 \chi_2\right] = M(\tau_1, \tau_2, \ldots) \quad (3.114)$$

In expressing the constant of integration M as function of the slow-scales $\tau_1, \tau_2, \ldots$, we recognize that the first integral of motion (3.114) refers only to the first-order dynamics, i.e., it is only constant correct to $O(1)$; mathematically, the slow variation of the first integral (3.114) is justified by the fact that the multiple scales of the problem are considered to be distinct and independent from each other. Hence, by (3.114) we allow slow variation of the dynamics, but at higher-order (slower) time scales. By the same reasoning, the equilibrium points, $\Phi(\tau_1, \tau_2, \ldots)$ of the first-order system (3.113) may be constant with respect to the first-order time scale, but may slowly vary with respect to higher-order (superslow) time scales; hence, the equilibrium points may depend on the higher-order superslow time scales $\tau_1, \tau_2, \ldots$. These equilibrium points of the slow flow are computed by solving the following algebraic equation:

$$\frac{j}{2}\Phi + \frac{\lambda}{2}\Phi - \frac{j}{2}|\Phi|^2\Phi = M(\tau_1, \tau_2, \ldots) \quad (3.115)$$

Clearly, if an equilibrium is stable it holds that

$$\Phi(\tau_1, \tau_2, \ldots) = \lim_{\tau_0 \to +\infty} \chi_2(\tau_0, \tau_1, \tau_2, \ldots) < \infty$$

whereas it holds that

$$\Phi(\tau_1, \tau_2, \ldots) = \lim_{\tau_0 \to -\infty} \chi_2(\tau_0, \tau_1, \tau_2, \ldots) < \infty$$

if that equilibrium is unstable. One can show that the first-order dynamical system (3.113) does not possess any limit sets besides equilibrium points [for instance by applying Bendixon's criterion (Guckenheimer and Holmes, 1982; Wiggins, 1990)].

Since we will carry the analysis only up to $O(\varepsilon)$, we omit from here on slow time scales of order higher than one and express the solution of (3.115) in the following polar form:

$$\Phi(\tau_1) = N(\tau_1) \exp(j\gamma(\tau_1)) \quad (3.116)$$

Upon substituting into (3.115) and separating real and imaginary terms, we reduce the computation of the equilibrium points of the slow flow to

$$\lambda^2 Z(\tau_1) + Z(\tau_1)[1 - Z(\tau_1)]^2 = 4|M(\tau_1)|^2 \quad (3.117)$$

where $Z(\tau_1) \equiv N^2(\tau_1)$. The number of solutions of equation (3.117) depends on $|M(\tau_1)|$ and λ. The function on the left-hand side can be either monotonous, or can have a maximum and a minimum. In the former case the change of $|M(\tau_1)|$ has no effect on the number of solutions and equation (3.117) provides a single positive solution. In the latter case, however, the change of $|M(\tau_1)|$ brings about a pair of saddle-node bifurcations, and hence multiple solutions.

In order to distinguish between the different cases, we check the roots of the derivative with respect to $Z(\tau_1)$ of the left-hand side of (3.117):

$$1 + \lambda^2 - 4Z + 3Z^2 = 0 \Rightarrow Z_{1,2} = [2 \pm \sqrt{1 - 3\lambda^2}]/3 \quad (3.118)$$

It follows that for $\lambda < 1/\sqrt{3}$ there exist two additional real roots and a pair of saddle-node bifurcations, whereas at the critical damping value $\lambda = 1/\sqrt{3}$ the two saddle-node bifurcation points coalesce forming the typical structure of a cusp. Extending these results to equation (3.117), if a single equilibrium exists, this equilibrium is stable with respect to the time scale τ_0. If three equilibrium points exist, two of them are stable nodes, and the third is an unstable saddle with respect to the time scale τ_0. Therefore, the $O(1)$ dynamics is attracted always to a stable node.

The characteristic rate of attraction of the dynamics near a node may be evaluated by linearizing equation (3.114), and considering the following perturbation of the dynamics near an equilibrium point:

$$\chi_2(\tau_0, \tau_1) = \Phi(\tau_1) + \delta(\tau_0), \quad |\delta| \ll |\Phi| \quad (3.119)$$

Upon substitution of (3.119) into (3.114) yields the following linearized equation,

$$\frac{\partial \delta}{\partial \tau_0} + \left[\frac{j}{2}\delta + \frac{\lambda}{2}\delta - j|\Phi|^2\delta - \frac{j}{2}\Phi^2\delta^*\right] = 0 \quad (3.120)$$

where asterisk denotes complex conjugate. Rewriting equation (3.120) as

$$\left(\frac{\partial}{\partial \tau_0} + \frac{j}{2} + \frac{\lambda}{2} - j|\Phi|^2\right)\delta = \frac{j}{2}\Phi^2\delta^* \quad (3.121)$$

taking its complex conjugate and combining the two equations, we derive an expression that explicitly computes the evolution of the perturbation $\delta(\tau_0)$ (note that Φ depends only on τ_1 and not on the time scale τ_0),

$$\left[\frac{\partial^2}{\partial \tau_0^2} + \lambda \frac{\partial}{\partial \tau_0} + \frac{1}{4}(1 + \lambda^2 - 4Z + 3Z^2)\right]\delta = 0 \Rightarrow$$

$$\delta = \delta_0 \exp[(-\lambda \pm j\omega)t/2] \quad (3.122)$$

where $\omega = \sqrt{3Z^2 - 4Z + 1}$. Solution (3.122) reveals that the linearized dynamics in the vicinity of the equilibrium points depends on λ and Z.

The following possible alternatives are now described. For relatively large values of damping above the critical value, $\lambda > 1/\sqrt{3}$, there exists a single stable node in the $O(1)$ dynamics. For $Z > 1$ or $Z < 1/3$ the attraction of the dynamics to that node is through oscillations [i.e., ω is real-underdamped cases], whereas for $1 > Z > 1/3$ the attraction is through a decaying motion [i.e., ω is imaginary – overdamped case).

For relatively small damping values, $\lambda < 1/\sqrt{3}$, the situation is more complex, since there exist two additional real equilibrium points given by (3.118). For $Z > 1$ or $Z < 1/3$ the attraction of the dynamics to the stable node is oscillatory (underdamped cases), whereas for $1 > Z > Z_1$ or $Z_2 > Z > 1/3$ the attraction is through a decaying motion (overdamped cases). For $Z_1 > Z > Z_2$ we obtain an unstable equilibrium, and the linearized model predicts exponential growth in the dynamics.

In summary, as Z slowly decreases due to its dependency on the slow-time scale τ_1, and depending on the damping value λ, the $O(1)$ dynamics undergoes qualitative changes (bifurcations). In particular, if $\lambda > 1/\sqrt{3}$ we anticipate the dynamics to remain always stable, since in that case there exists a single slowly-varying attracting manifold of the $O(1)$ averaged flow. However, if $\lambda < 1/\sqrt{3}$ the dynamics becomes unstable, in which case we expect that the $O(1)$ averaged flow will make a sudden transition from one attracting manifold to another for slowly decreasing Z. In order to study this complicated damped transition, one should investigate the slow evolution of the equilibrium of the $O(1)$ averaged flow $\Phi(\tau_1)$.

To this end, we consider the $O(\varepsilon)$ terms in the multiple-scale expansion (3.111–3.112):

$$2\frac{\partial^2 \chi_2}{\partial \tau_0 \partial \tau_1} + \frac{\partial}{\partial \tau_1}\left[\frac{j}{2}\chi_2 + \frac{\lambda}{2}\chi_2 - \frac{j}{2}|\chi_2|^2 \chi_2\right]$$
$$+ \frac{\partial}{\partial \tau_0}\left[\frac{\lambda}{2}\chi_2 - \frac{j}{2}|\chi_2|^2 \chi_2\right] + \frac{j}{4}[\lambda \chi_2 - j|\chi_2|^2 \chi_2] = 0 \quad (3.123)$$

We are interested in the behavior of the solution of the $O(\varepsilon)$ averaged flow in the neighborhood of a stable equilibrium point, or equivalently, in the neighborhood of the damped NNM invariant manifold $\Phi(\tau_1) = \lim_{\tau_0 \to +\infty} \chi_2(\tau_0, \tau_1)$. Therefore, by taking the limit $\tau_0 \to +\infty$ in equation (3.123) we obtain the following equation which describes the evolution of the dynamics at the slower time scale τ_1:

$$\frac{\partial}{\partial \tau_1}\left(\frac{j}{2}\Phi + \frac{\lambda}{2}\Phi - \frac{j}{2}|\Phi|^2 \Phi\right) + \frac{j}{4}(\lambda \Phi - j|\Phi|^2 \Phi) = 0 \quad (3.124)$$

In deriving this equation we take into account that on the slowly-varying, stable invariant manifold there is no dependence of the dynamics on τ_0, since $\Phi(\tau_1)$ was defined previously as the equilibrium point of the $O(1)$ averaged flow (3.113–3.114). Hence, the differential equation (3.124) describes the slow evolution of the stable equilibrium points of equation (3.113) (these are equilibrium points with respect to the fast time scale τ_0, but not with respect to the slow time scale τ_1 and to slow time scales of higher orders, which, however are omitted from the present analysis). The slowly varying equilibrium $\Phi(\tau_1)$ provides an $O(\varepsilon)$ approximation to the *damped NNM manifold* of the dynamics of the system (3.98); this is an *invariant manifold* of the damped dynamics and can be regarded as the analytical continuation for weak damping of the corresponding NNM of the underlying Hamiltonian system (Shaw and Pierre, 1991, 1993).

Rearranging equation (3.124) in the form

$$\left(\frac{j}{2} + \frac{\lambda}{2} - j|\Phi|^2\right) \frac{\partial \Phi}{\partial \tau_1} - j\Phi^2 \frac{\partial \Phi^*}{\partial \tau_1} = -\frac{j}{4}(\lambda\Phi - j|\Phi|^2 \Phi) \qquad (3.125)$$

and adding to it its complex conjugate, we obtain the following explicit expression for the slowly varying derivative of the equilibrium point of the $O(1)$ slow flow:

$$\frac{\partial \Phi}{\partial \tau_1} = \frac{-\lambda\Phi + j\left(|\Phi|^2 \Phi - 3|\Phi|^4 \Phi - \lambda^2 \Phi\right)}{2\left(1 + \lambda^2 - 4|\Phi|^2 + 3|\Phi|^4\right)} \qquad (3.126)$$

Using the polar representation, $\Phi(\tau_1) = N(\tau_1) \exp(i\gamma(\tau_1))$, and separating real and imaginary parts, equation (3.126) yields the following set of real differential equations governing the slow evolution of the magnitude and phase of the stable equilibrium points of the $O(1)$ averaged flow (i.e., of the stable damped NNM manifolds),

$$\frac{\partial N}{\partial \tau_1} = \frac{-\lambda N}{2\left(1 + \lambda^2 - 4Z + 3Z^2\right)}$$

$$\frac{\partial \gamma}{\partial \tau_1} = \frac{(Z - 3Z^2 - \lambda^2)}{2\left(1 + \lambda^2 - 4Z + 3Z^2\right)} \qquad (3.127)$$

where $Z(\tau_1) \equiv N^2(\tau_1)$. The first of equations (3.127) can be integrated exactly by quadratures to yield

$$(1 + \lambda^2) \ln Z(\tau_1) - 4Z(\tau_1) + (3/2)Z^2(\tau_1) = K - \lambda \tau_1 \qquad (3.128)$$

where K is a constant of integration [it actually depends on the higher-order time scales $\tau_2, \tau_3, \ldots$, but these are nor considered here as the analysis is restricted to $O(\varepsilon)$].

Expression (3.128) implicitly determines the evolution of $Z(\tau_1)$ and, consequently, of $N(\tau_1)$. The slow evolution of the phase $\gamma(\tau_1)$ is described by the second of equations (3.127), and may be computed by direct integration once $Z(\tau_1)$ is known; due to the implicit form of (3.128), however, this task cannot be performed analytically and requires a numerical solution.

Essential information concerning the qualitative behavior of the solution may be extracted from relation (3.127) even without explicitly solving it. Indeed, for sufficiently strong damping, $\lambda > 1/\sqrt{3}$, the denominator on the right-hand side

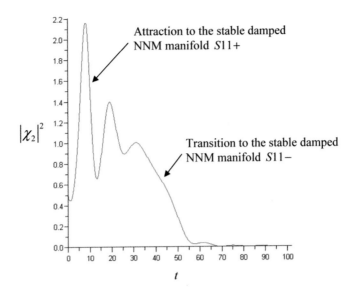

Fig. 3.51 Response of the averaged system (3.111) in the regime of 1:1 resonance capture, for $\varepsilon = 0.05$, $\lambda = 0.2 < 1/3$, and initial conditions given by $\chi_1(0) = 0.7 + 0j$, $\chi_2(0) = 0.7 + 0j$.

terms is always positive, and the first equation describes a monotonous decrease of $Z(\tau_1)$ towards zero with increasing τ_1. In other words, we conjecture that the slowly varying dynamics remains always on the in-phase damped NNM manifold $S11+$. By contrast, for relatively weak damping, $\lambda < 1/\sqrt{3}$, the velocity $\partial Z/\partial \tau_1$ is a negative quantity for $Z > Z_1$, but becomes divergent as the limit $Z \to Z_1$ is approached from above.

We cannot proceed to any statement regarding the sign of the velocity when the amplitude is in the range $Z_2 > Z > Z_1$, as the equilibrium point is unstable there; therefore, we infer that as Z decreases below the critical amplitude Z_1 the damped dynamics should be attracted to a NNM damped manifold distinct from $S11+$. This distinct manifold is a weakly nonlinear (linearized) branch of the damped NNM invariant manifold $S11-$. Of course, this conclusion is valid only for the averaged system (3.110–3.111), which was derived under the condition of 1:1 resonance capture. In the original system (3.98) attraction of the dynamics to other (i.e., different from 1:1) subharmonic or superharmonic resonance manifolds may take place, depending on the initial conditions and the system parameters. Similar averaging arguments could be used to study such more complex damped transitions.

The previous analytical findings are illustrated by performing numerical simulations of the averaged system (3.111) for parameters $\varepsilon = 0.05$, $\lambda = 0.2$, and initial conditions $\chi_1(0) = 0.7 + 0j$, $\chi_2(0) = 0.7 + 0j$. The time evolution of the square of the modulation of the envelope of the NES response, $|\chi_2|^2$, is depicted in Figure 3.51. Clearly, both the magnitude and frequency of the envelope modulation of the NES response tend to zero as the trajectory approaches the critical value

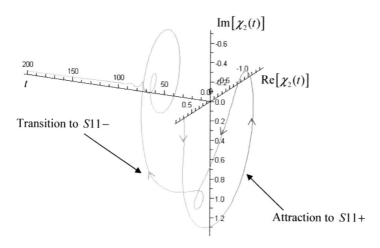

Fig. 3.52 Real and imaginary parts of the complex modulation χ_2 of the NES plotted against time, in the regime of 1:1 resonance capture for $\varepsilon = 0.05$, $\lambda = 0.2$, and initial conditions $\chi_1(0) = 0.7 + 0j$, $\chi_2(0) = 0.7 + 0j$.

$Z_1 = 0.979$. In the vicinity of this value, the trajectory jumps to the alternative stable attractor $S11-$. This point may be further illustrated using the three-dimensional plot depicted in Figure 3.52, where the real and imaginary parts of the complex envelope modulation of the NES, χ_2, are plotted in a parametric plot for increasing time. The damped trajectory of the envelope modulation of the NES starts from zero, gets attracted initially by the stable damped NNM manifold $S11+$, and then makes a transition (jump) to the weakly nonlinear, low-energy stable NNM manifold $S11-$.

In order to check the validity of the asymptotic approximations, we performed direct simulations of the original set (3.98) (i.e., of the exact system before averaging) with the same initial conditions used for the plots of Figures 3.51 and 3.52; the result is presented in Figure 3.53. It is clear from this figure that the damped dynamics is initially attracted by the damped NNM manifold $S11+$, as evidenced by the in-phase 1:1 resonant oscillations of the NES and the LO, with nearly unit frequency. With diminishing amplitude of the NES, the critical amplitude is reached close to $t \sim 50$ s, and a transition of the damped dynamics to a out-of-phase linearized low-energy regime $S11-$ takes place, with the motion localizing to the LO. This is in accordance with the predictions of the averaging analysis.

The next simulation illustrates the dynamics of the averaged system (3.110) for the case of low damping (see Figure 3.54). The system parameters are chosen as $\varepsilon = 0.05$, $\lambda = 0.03$, and the initial conditions as $\chi_1(0) = 0.9 + 0j$ and $\chi_2(0) = 0.9 + 0j$. Despite the low damping value, the qualitative behavior of the dynamics is similar to the previous case, although it takes much more time for the dynamics to escape away from the damped NNM invariant manifold $S11+$. It should be mentioned that, technically, the multiple-scale analysis developed above is not formally valid in this case, because the damping coefficient is of $O(\varepsilon)$ and

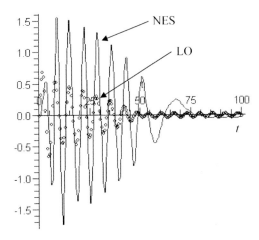

Fig. 3.53 Direct numerical simulation of the damped system (3.98) for parameters $\varepsilon = 0.05$, $\lambda_1 = 0$, $\lambda_2 = 0.01$, and initial conditions $x(0) = v(0) = \dot{x}(0) = 0$ and $\dot{x}(0) = 0.7$; the dynamics correspond to the analytical results of Figures 3.50 and 3.51.

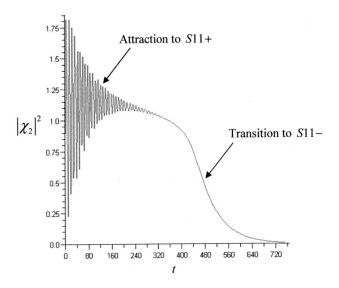

Fig. 3.54 Response of the averaged system (3.111) in the regime of 1:1 resonance capture, for $\varepsilon = 0.05$, $\lambda = 0.03$, and initial conditions given by $\chi_1(0) = 0.9 + 0j$ and $\chi_2(0) = 0.9 + 0j$.

not of $O(1)$ as assumed in the analysis. To check, however, the applicability of the approximation in this case, the original system (3.98) was again simulated for parameters and initial conditions corresponding to the ones of the averaged model. The result is presented in Figure 3.55. It is difficult to judge whether any real transition (jump) occurs at $t \sim 480$ s, but a gradual change of the NES frequency starts at this

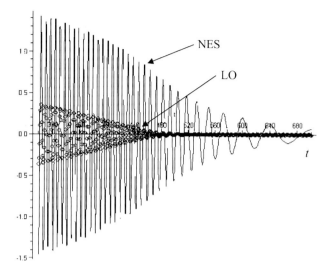

Fig. 3.55 Direct numerical simulation of the damped system (3.98) for parameters $\varepsilon = 0.05$, $\lambda_1 = 0$, $\lambda_2 = 0.0015$, and initial conditions $x(0) = v(0) = \dot{x}(0) = 0$ and $\dot{x}(0) = 0.9$; the dynamics correspond to the analytical result of Figure 3.54.

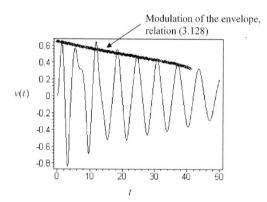

Fig. 3.56 Exact solution of the NES oscillation $v(t)$ – solid line, superimposed to the analytically predicted envelope modulation [computed from (3.128)] – dotted line, up to the point of transition away from the damped NNM manifold $S11+$.

time instant and reveals escape from the regime of 1:1 resonance capture, thereby confirming the analytic findings.

It is instructive to compare the result of the direct numerical simulation with the analytic expression (3.128) that computes approximately the modulation of the envelope of the response of the NES. The result of this comparison is presented in Figure 3.56. Expression (3.128) provides an accurate prediction for the modulation of the envelope of the NES response as long as the damped dynamics is in the

1:1 resonance capture regime, i.e., before the escape from the damped NNM manifold $S11+$. Still, the description of the response is not complete since the initial conditions should also be taken into account. The averaging approach successfully describes the process up to times of $O(1/\varepsilon)$, but is not suitable for later times since the limit $\tau_0 \to +\infty$ is irrelevant in this case, as the dynamics makes a transition away from the manifold $S11+$ at finite time.

Fortunately, this latter problem is even easier to address. Indeed, if one is interested only in the behavior of the system up to a time scale of $O(1)$, i.e., only during the initial transient regime of the motion, then to a first approximation it is possible to neglect all terms of $O(\varepsilon)$ from the problem; this is shown below. To this end, we reconsider the damped two-DOF system (3.98) in the form

$$\ddot{x} + \lambda_1 \dot{x} + \lambda_2 (\dot{x} - \dot{v}) + \omega_0^2 x + C(x - v)^3 = 0$$
$$\varepsilon \ddot{v} + \lambda_2 (\dot{v} - \dot{x}) + C(v - x)^3 = 0 \quad (3.129)$$

with $\lambda_1 = 0$, $\omega_0^2 = 1$, $\lambda_2 = \varepsilon \lambda$ and $C = \varepsilon$, $0 < \varepsilon \ll 1$. We introduce the change of variables, $y_1 = x + \varepsilon v$, $y_2 = x - v$, where y_1 describes the motion of the center of mass of the system, and y_2 the relative motion between the LO and the NES. System (3.129) is then transformed into the following form:

$$\ddot{y}_1 + \frac{y_1 + \varepsilon y_2}{1 + \varepsilon} = 0$$
$$\ddot{y}_2 + \frac{y_1 + \varepsilon y_2}{1 + \varepsilon} + (1 + \varepsilon) \lambda \dot{y}_2 + (1 + \varepsilon) y_2^3 = 0 \quad (3.130)$$

The important advantage of system (3.130) compared to (3.129) is that the highest derivatives are now multiplied by unity, and the perturbation parameter is shifted to the remaining terms. This permits the application of standard perturbation techniques (such as the methods of multiple scales or averaging) to the analysis of the dynamics. To a first approximation, we retain only terms of $O(1)$ in (3.130), rendering the resulting analytical transient approximations valid only up to times of $O(1)$, i.e., only in the initial, strongly nonlinear regime of the motion:

$$\begin{cases} \ddot{y}_1 + y_1 = 0 \\ \ddot{y}_2 + \lambda \dot{y}_2 + y_2^3 = -y_1 \end{cases} \quad \text{(Early-time approximation))} \quad (3.131)$$

More accurate approximation to the dynamics may be obtained by carrying the analysis beyond the $O(1)$ approximation, for example, by analyzing the transformed system (3.130) by the method of multiple scales or averaging. This, however, would recover the averaging results of the previous analysis which are valid up to times of $O(1/\varepsilon)$, so this option is not pursued further here.

We note that the damping term in the second of equations (3.131) appears now as an $O(1)$ quantity, so the approximation is justified only if $\lambda = O(1)$; in the following simulations this condition is satisfied. Besides, the implicit assumption is that $O(y_1) = O(y_2)$, i.e., that the amplitude of the oscillation of the center of mass is comparable to the amplitude of the relative oscillation between the LO and the NES. This assumption is correct only during the initial regime of the motion, as further evolution of the variables brings about diffentiation of the relative scaling between amplitudes, and the coupling term of order ε in the equation for v is not negligible anymore. Furthermore, equations (3.131) do not conserve energy in the absence of damping, which means that they are not suitable for describing the global dynamics of the system (3.129).

We wish to develop analytical approximations of the early-time transient responses modeled by the dynamical system (3.131), subject to the general initial conditions

$$y_1(0) = Y_1, \quad \dot{y}_1(0) = V_1$$
$$y_2(0) = Y_1 - Y_2, \quad \dot{y}_2(0) = V_1 - V_2) \tag{3.132}$$

where Y_1 and V_1 are the initial displacement and velocity of the LO, respectively, and Y_2 and V_2 the corresponding initial conditions for the NES. Hence, correct up to a time scale of $O(1)$, the system decomposes approximately to an unforced, undamped LO and a strongly damped and strongly nonlinear oscillator forced by the linear one; in essence, the approximately linear oscillation of the center of mass drives the strongly nonlinear relative oscillation between the LO and the NES. As in the previous analysis carried out in this section we focus only on the early-time response under the condition of 1:1 transient resonance capture (TRC). In terms of the approximate system (3.131), this means that the relative displacement y_2 is assumed to perform fast oscillations with frequency nearly equal to unity, possibly modulated by a slowly-varying envelope. This paves the way for a slow-fast partition of the early-time dynamics.

Solving the first of equations (3.131), we may reduce the approximate system to a single nonlinear differential equation:

$$y_1 = Y_1 \cos t + V_1 \sin t$$
$$\ddot{y}_2 + \lambda \dot{y}_2 + y_2^3 = -Y_1 \cos t - V_1 \sin t,$$
$$y_2(0) = Y_1 - Y_2, \quad \dot{y}_2(0) = V_1 - V_2 \tag{3.133}$$

Restricting the analysis to the subset of initial conditions that correspond to the domain of attraction of the 1:1 resonance manifold (i.e., that provide the conditions for 1:1 TRC), we introduce the following slow flow partition of the dynamics:

$$\psi(t) \equiv y_2(t) + j\dot{y}_2(t) = \varphi(t) e^{jt} \tag{3.134}$$

where e^{jt} represents the fast oscillation of the system and $\varphi(t)$ the corresponding slow modulation. Clearly, the original variables can be recovered using the relations

$$y_2(t) = [\psi(t) - \psi^*(t)]/2j \quad \text{and} \quad \dot{y}_2(t) = [\psi(t) + \psi^*(t)]/2 \tag{3.135}$$

where the asterisk denotes complex conjugate. Introducing the expressions (3.134) and (3.135) into (3.133) we obtain

$$y_1 = Y_1 \cos t + V_1 \sin t$$

$$\dot{\varphi}e^{jt} + j\varphi e^{jt} + \frac{(\lambda-1)}{2}(\varphi e^{jt} + \varphi^* e^{-jt}) + \frac{j}{8}(\varphi e^{jt} - \varphi^* e^{-jt})^3$$

$$= -Y_1 \frac{e^{jt} + e^{-jt}}{2} - V_1 \frac{e^{jt} - e^{-jt}}{2j} \tag{3.136}$$

with initial condition $\varphi(0) = (V_1 - V_2) + j(Y_1 - Y_2)$.

To explore the slow flow dynamics (3.136) we perform time averaging with respect to the fast frequency, and obtain the following reduced, early-time slow flow system:

$$\dot{\varphi} + \frac{(\lambda+j)}{2}\varphi - \frac{3j}{8}|\varphi|^2\varphi = -\frac{Y_1}{2} - \frac{V_1}{2j}, \quad \varphi(0) = (V_1 - V_2) + j(Y_1 - Y_2) \tag{3.137}$$

This complex modulation equation governs approximately the slow dynamics of the early-time dynamics in the neighborhood of the 1:1 resonance manifold. To derive a set of real modulation equations, we employ the polar form representation, $\varphi(t) = N(t)e^{j\delta(t)}$, and set separately real and imaginary parts equal to zero to derive a set of two real modulation equations governing the amplitude and the phase:

$$\dot{N} + (\lambda/2)N = -(Y_1/2)\cos\delta + (V_1/2)\sin\delta$$

$$N\dot{\delta} + (N/2) - (3N^3/8) = (Y_1/2)\sin\delta + (V_1/2)\cos\delta$$

$$N(0) = \sqrt{(Y_1-Y_2)^2 + (V_1-V_2)^2},$$

$$\tan\delta(0) = (Y_1-Y_2)/(V_1-V_2) \tag{3.138}$$

From the physical viewpoint, the amplitude $N(t)$ may be associated with a characteristic amplitude of the early-time nonlinear oscillations.

The 1:1 damped invariant NNM manifold of the early-time dynamics corresponds to the set of equilibrium points of the slow flow (3.138) up to time scale of $O(1)$. In order to determine this set we impose stationarity conditions $\dot{N} = \dot{\delta} = 0$ yielding the following relations:

$$N^6 - (8/3)N^4 + (16/9)(1+\lambda^2)N^2 - (16/9)(Y_1^2 + V_1^2) = 0$$

$$\cos\delta = [V_1 N(1 - 3N^2/4) - \lambda Y_1 N]/(Y_1^2 + V_1^2)$$

$$\sin\delta = [Y_1 N(1 - 3N^2/4) + \lambda V_1 N]/(Y_1^2 + V_1^2) \tag{3.139}$$

The stability of an equilibrium point is specified by the nature of the eigenvalues of the Jacobian matrix of the linearization of system (3.138) evaluated at that equilibrium point:

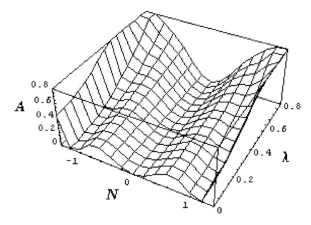

Fig. 3.57 Early-time dynamics: surface of equilibrium points $A = f(N, \lambda)$ as solutions of the first of equations (3.139).

$$\mu_{1,2} = (1/2)\left[-\lambda \pm \sqrt{4(9N/8 - 1/2N)(N/2 - 3N^3/8)}\right] \qquad (3.140)$$

Bifurcations of equilibrium points can be studied by considering the topology of the two-dimensional surface $A = f(N, \lambda)$, where $A = Y_1^2 + V_1^2$; this is depicted in Figure 3.57. This surface is defined by the first of equations (3.139), and all equilibrium points lie on it. The folding lines on this surface form the boundaries separating the parameter regions where one or three equilibrium points exist. These are defined by the following equation:

$$\frac{\partial}{\partial N}\left[N^6 - (8/3)N^4 + (16/9)(1 + \lambda^2)N^2 - (16A/9)\right] = 0 \quad \text{(Folding curves)}$$
(3.141)

A projection of the fold to the plane (λ, A) is obtained by eliminating N between equation (3.141) and the first of equations (3.139). The point on the plane (λ, A) where the two folding curves intersect is computed as $(\lambda_{\text{deg}} = 1/3^{1/2}, A_{\text{deg}} = 0.39506)$, and can be regarded as the most degenerate point of the surface of equilibrium points. In Figure 3.58 we depict the two folding curves projected onto the plane (λ, A) with the degenerate point of intersection also indicated. In the region between the two folding curves, the early-time, slow flow dynamical system (3.138) possesses three equilibrium points, whereas in the complementary region only one. Qualitative changes in the dynamics are anticipated as the folding curves are crossed transversely.

It should be mentioned that the equilibrium points discussed above are the only limit sets of the equation (3.137). This fact may be rigorously proved with the help of Bendixon's criterion (Guckenheimer and Holmes, 1982). That is why the classification of phase trajectories on the basis of the equilibrium points to which they are attracted is justified.

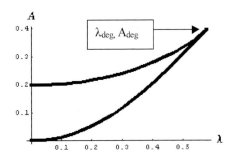

Fig. 3.58 Early-time dynamics: projection of the fold of the surface $A = f(N, \lambda)$ onto the plane (λ, A).

In order to study the evolution of the trajectories of the dynamical system in phase space, we consider again the energy-like quantity $N(t)$, and provide the following alternative expression related to the responses of the original system (3.129):

$$N(t) = \sqrt{(\dot{x}(t) - \dot{v}(t))^2 + (x(t) - v(t))^2} \qquad (3.142)$$

The same quantity was defined previously by the polar transformation of the slow complex amplitude,

$$\phi(t) = N(t) e^{j\delta(t)} \qquad (3.143)$$

and its temporal (slow) evolution is governed by the first of equations (3.138). Hence, it is possible to compare directly the dynamics of the exact system (3.129) and the averaged dynamics governed by the early-time slow modulation equations (3.137) or (3.138). To study quantitative changes in the dynamics of the exact system associated with bifurcations of equilibrium points of the reduced early-time dynamical system (3.138), we consider two case studies corresponding to different values of damping and initial conditions.

First, we consider system (3.129) with damping $\lambda_2 = \lambda = 0.1$, and initial conditions corresponding to $A = 0.1$. The additional damping coefficient is chosen to be zero in the following computations, i.e., $\lambda_1 = 0$. This corresponds to a point inside the area defined by the folding curves in the (λ, A) plane (see Figure 3.58), which means that the reduced early-time system (3.138) possesses three equilibrium positions. These are computed as:

$$(\delta, N) = (0.109375, 0.345185) \quad \text{(Lower Focus)}$$
$$(\delta, N) = (0.306884, 0.955293) \quad \text{(Middle Saddle)}$$
$$(\delta, N) = (2.755332, 1.278643) \quad \text{(Upper Focus)}$$

Taking into account the previously introduced coordinate transformations, this specific case corresponds to the following two-parameter set of initial conditions of the original dynamical system (3.129)

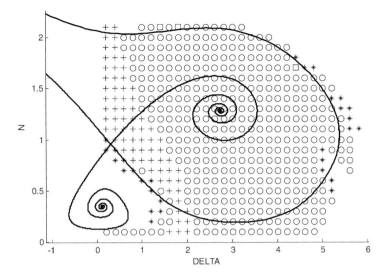

Fig. 3.59 Domains of attraction of the early-time averaged system (3.138) with $\lambda = 0.1$ and $A = 0.1$, superimposed to a grid of initial conditions $(\delta(0), N(0))$ with each symbol indicating the level of efficiency of TET realized in the exact damped system (3.129): (+) EDM > 70%, (o) 50% < EDM < 70%, (□) 30% < EDM < 50%, (*) 10% < EDM < 30%.

$$Y_1 = 0.0, \quad V_1 = 0.316228,$$
$$Y_2 = Y_1 - N(0)\cos\delta(0), \quad V_2 = V_1 - N(0)\cos\delta(0) \quad (3.144)$$

with parameters $\delta(0)$ and $N(0)$. The exact system (3.129) was integrated for $\varepsilon = 0.01$ using each time a different initial point ($\delta(0), N(0)$) on the plane (δ, N). For each simulation we computed the corresponding energy dissipation measure – EDM, i.e., the percentage of initial energy of the system that is eventually dissipated by the damper of the NES damper. *Our effort was to relate the effectiveness of TET in the original system (3.129), to the domains of attraction of the stable equilibrium points of the early-time averaged system (3.138)*. In Figure 3.59 we depict the domains of attraction of the upper and lower foci of the reduced system, superimposed to a grid of initial points ($\delta(0), N(0)$). The different symbols of the grid points are related to the percentage of total initial energy eventually dissipated by the NES for the corresponding initial condition.

The picture clearly demonstrates that *most efficient TET in the exact system (3.129) occurs if the dynamics of the early-time reduced system (3.138) is initiated inside or below the basin of attraction of the upper focus*. A worthwhile caution is that the depicted basin of attraction is only approximate since it is computed only up to a time scale of $O(1)$, and, hence, is valid only in the initial high-energy regime of the dynamics [since in the transformed system (3.130) only $O(1)$ terms were retained in the analysis]. The following numerical simulations support the above-mentioned conclusion.

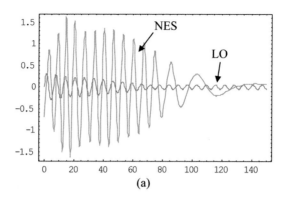

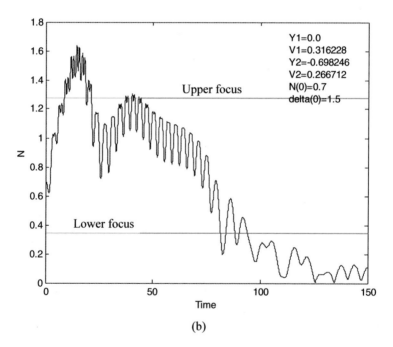

Fig. 3.60 Transient response of the exact system (3.129) for initial conditions, $Y_1 = 0.0$, $V_1 = 0.316228$, $Y_2 = -0.698246$, $V_2 = 0.266712$: (a) LO and NES displacements; (b) evolution of $N(t)$ from the early-time averaged system (3.138).

In Figure 3.60a we depict the exact responses of the LO and the NES for an initial condition inside the basin of attraction of the upper focus of the averaged system ($N(0) = 0.7$, $\delta(0) = 1.5$), whereas in Figure 3.60b we depict the corresponding temporal evolution $N(t)$ computed by integrating the reduced system (3.138). The results indicate that around $t = 20$ s the dynamics is in the domain of attraction of the upper focus. At $t = 50$–60 s the trajectory escapes this regime, and this coincides with a rather abrupt decrease of the amplitudes of oscillation. It is interesting to note that the LO is continuously oscillating with unit frequency, whereas the NES is oscillating with a lower frequency than that. This result may be understood in terms of the previously discussed invariant manifold approach; that is, an abrupt change of the dynamical regime is caused by the breakdown of the invariant manifold by the saddle-node bifurcation described previously. In Figure 3.60 one can also clearly distinguish the crossover between the initial transient and the slow evolution of the invariant manifold. Up to $t = 50$ s the oscillations of the dynamical flow around the upper focus are clear, whereas afterwards a rapid escape of the dynamics away from the domain of attraction of the upper focus takes place.

In Figures 3.61a, b the corresponding plots for an initial condition inside the basin of attraction of the lower focus ($N(0) = 0.5$, $\delta(0) = 0.4$) are depicted. Around $t = 50$ s the dynamics is attracted by the lower focus, and the two oscillators are oscillating with approximately unit frequency. In this case, the amount of energy transferred from the LO to the NES is small. From the viewpoint of invariant manifolds, this case corresponds to the situation without bifurcation. One cannot expect efficient TET in this case, since the NES is almost not excited.

The case when the NES is initially at rest, which is important from a practical viewpoint, is now considered. This corresponds to excitation of an impulsive orbit (IO). In order to study the EDM (the percentage of energy eventually dissipated by the NES) for given initial velocities of the LO and for varying damping values λ, we performed an additional series of numerical simulations. The results are depicted in Figure 3.62, superimposed to the folding boundary curves of Figure 3.58. The plot shows that most efficient TET occurs when the dynamics is initiated above and close to the upper folding boundary curve. This result can be related to previous results based on the damped NNM manifold approach, and demonstrates that *the most efficient TET is realized when the dynamics is attracted to the stable damped NNM manifold, close to the point of bifurcation of that manifold*. Otherwise, if the dynamics is attracted relatively far from the bifurcation point, it undergoes a few cycles of oscillation around the stable focus before breaking down (these cycles are, in fact, nonlinear beats). *Therefore we conclude that with the NES initially at rest, TET is most efficient in the region close (but above) the upper folding boundary curve in the (λ, A) plane.*

This conclusion is supported by the simulations of Figures 3.63a, b depicting the transient responses of exact system (3.129) for initial conditions ($Y_1 = Y_2 = V_2 = 0$, $V_1 = \sqrt{0.25}$), and parameters $\lambda = 0.25$ and $\varepsilon = 0.01$. In this case, the EDM is over 50%. The plots demonstrate that around $t = 35$ s the dynamics of the system is attracted by the stable focus of the averaged system (corresponding to $\delta = 2.427$ and $N = 1.3105$), with the amplitude of LO decreasing smoothly up to $t = 100$ s. The amplitude of the NES increases up to $t = 45$ s and then decreases abruptly.

Figure 3.64 depicts the response of system (3.129) for initial conditions $Y_1 = Y_2 = V_2 = 0$, $V_1 = \sqrt{0.065}$, and $\lambda = 0.3$, $\varepsilon = 0.01$. The selected initial conditions and damping value correspond to an initial point lying below the lower folding boundary curve of the (λ, A) plane. The EDM is below 20% in this case. The plots

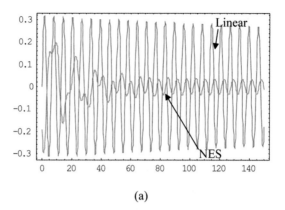

(a)

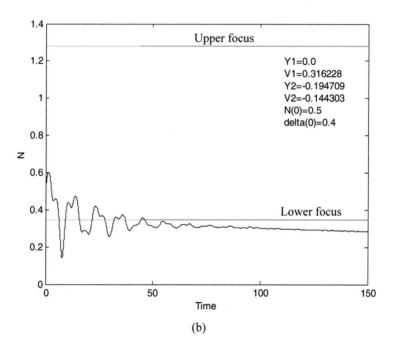

(b)

Fig. 3.61 Transient response of the exact system (3.129) for initial conditions, $Y_1 = 0.0$, $V_1 = 0.316228$, $Y_2 = -0.194709$, $V_2 = -0.144303$: (a) LO and NES displacements; (b) evolution of $N(t)$ from the early-time averaged system (3.138).

demonstrate that around $t = 10$ s the dynamics is attracted by the basin of attraction of the focus corresponding to $\delta = 0.9055$ and $N = 0.2556$. The amplitude of the LO decreases smoothly, while the amplitude of the NES remains almost constant. It is interesting to note that both oscillators are oscillating with unit frequency, implying the continuity (i.e., lack of bifurcation) of the invariant manifold.

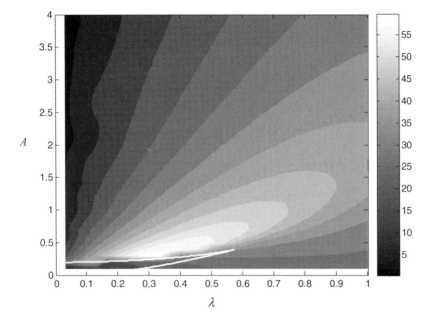

Fig. 3.62 Efficiency of TET as expressed by EDM (%), for varying damping values and initial conditions, with the NES being initially at rest; the fold predicted by the early-time averaging analysis is also shown.

The results presented above demonstrate that the problem of identifying appropriate initial conditions for enhanced TET may be reduced to the problem of predicting of the domains of attraction for a limited number of equilibrium points of the early-time averaged slow flow. This latter approximation does not coincide with the damped NNM manifold approach discussed in the beginning of this section, and the resulting two-dimensional slow phase plane (N, δ) does not coincide with that of the damped NNM manifold at later stages of the dynamical process. Nevertheless, equilibrium points in this slow phase plane obviously correspond to damped NNM manifolds, which provide a direct connection between these two approaches. In other words, the approach developed above enables the determination of the specific equilibrium point eventually reached by the dynamics of the system for the majority of initial conditions. Once this question is answered, the dynamics and efficiency of TET may be assessed using the damped NNM manifold framework. Consequently, the combination of the two methods leads to the analytical modeling of TET dynamics over its entire time span, and answers the question of robustness of TET to changes in initial conditions. The numerical results presented in this section clearly support the predictions of these analytical methodologies.

In summary, in this section we analyzed the damped dynamics of the essentially nonlinear two-DOF system (3.98) or (3.129) under conditions of 1:1 resonance capture. The resulting fundamental TET was studied by considering the damped NNM

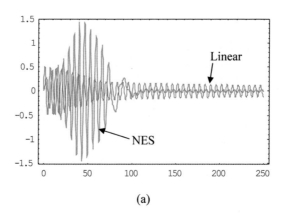

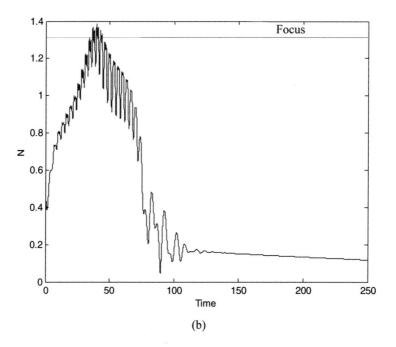

Fig. 3.63 Transient response of the exact system (3.129) for initial conditions, $Y_1 = 0.0$, $V_1 = \sqrt{0.25}$, $Y_2 = 0.0$, $V_2 = 0.0$: (a) LO and NES displacements; (b) evolution of $N(t)$ from the early-time averaged system (3.138).

manifolds of the slow flow, and by analyzing the attraction of the dynamics on these manifolds, as well as by studying damped transitions between damped NNM manifolds. More importantly, we demonstrated that the rate of energy dissipation by the NES, i.e., TET efficiency, is closely related to the bifurcation structure of the NNM

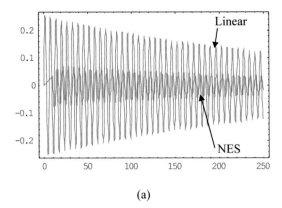

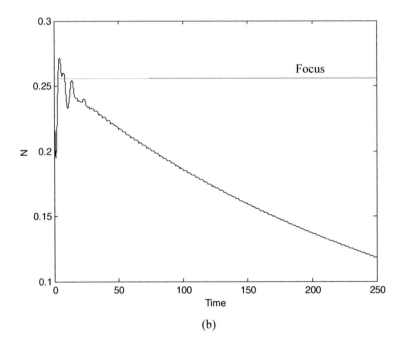

Fig. 3.64 Transient response of the exact system (3.129) for initial conditions, $Y_1 = 0.0$, $V_1 = \sqrt{0.065}$, $Y_2 = 0.0$, $V_2 = 0.0$: (a) LO and NES displacements; (b) evolution of $N(t)$ from the early-time averaged system (3.138).

invariant manifolds. Indeed, it was found numerically that with the NES initially at rest (i.e., when an impulsive orbit is excited), optimal TET is realized when the damped dynamics is attracted by a stable in-phase damped NNM invariant manifold, close to the point of bifurcation of that manifold, or equivalently, close (but above) the upper folding boundary curve in the (λ, A) plane depicted in Figure 3.62; this folding curve was computed by performing an analysis of the early-time dynamics. This naturally leads us to the more detailed study of the conditions for optimal fundamental TET in system (3.98), which is performed in the next section.

3.4.2.5 Conditions for Optimal Fundamental TET

In Section 3.3.5 we discussed some topological features of the Hamiltonian dynamics of the two-DOF system (3.98) with no damping terms. Focusing in the intermediate-energy region close to the 1:1 resonance manifold of the Hamiltonian system we studied the topological changes of intermediate-energy impulsive orbits (IOs) for varying energy (see Figure 3.39). Specifically, we found that above the critical value of energy-like variable $r = r_{cr}$ (see Section 3.3.5) the topology of intermediate-energy IOs changes drastically, as these make much larger excursions into phase space, resulting in continuous strong energy exchanges between the LO and the nonlinear attachment in the form of strong nonlinear beats. We also mentioned in Section 3.3.5 that this critical energy of the Hamiltonian system may be directly related to the energy threshold required for TET in the weakly damped system (as discussed in Section 3.2 and Figure 3.4).

In this section we study the intermediate-energy dynamics of the weakly damped system (3.98),

$$\ddot{x} + \lambda_1 \dot{x} + \lambda_2(\dot{x} - \dot{v}) + \omega_0^2 x + C(x - v)^3 = 0$$
$$\varepsilon \ddot{v} + \lambda_2(\dot{v} - \dot{x}) + C(v - x)^3 = 0 \quad (3.98)$$

in an effort to formulate conditions for optimal fundamental TET; as usual we assume that $0 < \varepsilon \ll 1$. It follows that our study will be necessarily restricted to the neighborhood of the 1:1 resonance manifold of the underlying Hamiltonian system, and the damped dynamics will be studied under the condition of 1:1 resonance capture. However, the ideas and techniques presented here can be extended to study optimal conditions for the more general case of m:n subharmonic TET.

To initiate our analysis, we set $\omega_0^2 = 1$ in (3.98), and consider the following *ansatz* for the damped responses close to the 1:1 resonance manifold of the Hamiltonian system (i.e., for $\omega \approx 1$):

$$x(t) \approx \frac{a_1(t)}{\omega} \cos[\omega t + \alpha(t)], \quad v(t) \approx \frac{a_2(t)}{\omega} \cos[\omega t + \beta(t)] \quad (3.145)$$

Substituting (3.145) into (3.98) and averaging out all frequency components with frequencies higher than ω, we derive a system of four modulation equations governing the slow evolution of the amplitudes $a_1(t)$, $a_2(t)$ and phases $\alpha(t)$, $\beta(t)$ of the two oscillators; this defines the slow flow of system (3.98) in the neighborhood of the 1:1 resonance manifold.

In Section 3.3.5 we found that the slow flow of the corresponding undamped system is fully integrable and can be reduced to the sphere $(R^+ \times S^1 \times S^1)$. Motivated by these results, we introduce the phase difference $\phi = \alpha - \beta$, the energy-like variable $r^2 = a_1^2 + (\sqrt{\varepsilon}a_2)^2$, and the angle $\psi \in [-\pi/2, \pi/2]$ defined by the relation $\tan[\psi/2 + \pi/4] = a_1/\sqrt{\varepsilon}a_2$. Enforcing the condition of weak damping by rescaling the damping coefficients according to $\lambda_1 \to \varepsilon\lambda_1$, $\lambda_2 \to \varepsilon\lambda_2$, and expressing the slow flow equations in terms of the new variables, we reduce the slow flow of the damped dynamics to the sphere $(r, \phi, \psi) \in (R^+ \times S^1 \times S^1)$:

$$\dot{r} = -\frac{r}{2}\left\{\varepsilon\lambda_1(1+\sin\psi) + \varepsilon\lambda_2\left[(1+\varepsilon) - (1-\varepsilon)\sin\psi - 2\varepsilon^{1/2}\cos\psi\cos\varphi\right]\right\}$$

$$\dot{\psi} = \frac{-3Cr^2}{8\varepsilon^{3/2}}\left[(1+\varepsilon) - (1-\varepsilon)\sin\psi - 2\sqrt{\varepsilon}\cos\psi\cos\varphi\right]\sin\varphi$$

$$- \frac{\varepsilon\lambda_1}{2}\cos\psi + \frac{\lambda_2}{2}\left[(1-\varepsilon)\cos\psi - 2\varepsilon^{1/2}\sin\psi\cos\varphi\right]$$

$$\dot{\varphi} = \frac{1}{2} - \frac{3Cr^2}{16\varepsilon^2}\left[(1+\varepsilon) - (1-\varepsilon)\sin\psi - 2\sqrt{\varepsilon}\cos\psi\cos\varphi\right]$$

$$\times \left[(1-\varepsilon) - 2\varepsilon^{1/2}\frac{\sin\psi\cos\varphi}{\cos\psi}\right] - \varepsilon^{1/2}\lambda_2\frac{\sin\varphi}{\cos\psi} \tag{3.146}$$

We note that when $\lambda_1 = \lambda_2 = 0$ the slow flow reduces to the integrable system (3.89) on a two-torus possessing the first integral (3.90). For non-zero damping, however, the slow flow dynamics is non-integrable and the dimensionality of the system (3.146) cannot be further reduced.

In Figure 3.65 projections of damped IOs to the three-dimensional space $(r, \phi, \psi) \in (R^+ \times S^1 \times S^1)$ are depicted for three different initial energy levels; these results were obtained by direct numerical simulations of the damped system (3.98) subject to initial conditions corresponding to IOs, and can be directly compared to the plots of Figure 3.37 which depict isoenergetic projections of the underlying Hamiltonian dynamics. In the damped case, however, instead of the equilibrium points corresponding to NNMs on branches $S11\pm$ we get in-phase and out-of-phase damped NNM invariant manifolds (Shaw and Pierre, 1991, 1993). For the case of large initial energy there is an initial transient phase (denoted as Stage I in Figure 3.65b) as the orbit gets attracted to the damped NNM manifold $S11+$; this is followed by the slow evolution of the damped motion along $S11+$ as energy decreases due to damping dissipation, with the motion predominantly localized to the NES as evidenced by the fact that $\psi(t) \approx -\pi/2$ (Stage II in Figure 3.65b). Finally, the damped NNM $S11+$ becomes unstable, and the dynamics makes a final transition to the weakly nonlinear (linearized) NNM manifold $S11-$; the resulting out-of-phase oscillations are localized predominantly to the LO, as evidenced by the fact that $\lim_{t\to\infty}\psi(t) = \pi/2$ (Stage III in Figure 3.65b). TET in this case occurs predominantly during Stage I (TET through nonlinear beat) and Stage II (fundamental TET).

For lower initial energy (i.e., in the intermediate energy level), the initial transients of the dynamics during the attraction to $S11+$ possess larger amplitudes

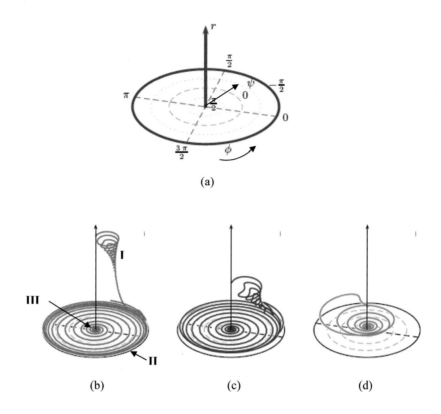

Fig. 3.65 Phase space projection of damped IOs for $\varepsilon = 0.1$, $C = 2/15$, and $\lambda_1 = \lambda_2 = 0.1$: (a) projection definition, (b) $r(0) = 2.0$, (c) $r(0) = 1.0$, (d) $r(0) = 0.5$.

(Stage I, Figure 3.65c), leading to an increase of the resulting TET due to nonlinear beats; in later times, Stages II and III of the dynamics are similar to the corresponding ones of the higher-energy case. Compared to the previous case, TET is enhanced, especially during the initial transients of the motion where the LO and the NES undergo larger-amplitude nonlinear beats. Qualitatively different dynamics is observed when the initial energy is further decreased; this can be noted from the projection of Figure 3.65d, where the low-energy motion rapidly localizes to the LO as the dynamics gets directly attracted by the weakly nonlinear branch of the damped NNM manifold $S11-$, and, as a result, TET drastically diminishes. In essence, for this low energy value only Stage III of the dynamics is realized.

An analytical study of the stability of the damped NNM manifolds $S11\pm$, which, as we showed, affects the damped transitions of system (3.98) and the resulting TET, is carried out in Quinn et al. (2008). In that work a detailed study of TET efficiency as judged by the time required by the NES to passively absorb and dissipate a significant amount of initial energy of the LO is performed as well. A representative

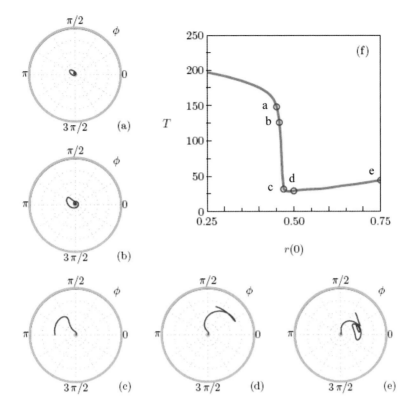

Fig. 3.66 Damped IO simulations at various initial energy levels: projections of the damped motions onto the unit disk for (a) $r(0) = 0.45$, (b) $r(0) = 0.46$, (c) $r(0) = 0.47$, (d) $r(0) = 0.50$, (e) $r(0) = 0.75$; (f) time T required for decay of $r(0)$ by a factor of e^{-1} as a function of $r(0)$, circles refer to the projections (a–e).

result of this study is presented in Figure 3.66, depicting the time T required for the initial value of the energy-like variable, $r(0)$, to decay by a factor of e, when the motion is initiated on an IO:

$$r(T) = r(0)\, e^{-1} \qquad (3.147)$$

We note that for a classical viscously damped SDOF linear oscillator with damping constant $\varepsilon\lambda$, the corresponding time interval T would be equal to $\varepsilon\lambda/2$. The numerical results depicted in Figure 3.66 were derived for parameters $\varepsilon = 0.05$, $\lambda_1 = 0$, $\lambda_2 = 0.2$ and $C = 2/15$, and the damped IOs in Figures 3.66a–e are only depicted in the time interval $0 < t < T$. We note that as we increase $r(0)$ from 0.46 to 0.47 there is a drastic reduction in T, signifying drastic enhancement of TET efficiency. This is associated with a sudden 'excursion' of the damped IO in the projection of the phase space, as the dynamics makes a transition from a motion that is predominantly localized to the LO (Figures 3.66a, b) to a motion where large relative motion

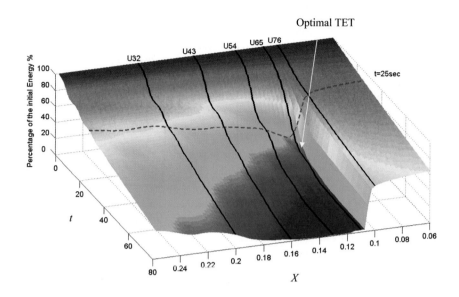

Fig. 3.67 Percentage of energy dissipated in system (3.148) when intermediate-energy damped IOs are excited ($\varepsilon = 0.05$, $C = 1$ and $\varepsilon\lambda = 0.005$): solid lines correspond to excitation of specific periodic IOs, and the dashed line indicates the energy remaining in the system at $t = 25$ s.

between the LO and the NES takes place (see Figure 3.66c); this, in turn leads to enhanced TET through nonlinear beats.

It is interesting to note that the sudden jump in TET efficiency in Figure 3.66 occurs in the intermediate-energy regime, in the neighborhood of the 1:1 resonance manifold of the Hamiltonian system. In this regime of the dynamics the slow flow model (3.146) is valid, so it can be used to study the conditions for optimal TET efficiency. An analytic study of the conditions for optimal TET through the excitation of intermediate-energy IOs is carried out in Sapsis et al. (2008), and elements of this study will be reproduced here. Although the study is carried out under the assumption of 1:1 resonance capture, and is based on CX-A, the analysis of the resulting averaged slow flow is different than that carried out for the underlying Hamiltonian system.

Hence, we reconsider the two-DOF system (3.98) with $\lambda_1 = \lambda_2 = \varepsilon\lambda$ and $\omega_0^2 = 1$,

$$\ddot{x} + \varepsilon\lambda\dot{x} + \varepsilon\lambda(\dot{x} - \dot{v}) + x + C(x - v)^3 = 0$$
$$\varepsilon\ddot{v} + \varepsilon\lambda(\dot{v} - \dot{x}) + C(v - x)^3 = 0 \qquad (3.148)$$

with initial conditions corresponding to excitation of an impulsive orbit (IO), $v(0) = \dot{v}(0) = x(0) = 0$ and $\dot{x}(0) = X$, and $0 < \varepsilon \ll 1$. In Figure 3.67 we depict the dissipation of instantaneous energy in this system with $\varepsilon = 0.05$, $C = 1$

and $\varepsilon\lambda = 0.005$ (these parameter values will be used in the remainder of this section, unless stated otherwise), when damped IOs are excited. In accordance with previous findings of this Chapter, we find that strong energy dissipation, i.e., strong TET, is realized in the intermediate energy region and more specifically in the neighborhood of the 1:1 resonance manifold of the underlying Hamiltonian system (we note that the FEP of the corresponding undamped system with the positions of periodic IOs indicated, is depicted at Figure 3.20).

Moreover, *optimal TET*, as judged by the strongest energy dissipation in the least possible time in the plot of Figure 3.67, is realized for initial impulses X (i.e., initial energies) in the range between the periodic IOs $U65$ and $U76$; from the FEP of Figure 3.20, we note that these periodic IOs are close to the energy level of a saddle-node bifurcation of the linearized and strongly nonlinear components of the backbone branch $S11-$. At this energy level, an unstable hyperbolic periodic orbit is generated on the strongly nonlinear component of $S11-$. As shown below, it is the homoclinic orbit of this hyperbolic periodic orbit that affects the topology of nearby IOs and defines conditions for optimal TET in the weakly damped system. This observation is in accordance with the discussion of Section 3.3.5 and the results depicted in Figures 3.39 and 3.66, indicating that above a critical energy level the topology of the IOs changes drastically, with IOs making large excursions in phase space (actually, this critical energy level in the Hamiltonian system may be defined as the energy where with the IO coincides with the homoclinic orbit – Figure 3.39d). Finally, we note that these observations are also in accordance with the findings of the approach based on damped NNM invariant manifolds (see Section 3.4.2.4), where it was noted that the most efficient TET is realized when the damped dynamics is attracted to a stable damped NNM manifold, close to the point of bifurcation of that manifold.

The analytical study of conditions for optimal fundamental TET is carried out by applying the CX-A technique to system (3.148) under condition of 1:1 internal resonance between the LO and the NES. Moreover, only intermediate-energy IOs are considered, focusing to those lying close to the 1:1 resonance manifold with dominant (fast) frequency $\omega \approx 1$ (see the FEP of Figure 3.20). Applying the usual complexification,

$$\psi_1(t) = \dot{v}(t) + jv(t) \equiv \phi_1(t)\,e^{jt}, \quad \psi_2(t) = \dot{x}(t) + jx(t) \equiv \phi_2(t)\,e^{jt}$$

and performing averaging with respect to the fast term e^{jt}, we derive the following set of complex modulation equations,

$$\dot{\phi}_1 + (j/2)\phi_1 + (\lambda/2)(\phi_1 - \phi_2) - (3jC/8\varepsilon)|\phi_1 - \phi_2|^2(\phi_1 - \phi_2) = 0$$
$$\dot{\phi}_2 + (\varepsilon\lambda/2)(2\phi_2 - \phi_1) + (3jC/8)|\phi_1 - \phi_2|^2(\phi_1 - \phi_2) = 0 \quad (3.149)$$

with initial conditions $\phi_1(0) = 0$ and $\phi_2(0) = X$. Introducing the new complex variables,

$$\left.\begin{array}{l} u = \phi_1 - \phi_2 \\ w = \varepsilon\phi_1 + \phi_2 \end{array}\right\} \Leftrightarrow \left\{\begin{array}{l} \phi_1 = (u + w)/(1 + \varepsilon) \\ \phi_2 = (w - \varepsilon u)/(1 + \varepsilon) \end{array}\right. \quad (3.150)$$

we express system (3.149) as

$$\dot{u} + \frac{(1+\varepsilon)\lambda}{2}u - \frac{3(1+\varepsilon)jC}{8\varepsilon}|u|^2 u + j\frac{u+w}{2(1+\varepsilon)} - \varepsilon\lambda\frac{w-\varepsilon u}{2(1+\varepsilon)} = 0$$

$$\dot{w} + j\varepsilon\frac{u+w}{2(1+\varepsilon)} + \varepsilon\lambda\frac{w-\varepsilon u}{2(1+\varepsilon)} = 0 \qquad (3.151)$$

with initial conditions $u(0) = -X$ and $w(0) = X$. Hence, we have reduced the problem of studying intermediate-energy damped IOs of the initial system of coupled oscillators (3.148) to the above system of first-order complex modulation equations governing the slow flow close to the 1:1 resonance manifold. These equations are valid only for small- and moderate-energy IOs, i.e., for initial conditions $X < 0.5$ (see Figure 3.67), since above this level the fast frequency of the response depends significantly on the energy level and the assumption $\omega \approx 1$ is violated.

Since we are interested in the study of optimal energy dissipation by the NES, we shall now derive expressions for the various energy quantities in terms of the complex modulations u and w. These expressions will be further exploited in an effort to study conditions on u and w that optimize TET. Thus, for computing the instantaneous total energy stored in the LO we derive the expression:

$$E_L(t) \equiv \frac{1}{2}[x^2(t) + \dot{x}^2(t)]$$

$$\approx \frac{1}{2}[(\text{Im}[\phi_2 e^{jt}])^2 + (\text{Re}[\phi_2 e^{jt}])^2] = \frac{1}{2}|\phi_2|^2 = \frac{|w-\varepsilon u|^2}{2(1+\varepsilon)^2}$$
(3.152)

The instantaneous energy stored in the NES is approximately evaluated as:

$$E_{\text{NL}}(t) = \frac{1}{2}\left\{\varepsilon\dot{v}^2(t) + \frac{C}{2}[x(t) - v(t)]^4\right\}$$

$$\approx \frac{1}{2}\left\{\varepsilon(\text{Im}[\phi_1 e^{jt}])^2 + \frac{C}{2}(\text{Re}[ve^{jt}])^4\right\}$$

$$= \frac{1}{2}\left\{\varepsilon\left(\text{Im}\left[\frac{u+w}{1+\varepsilon}e^{jt}\right]\right)^2 + \frac{C}{2}(\text{Re}[ue^{jt}])^4\right\} \qquad (3.153)$$

Finally, the most important energy measure as far as our analysis is concerned will be the energy dissipated by the damper of the NES, approximated as:

$$E_{\text{DISS}}(t) = \int_0^t \varepsilon\lambda\,[\dot{x}(t) - \dot{v}(t)]^2\,dt \approx \varepsilon\lambda\int_0^t (\text{Re}[ue^{jt}])^2\,dt$$

$$= \varepsilon\lambda \int_0^t \left\{(\text{Re }[u])^2 \cos^2 t + (\text{Im }[u])^2 \sin^2 t - \text{Re }[u]\,\text{Im }[u] \sin 2t\right\} dt$$

$$= \varepsilon\lambda \int_0^t \left\{(\text{Re }[u])^2 \frac{1+\cos 2t}{2} + (\text{Im }[u])^2 \frac{1-\cos 2t}{2} - \text{Re }[u]\,\text{Im }[u] \sin 2t\right\} dt \tag{3.154}$$

Omitting terms with fast frequencies greater than unity from the integrand (this is consistent with our analysis based on averaging with respect to the fast frequency equal to unity), the above integral can be approximated by the following simple expression:

$$E_{\text{DISS}}(t) \approx \frac{\varepsilon\lambda}{2} \int_0^t \left\{(\text{Re }[u])^2 + (\text{Im }[u])^2\right\} dt = \frac{\varepsilon\lambda}{2} \int_0^t |u(t)|^2 dt \tag{3.155}$$

Hence, *within the approximations of the analysis, the energy dissipated by the NES is directly related to the modulus of u(t) which characterizes the relative response between the LO and the NES.* It follows, that enhanced TET in system (3.148) is associated with the modulus $|u(t)|$ attaining large amplitudes, especially during the initial phase of motion where the energy is at its highest.

Returning to the slow flow (3.151), the second modulation equation can be solved explicitly as follows:

$$w(t) = X \exp\left(-\frac{\varepsilon(j+\lambda)t}{2}\right) + \frac{\varepsilon(\varepsilon\lambda - j)}{2(1+\varepsilon)} \int_0^t \exp\left(-\frac{\varepsilon}{2}(j+\lambda)(t-\tau)\right) u(\tau)\,d\tau \tag{3.156}$$

which, upon substitution into the first modulation equation yields:

$$\dot{u} - \frac{3jC(1+\varepsilon)}{8\varepsilon} |u|^2 u + \frac{j+\lambda\left[\varepsilon^2 + (1+\varepsilon)^2\right]}{2(1+\varepsilon)} u$$

$$= \frac{\varepsilon\lambda - j}{2(1+\varepsilon)} X \exp\left(-\frac{\varepsilon(j+\lambda)t}{2}\right)$$

$$+ \varepsilon\left[\frac{(\varepsilon\lambda - j)}{2(1+\varepsilon)}\right]^2 \int_0^t \exp\left(-\frac{\varepsilon}{2}(j+\lambda)(t-\tau)\right) u(\tau)\,d\tau, \quad u(0) = -X \tag{3.157}$$

This complex integro-differential equation governs the slow flow of a damped IO in the intermediate-energy regime, as it is equivalent to system (3.151). It follows that the above dynamical system provides information on the slow evolution of the damped dynamics close to the 1:1 resonance manifold.

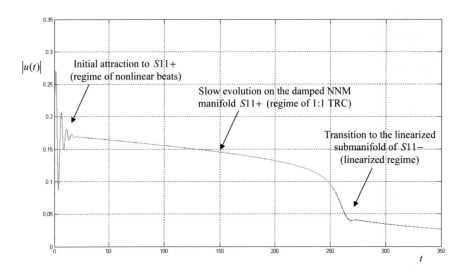

Fig. 3.68 Slow flow (3.151) of a damped IO in the intermediate-energy regime of Figure 3.67.

In Figure 3.68 we present a typical solution of (3.151) depicting the slow flow of a damped IO in the upper intermediate-energy regime of Figure 3.67. The initial 'wiggles' in the slow flow represent the initial attraction of the IO dynamics by the damped NNM manifold $S11+$, and correspond to initial nonlinear beats in the full response. Although short in duration, the energy dissipated by the NES in the initial regime of nonlinear beats can be quite significant as discussed below.

In Figure 3.69 we examine the dynamics of the averaged system (3.151) [or equivalently (3.157)] over the entire intermediate-energy regime of damped IOs. Starting from relatively high energies (i.e., the highest value of impulsive magnitude X, Figure 3.69a), the initial regime of nonlinear beats (corresponding to the attraction of the dynamics to the stable damped NNM invariant manifold $S11+$) leads to strong energy exchanges between the LO and the NES; as the dynamics settles to $S11+$ the energy exchanges diminish and slow energy dissipation is noted in both oscillators; finally the dynamics makes the transition to the linearized damped NNM submanifold $S11-$ at the later stage where nearly the entire energy of the system has been dissipated. We conclude that in the upper region of the intermediate-energy regime TET is relatively weak as the impulsively excited LO retains most of its energy throughout the oscillation. As the impulsive energy decreases (see Figures 3.69b, c) the initial regime of nonlinear beats expands and stronger energy exchanges between the impulsively forced LO and NES take place; moreover, the dynamics instead of settling to $S11+$, proceeds to make a transition to the weakly nonlinear branch of $S11-$. These features of the slow dynamics enhance TET in the system, as judged by the efficient dissipation of energy in both oscillators. Overall, optimal energy dissipation, and hence optimal TET, is realized

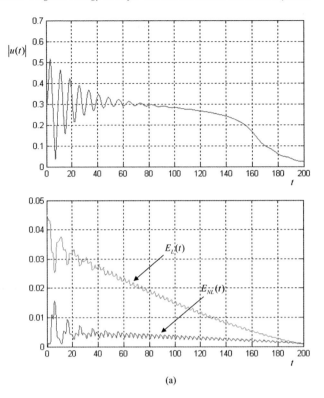

Fig. 3.69 Slow flow (3.151) or (3.157) of damped IOs in the intermediate-energy regime: (a) $X = 0.30$ (upper regime).

in Figure 3.69d, where the initial regime of beats is replaced by *a (super)slow oscillation during which the entire energy of the LO gets transferred to the NES over a single half-cycle*; some of this energy gets 'backscattered' to the LO at a later stage of the motion, during some low-amplitude nonlinear beats, but the major amount of energy gets dissipated during the initial half-cycle energy transfer where the energy of the system is at its highest; this provides the condition for optimal TET in this system, and corresponds to the 'ridge' in Figure 3.67 at $X \approx 0.11$. A slight decrease of the impulsive magnitude X changes qualitatively the slow dynamics, as both oscillators now settle into linearized responses and negligible TET takes place; in this case the slow dynamics gets directly attracted to the weakly nonlinear branch of $S11-$.

Hence, the slow dynamics of the damped IOs in the intermediate-energy regime is quite complex. Indeed, based on the qualitative features of the damped IO dynamics we may divide the intermediate-energy regime of Figure 3.67 into three distinct subregimes; these can be distinguished by the features of the slow flow dynamics (3.157) during the initial, highly energetic stage of the impulsive motion where most TET is realized. In the *upper subregime* corresponding to higher impulsive magni-

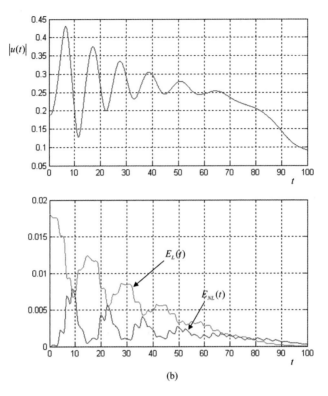

Fig. 3.69 Slow flow (3.151) or (3.157) of damped IOs in the intermediate-energy regime: (b) $X = 0.19$.

tudes (see Figures 3.69a–c) TET through nonlinear beats takes place. The *middle subregime* (see Figure 3.69d) is the regime of optimal TET, and is governed by the most complex dynamics, since the initial slow flow dynamics consists of a single 'super-slow' half-cycle during which the entire energy of the LO gets transferred to the NES. Hence, it appears that the initial nonlinear beats realized in the upper subregime degenerate to a single 'super-slow' half-cycle of the slow flow as the middle subregime is approached. As shown in the following analysis, the dynamical mechanism that leads to this 'super-slow' degeneration of the slow dynamics in Figure 3.69d is the homoclinic orbit of the unstable damped NNM on $S11-$ that is generated by the saddle-node bifurcation at the critical energy level between the periodic IOs $U65$ and $U76$ in the FEP of Figure 3.20. Finally, the *lower subregime* is characterized by linearized motion predominantly localized to the LO, with complete absence of nonlinear beats and negligible TET.

We note that this disussion can be directly related to the analysis presented in Section 2.3 where the dynamics of a two-DOF system of a different configuration (with a grounded NES) was studied asymptotically in the neighborhood of the 1:1 resonance manifold of the dynamics. Indeed, the homoclinic orbit of the unstable

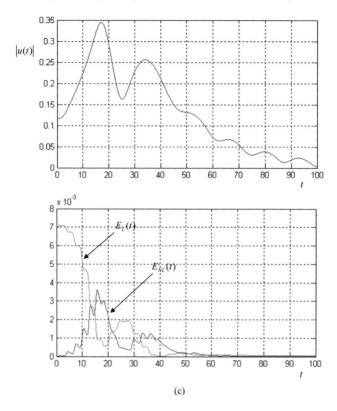

Fig. 3.69 Slow flow (3.151) or (3.157) of damped IOs in the intermediate-energy regime: (c) $X = 0.12$.

undamped NNM on $S11-$ of the Hamiltonian system (3.148) studied in the present section, is similar to the homoclinic loop appearing in Figure 2.10 of the Hamiltonian system (2.31). As shown in Section 2.3, when sufficiently weak damping is added to the system [refer to condition $\mu > \nu$ in equation (2.47)] the Hamiltonian homoclinic loop is perturbed (to first order as in Figure 2.11a, and to second order as schematically shown in Figure 2.13). Hence, following a similar reasoning, we can relate the results regarding TET efficiency of this section to the damped system (2.41) with grounded NES, by relating the dynamics in the neighborhood of the perturbed homoclinic orbit of that system to TET efficiency.

The previous discussion and results provide ample motivation for focusing in the initial, highly energetic regime of the slow flow dynamics (3.151) [or equivalently (3.157)], as this represents the most critical stage for TET. Hence, we consider the modulation equation (3.157) and *restrict the analysis to the initial stage of the dynamics*. Mathematically, we will be interested in the dynamics up to times of $O(1/\varepsilon^{1/2})$, and for initial conditions (impulses) of order $X = O(\varepsilon^{1/2})$. Under these assumptions we consider the integral term on the right-hand side of (3.157)

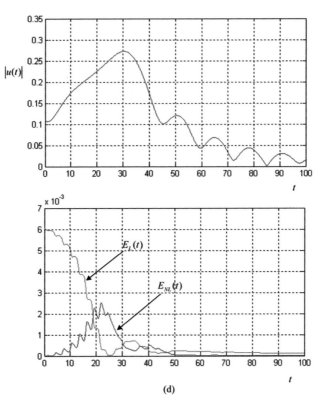

Fig. 3.69 Slow flow (3.151) or (3.157) of damped IOs in the intermediate-energy regime: (d) $X = 0.11$ (optimal TET).

and express it as follows:

$$I \equiv \varepsilon \left[\frac{(\varepsilon\lambda - j)}{2(1+\varepsilon)}\right]^2 \int_0^t \exp\left(-\frac{\varepsilon}{2}(j+\lambda)(t-\tau)\right) u(\tau)\, d\tau$$

$$= \varepsilon \left[\frac{1}{2(1+\varepsilon)}\right]^2 \int_0^t \exp\left(-\frac{\varepsilon}{2}(j+\lambda)(t-\tau)\right) u(\tau)\, d\tau + O(\varepsilon^2)$$

When $t = O(\varepsilon^{-1/2})$, we have also that $(\tau - t) = O(\varepsilon^{-1/2})$; it follows that by expanding the exponential in the integrand in Taylor series in terms of ε, the integral I can be approximated as

$$I \approx \varepsilon \left[\frac{1}{2(1+\varepsilon)}\right]^2 \int_0^t u(\tau)\, d\tau + O(\varepsilon^{3/2})$$

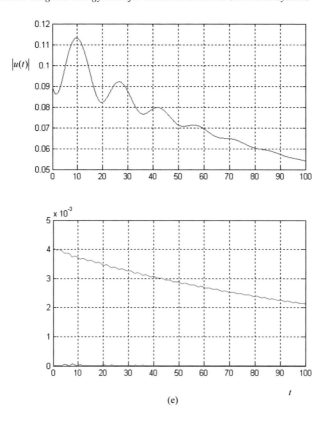

Fig. 3.69 Slow flow (3.151) or (3.157) of damped IOs in the intermediate-energy regime: (e) $X = 0.09$ (lower regime).

or, by invoking the mean value theorem of integral calculus, as

$$I \approx 2^{-2}(1+\varepsilon)^{-2}\varepsilon t\, u(t_0)$$

for some t_0 in the interval $0 < t_0 < t$. Given that $t = O(\varepsilon^{-1/2})$ and $u(t_0) = O(X) = O(\varepsilon^{1/2})$, we prove that for times smaller than $O(\varepsilon^{-1/2})$, the integral is ordered as $I = O(\varepsilon)$, and hence is a small quantity.

Taking this result into account, and introducing the variable transformations

$$u = \varepsilon^{1/2} z \quad \text{and} \quad X = \varepsilon^{1/2} Z$$

to account for the scaling of the initial condition (impulse) $X = O(\varepsilon^{1/2})$, we express the modulation equation (3.157) in the form

$$\dot{z} - \frac{3jC}{8}|z|^2 z + \frac{j+\lambda}{2} z = -\frac{jZ}{2} + O(\varepsilon, \varepsilon^{1/2}\lambda), \quad z(0) = -Z, \quad t \text{ up to } O(\varepsilon^{-1/2}) \tag{3.158}$$

where the variable z and initial condition Z are assumed to $O(1)$ quantities, unless otherwise noted. Finally, introducing the rescalings

$$z \to \left(\frac{4}{3C}\right)^{1/2} z, \quad w \to \left(\frac{4}{3C}\right)^{1/2} w \qquad (3.159a)$$

the new notation,

$$B = -\left(\frac{3C}{4}\right)^{1/2} Z \qquad (3.159b)$$

and the additional scaling for the damping coefficient, $\lambda = \varepsilon^{1/2}\hat{\lambda}$, the system is brought into the following final form,

$$\dot{z} - \frac{j}{2}|z|^2 z + \frac{j + \varepsilon^{1/2}\hat{\lambda}}{2} z = \frac{jB}{2} + O(\varepsilon), \quad z(0) = B, \quad t \text{ up to } O(\varepsilon^{-1/2}) \quad (3.160)$$

and all quantities other than the small parameter ε are assumed to be $O(1)$ quantities. The complex modulation equation (3.160) provides an approximation to the initial slow flow dynamics, and is valid formally only up to times of $O(\varepsilon^{-1/2})$.

In Figure 3.70 we compare the initial approximation of the slow flow (3.158) or (3.160) and the full slow flow (3.151) or (3.157), by computing the predicted energy dissipated in the intermediate-energy regime of damped IOs by the two approximations. This comparison clearly validates the slow flow approximation (3.160) in the intermediate-energy level of interest in this study.

Introducing the polar transformation, $z = Ne^{j\delta}$, substituting into (3.160) and separating real and imaginary parts, this system can be expressed in terms of the following two real modulation equations:

$$\dot{N} + \frac{\varepsilon^{1/2}\hat{\lambda}}{2} N = \frac{B}{2} \sin \delta + O(\varepsilon), \quad N(0) = B$$

$$\dot{\delta} + \frac{1}{2} - \frac{1}{2}N^2 = \frac{B}{2N} \cos \delta + O(\varepsilon), \quad \delta(0) = 0 \qquad (3.161)$$

These equations govern the slow evolutions of the amplitude N and phase δ of the complex modulation z of the IO, during the initial (high-energy) regime of the dynamics.

In Figure 3.71 we depict the initial regime of slow flow dynamics (3.160–3.161) for $\varepsilon = 0.05$, $\hat{\lambda} = 0.4472$ and three different normalized impulses (initial conditions) B. For B above the critical level $B_{cr}(\hat{\lambda} = 0.4472) \approx 0.3814$, the slow flow model (3.160) predicts large excursion of the damped IO in phase space. In fact, after executing relatively large-amplitude transients, the orbit is being ultimately attracted by the stable in-phase damped NNM $S11+$ (which, within the order of approximation of the present analysis, appears as a fixed point, although as shown in previous sections in actuality it 'drifts' slowly, i.e., it depends on higher order time scales); these initial transients correspond to the nonlinear beats (the 'wiggles') observed in the initial stage of the full slow flow model (3.157) in the upper subregime

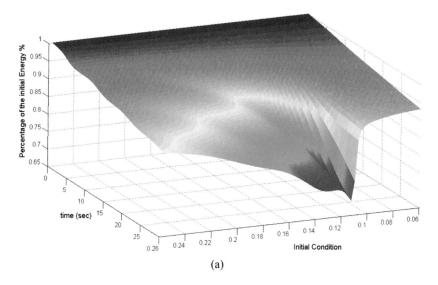

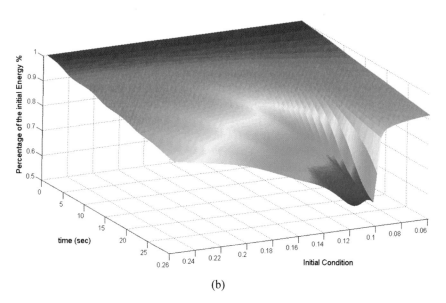

Fig. 3.70 Percentage of energy dissipated when intermediate-energy damped IOs are excited ($\varepsilon = 0.05$, $C = 1$ and $\lambda \varepsilon^{1/2} \hat{\lambda} = 0.1$): (a) full slow slow (3.151) or (3.157), (b) approximation of the slow flow in the initial stage of the dynamics, (3.158) or (3.160).

of the intermediate-energy regime (see Figures 3.68 and 3.69a–c). Note, that since the model (3.160–3.161) is valid only for the initial stage of the slow flow dynamics, it cannot predict the eventual transition of the dynamics from $S11+$ to $S11-$ in the later, low-energy (linearized) stage of the oscillation (nor the slow 'drift' of the dynamics on these damped invariant manifolds).

For B below the critical level $B_{cr}(\hat{\lambda})$, there is a significant qualitative change in the dynamics as the IO executes small-amplitude oscillations, and the dynamics is being attracted to the weakly nonlinear out-of-phase damped NNM $S11-$; this corresponds to the weakly nonlinear dynamics realized in the lower subregime of the intermediate-energy range (see Figure 3.69e). It follows, that the critical orbit that separates these two qualitatively different regimes of the dynamics is a *'perturbed homoclinic orbit'* realized for $B = B_{cr}(\hat{\lambda})$. This special orbit is formed by one of the branches resulting from the 'break-up' of the Hamiltonian homoclinic loop when weak damping is added to the system. We recall that the homoclinic loop of the unstable undamped NNM $S11-$ is generated due to a saddle-node bifurcation that occurs at an energy level between the periodic IOs on branches $U65$ and $U76$ in the FEP of Figure 3.20. The damped perturbed homoclinic orbit appears as the initial 'super-slow' half-cycle in the plot of Figure 3.69d, and corresponds to the case of *optimal TET* in the system. In Figure 3.71 we depict the portion of this damped homoclinic orbit corresponding to the solution of the slow flow dynamical systems (3.160–3.161) for the given initial condition $z(0) = B$; we note that these are peculiar forms of dynamical systems, as the initial conditions appear also as excitation terms on their right-hand sides. In what follows, the damped perturbed homoclinic orbit will be analytically studied, in an effort to analytically model the optimal TET regime depicted in Figure 3.69d. This analysis is analogous to, but different from the analytical study performed in Section 2.3 concerning the 'break-up' of the homoclinic orbit (depicted in Figure 2.10) of system (2.31) with grounded NES when damping was added.

Reconsidering system (3.160–3.161), we seek its solution in the following regular perturbation series form:

$$z(t) = z_0(t) + \varepsilon^{1/2}\hat{\lambda}\, z_1(t) + O(\varepsilon), \quad B = B_0 + \varepsilon^{1/2}\hat{\lambda} B_1 + O(\varepsilon) \quad (3.162)$$

Substituting into (3.160) and considering only $O(1)$ terms we derive the following system at the first order of approximation,

$$\dot{z}_0 - \frac{j}{2}|z_0|^2 z_0 + \frac{j}{2}z_0 = \frac{jB_0}{2}, \quad z_0(0) = B_0 \quad (3.163a)$$

or in terms of the polar transformation $z_0 = N_0 e^{j\delta_0}$,

$$\dot{N}_0 = \frac{B_0}{2}\sin\delta_0, \quad N_0(0) = B_0$$

$$\dot{\delta}_0 + \frac{1}{2} - \frac{1}{2}N_0^2 = \frac{B_0}{2N_0}\cos\delta_0, \quad \delta_0(0) = 0 \quad (3.163b)$$

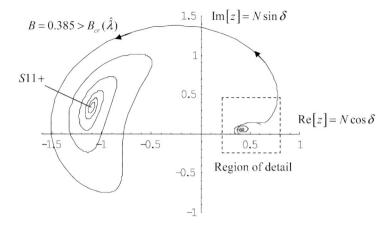
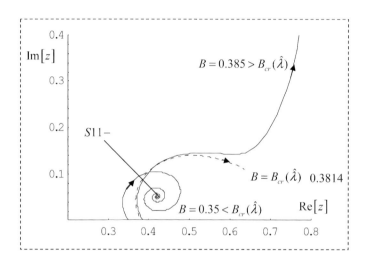

Fig. 3.71 Parametric plots for $\hat{\lambda} = 0.4472$ Im[z] against Re[z] with t being the parametrizing variable: initial regime of slow flow dynamics of intermediate-energy damped IOs for different normalized impulses B [slow flow (3.160) or (3.161)].

We note that there exist no damping terms in this first order of approximation, as these terms enter into the problem at the next order of approximation.

It can be proved that the undamped slow flow possess the following Hamiltonian (first integral of the motion):

$$\frac{j}{2}|z_0|^2 - \frac{j}{4}|z_0|^4 - \frac{jB_0}{2}z_0^* - \frac{jB_0}{2}z_0 = h \qquad (3.164)$$

where the asterisk denotes complex conjugate. This relation reduces (3.163b) to the following one-dimensional slow flow:

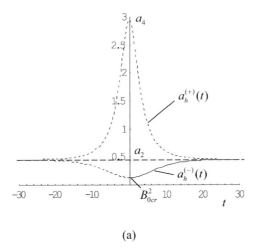

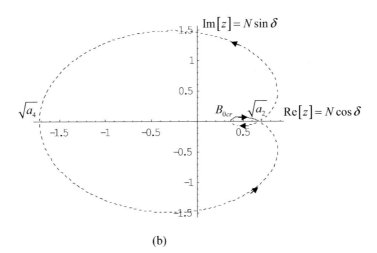

Fig. 3.73 Homoclinic orbits (3.167a, b) and (3.168a, b): (a) $a_h^{(\pm)}(t)$, (b) parametric plot of Im[z] against Re[z] with t being the parametrizing variable; the solid line represents the homoclinic solution of the slow-flow problem (3.165).

$$z_{1h}(t) = z_{1HS}(t) + z_{1PI}(t) \qquad (3.170)$$

i.e., it is expressed as a superposition of the general homogeneous solution $z_{1HS}(t)$ and of a particular integral $z_{1PI}(t)$. Key in solving the problem, is the computation of two linearly independent homogeneous solutions of (3.169), since then, a particular integral may be systematically computed by either solving the differential equation satisfied by the Wronskian of the homogeneous solutions, or through the method of variation of parameters.

We can easily prove (by simple substitution into the complex homogeneous equation) that one homogeneous solution of (3.169) can be computed in terms of the time derivative of the $O(1)$ homoclinic solution as $z_{1HS}^{(1)}(t) = \Lambda \dot{z}_{0h}(t)$, $\Lambda \in R$. At this point we decompose the complex solution into real and imaginary parts:

$$z_{1h}(t) = x_{1h}(t) + jy_{1h}(t), \quad z_{0h}(t) = x_{0h}(t) + jy_{0h}(t) \tag{3.171}$$

Then the first homogeneous solution of (3.169) is expressed as

$$\left. \begin{array}{l} x_{1HS}^{(1)}(t) = \dfrac{2}{B_{0\,cr}} \dot{x}_{0h}(t) \\[2mm] y_{1HS}^{(1)}(t) = \dfrac{2}{B_{0\,cr}} \dot{y}_{0h}(t) \end{array} \right\}$$

$$\Rightarrow z_{1HS}^{(1)} = \dfrac{2}{B_{0\,cr}} [\dot{x}_{0h}(t) + j\dot{y}_{0h}(t)] \quad \text{(First homogeneous solution)} \tag{3.172}$$

where the real constant Λ was selected so that the first homogeneous solution satisfies the initial conditions $x_{1HS}^{(1)}(0) = 0$, $y_{1HS}^{(1)}(0) = +1 \Rightarrow z_{1HS}^{(1)}(0) = j$. In addition, the homogeneous solution (3.173) satisfies the limiting conditions $\lim_{t \to +\infty} x_{1HS}^{(1)}(t) = 0$ and $\lim_{t \to +\infty} y_{1HS}^{(1)}(t) = 0$.

To compute a second linearly independent homogeneous solution of (3.169) it is convenient to carry the entire analysis to the real domain, by decomposing (3.169) into the following set of two real quasi-linear coupled ordinary differential equations with non-homogeneous terms:

$$\begin{Bmatrix} \dot{x}_{1h} \\ \dot{y}_{1h} \end{Bmatrix} + \begin{bmatrix} x_{0h}\,y_{0h} & (x_{0h}^2 + 3y_{0h}^2 - 1)/2 \\ -(3x_{0h}^2 + y_{0h}^2 - 1)/2 & -x_{0h}\,y_{0h} \end{bmatrix} \begin{Bmatrix} x_{1h} \\ y_{1h} \end{Bmatrix}$$

$$= \begin{Bmatrix} -x_{0h}/2 \\ (B_{1\,cr} - y_{0h})/2 \end{Bmatrix} \tag{3.173}$$

Note that problem (3.173) governs the $O(\varepsilon^{1/2})$ perturbation of the $O(1)$ homoclinic solution (3.167a, b), and the real constant $B_{1\,cr}$ on the right-hand side denotes the $O(\varepsilon^{1/2})$ correction to $B_{0\,cr}$ in (3.162). We seek a second homogeneous solution of (3.173) satisfying the initial conditions, $x_{1HS}^{(2)}(0) = -1$, $y_{1HS}^{(2)}(0) = 0$. Accordingly, we consider the following relation satisfied by the Wronskian of (3.173):

$$W(t) = x_{1HS}^{(1)}(t)y_{1HS}^{(2)}(t) - x_{1HS}^{(2)}(t)y_{1HS}^{(1)}(t) \tag{3.174a}$$

From the theory of ordinary differential equations the Wroskian then satisfies the following relation:

$$\dot{W}(t) = 0 \Rightarrow W(t) = W(0) = 1 \tag{3.174b}$$

which provides a means for computing the second homogeneous through the relation,

$$x^{(1)}_{1HS}(t)y^{(2)}_{1HS}(t) - x^{(2)}_{1HS}(t)y^{(1)}_{1HS}(t) = 1 \Rightarrow x^{(2)}_{1HS}(t) = \frac{x^{(1)}_{1HS}(t)y^{(2)}_{1HS}(t) - 1}{y^{(1)}_{1HS}(t)} \tag{3.175}$$

When this expression is substituted into the second of equations (3.173) with the non-homogeneous term dropped, yields the following first-order quasi-linear differential equation governing $y^{(2)}_{1HS}$,

$$\dot{y}^{(2)}_{1HS} + \left[a_{21} \frac{x^{(1)}_{1HS}}{y^{(1)}_{1HS}} + a_{22} \right] y^{(2)}_{1HS} = \frac{a_{21}}{y^{(1)}_{1HS}}, \quad y^{(2)}_{1HS}(0) = 0 \tag{3.176}$$

with $a_{11} = x_{0h}y_{0h}$, $a_{12} = (x_{0h}^2 + 3y_{0h}^2 - 1)/2$, $a_{21} = -(3x_{0h}^2 + y_{0h}^2 - 1)/2$, and $a_{22} = -x_{0h}y_{0h}$. The solution of (3.176) provides the second linearly independent homogeneous solution of (3.169), which is computed explicitly as follows:

$$\left. \begin{aligned} x^{(2)}_{1HS}(t) &= \frac{x^{(1)}_{1HS}(t)y^{(2)}_{1HS}(t) - 1}{y^{(1)}_{1HS}(t)} \\ y^{(2)}_{1HS}(t) &= \int_0^t \frac{a_{21}(\tau)}{y^{(1)}_{1HS}(\tau)} \exp\left\{ -\int_\tau^t \left[a_{21}(s)\frac{x^{(1)}_{1HS}(s)}{y^{(1)}_{1HS}(s)} + a_{22}(s) \right] ds \right\} d\tau \\ \Rightarrow z^{(2)}_{1HS} &= x^{(2)}_{1HS}(t) + jy^{(2)}_{1HS}(t) \quad \text{(Second homogeneous solution)} \end{aligned} \right\}$$
$$\tag{3.177}$$

As mentioned previously, the second homogeneous solution satisfies the initial conditions $x^{(2)}_{1HS}(0) = -1$, $y^{(2)}_{1HS}(0) = 0 \Rightarrow z^{(2)}_{1HS}(0) = -1$, and, contrary to (3.172) it diverges with time, since it holds that $\lim_{t \to +\infty} x^{(2)}_{1HS}(t) = +\infty$ and $\lim_{t \to +\infty} y^{(2)}_{1HS}(t) = +\infty$.

Making use of the two linearly independent homogeneous (3.172) and (3.177) we may compute a first particular integral by the method of variation of parameters. Indeed, by expressing the real and imaginary parts of the particular integral $z_{1PI}(t) = x_{1PI}(t) + jy_{1PI}(t)$, in the form

$$\begin{Bmatrix} x_{1PI}(t) \\ y_{1PI}(t) \end{Bmatrix} = c_1(t) \begin{Bmatrix} x^{(1)}_{1HS}(t) \\ y^{(1)}_{1HS}(t) \end{Bmatrix} + c_2(t) \begin{Bmatrix} x^{(2)}_{1HS}(t) \\ y^{(2)}_{1HS}(t) \end{Bmatrix} \tag{3.178}$$

and evaluating the real coefficients $c_1(t)$ and $c_2(t)$ by substituting into (3.173), we obtain the following explicit solution of problem (3.169) which provides the $O(\varepsilon^{1/2})$ perturbation of the homoclinic orbit:

$$\begin{Bmatrix} x_{1h}(t) \\ y_{1h}(t) \end{Bmatrix}$$

$$= \left[\Lambda_1 + \int_0^t \left\{ -\frac{x_{0h}(\tau)}{2} y_{1HS}^{(2)}(\tau) - \left[\frac{B_{1cr} - y_{0h}(\tau)}{2} \right] x_{1HS}^{(2)}(\tau) \right\} d\tau \right] \begin{Bmatrix} x_{1HS}^{(1)}(t) \\ y_{1HS}^{(1)}(t) \end{Bmatrix}$$

$$+ \left[\Lambda_2 + \int_0^t \left\{ \frac{x_{0h}(\tau)}{2} y_{1HS}^{(1)}(\tau) + \left[\frac{B_{1cr} - y_{0h}(\tau)}{2} \right] x_{1HS}^{(1)}(\tau) \right\} d\tau \right] \begin{Bmatrix} x_{1HS}^{(2)}(t) \\ y_{1HS}^{(2)}(t) \end{Bmatrix}$$

(3.179)

This analytical expression contains three yet undetermined real constants, namely, the coefficients Λ_1, Λ_2, and the correction to the initial condition for motion on the homoclinic orbit, $B_{1\,cr}$. By imposing the initial condition of (3.169), $z_{1h}(0) = B_{1\,cr} \Rightarrow x_{1h}(0) = B_{1\,cr}$, $y_{1h}(0) = 0$, we compute the two coefficients as follows:

$$\Lambda_1 = 0 \quad \text{and} \quad \Lambda_2 = -B_{1\,cr} \quad (3.180)$$

Then, taking into account that the components of the second homogeneous solution $x_{1HS}^{(2)}(t)$ and $y_{1HS}^{(2)}(t)$ in the second additive term of (3.179) diverge as $t \to +\infty$, and in order to obtain bounded solutions for $x_{1h}(t)$ and $y_{1h}(t)$ as $t \to +\infty$, we require that

$$-B_{1\,cr} + \int_0^{+\infty} \left\{ \frac{x_{0h}(\tau)}{2} y_{1HS}^{(1)}(\tau) + \left[\frac{B_{1\,cr} - y_{0h}(\tau)}{2} \right] x_{1HS}^{(1)}(\tau) \right\} d\tau = 0 \quad (3.181a)$$

This evaluates $B_{1\,cr}$ according to the following expression:

$$B_{1\,cr} = \frac{\int_0^{+\infty} \left[x_{0h}(\tau) y_{1HS}^{(1)}(\tau) - y_{0h}(\tau) x_{1HS}^{(1)}(\tau) \right] d\tau}{2 - \int_0^{+\infty} x_{1HS}^{(1)}(\tau) d\tau} \quad (3.181b)$$

This completes the solution of the problem (3.169) and computes the perturbation of the homoclinic orbit in the damped system (3.160–3.161) with $O(\varepsilon^{1/2})$ damping. In summary, the analytic approximation of the perturbed homoclinic orbit is given by

$$z_h(t) = z_{0h}(t) + \varepsilon^{1/2} \hat{\lambda} z_{1h}(t) + O(\varepsilon), \quad B_{cr}(\hat{\lambda}) = B_{0\,cr} + \varepsilon^{1/2} \hat{\lambda} B_{1\,cr} + O(\varepsilon)$$

(3.182)

where $z_{0h}(t) = \sqrt{a_h^{(-)}(t)} \exp[\delta_h^{(-)}(t)]$ and $a_h^{(-)}(t)$, $\delta_h^{(-)}(t)$ are computed by (3.167a, b); $z_{1h}(t) = x_{1h}(t) + j y_{1h}(t)$, where $x_{1h}(t)$ and $y_{1h}(t)$ are computed by (3.179), (3.180) and (3.182b); $B_{0\,cr} \approx 0.36727$; and $B_{1\,cr}$ is computed by (3.181b).

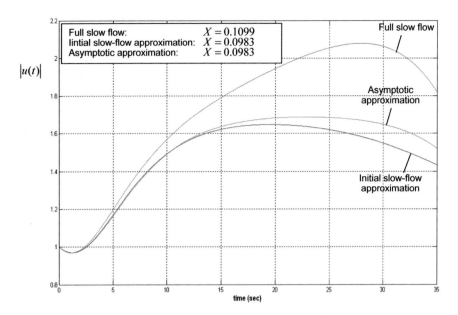

Fig. 3.74 Slow flow response in the regime of optimal TET ('super-slow' half-cycle of TET), for $\varepsilon = 0.05$, $C = 1$ and $\lambda = \varepsilon^{1/2}\hat{\lambda} = 0.1$; comparison of full slow slow (3.151) or (3.157), of the approximation of the slow flow in the initial stage of the dynamics (3.158) or (3.160), and of the asymptotic solution (3.182).

For $\varepsilon = 0.05$ and $\hat{\lambda} = 0.4472$ we estimate the initial condition as $B_{cr}(\hat{\lambda} = 0.4472) \approx 0.3806$, which compares to the numerical value of 0.3814 derived from the numerical integration of the initial approximation of the slow flow (3.160–3.161) (see Figure 3.71). Taking into account the previous coordinate transformations and rescalings for B, the previous analytical result leads to an estimated initial condition (impulse) of $X = 0.0983$ for optimal TET (i.e., for the excitation of the damped homoclinic orbit), compared to the numerical result of $X = 0.1099$ derived from the full averaged slow flow (3.157) (see Figure 3.69d); we note that the error is of $O(\varepsilon = 0.05)$ and compatible to our previous asymptotic derivations.

In Figure 3.74 we provide a comparison of the three approximate models for the slow flow dynamics in the regime of optimal TET; the asymptotic analysis correctly predicts the half-cycle 'super-slow' transfer of energy from the LO to the NES in the initial regime of the motion, although it underestimates the maximum amplitude of the response during this half-cycle; this can be explained by the fact that the slow flow approximation (3.158) or (3.160) is only valid in the initial regime of the motion.

This completes the analytical study of the regime of optimal TET in system (3.148) when intermediate-energy damped IOs are excited. In summary, in the weakly damped system, optimal TET is realized for initial energies where the excited damped IOs are in the neighborhood of the homoclinic orbit of the unstable out-of-phase damped NNM $S11-$; in the underlying Hamiltonian system this unstable NNM is generated at a critical energy through a saddle-node bifurcation. We studied analytically the perturbation of the homoclinic orbit in the weakly damped system, which introduces an additional 'super slow' time scale in the averaged dynamics and leads to optimal TET from the LO to the NES in a single 'super-slow' half cycle. At higher energies, this 'super-slow' half cycle is replaced by strong nonlinear beats (these are generated due to the attraction of the dynamics to the stable in-phase damped NNM $S11+$), which yield significant but non-optimal TET through nonlinear beats. At lower energies than the one corresponding to the optimal TET regime, the dynamics is attracted by the stable, weakly nonlinear (linearized), out-of-phase damped NNM $S11-$ and TET is negligible.

The above-mentioned conclusions are valid for the weakly damped system (3.148), under the assumption of sufficiently small ε, i.e., of for lightweight NESs and systems with strong mass asymmetries. In Figures 3.75–3.77 we study TET in system (3.148) for excitation of intermediate-energy damped IOs over a wider range of mass asymmetry ε and damping $\varepsilon\lambda$; these plots were derived by direct numerical integrations of the differential equations of motion, and monitoring the instantaneous energy of the system versus time. Numerical results indicate that, by increasing ε (i.e., by decreasing the mass asymmetry) and the damping coefficient $\varepsilon\lambda$, the capacity of the NES for optimal TET deteriorates. This is due to the fact that by increasing the inertia of the NES the amplitude of the relative response between the LO and the NES decreases, which hinters the capacity of the damper of the NES to effectively dissipate energy. Moreover, by increasing damping in the system, the damper of the LO dissipates an increasingly higher portion of the vibration energy which leads to deterioration of TET; this markedly slows energy dissipation in the system, as judged by comparing the time intervals required for energy dissipation in the plots of Figure 3.77 and the corresponding time intervals in the regimes of optimal TET in the plots of Figures 3.75 and 3.76.

3.5 Multi-DOF (MDOF) Linear Oscillators with SDOF NESs: Resonance Capture Cascades and Multi-frequency TET

Up to now we examined TET in a two-DOF system consisting of SDOF damped linear oscillator (LO) coupled to an essentially nonlinear attachment, acting, in essence, as nonlinear energy sink (NES). In this section, we extend the analysis to MDOF LOs with SDOF essentially nonlinear boundary attachments. The main result reported in this section is that the SDOF NES can interact with (and extract energy from) multiple linear modes of the linear system to which it is attached, due to *resonance capture cascades* (RCCs). Indeed, we will show that through RCCs the NES can passively extract broadband vibration energy from the linear system (i.e., over wide frequency ranges), through *multi-frequency TET*.

What enables a SDOF NES to interact with multiple linear modes over arbitrary frequency ranges is its essential stiffness nonlinearity, which enables it to engage in

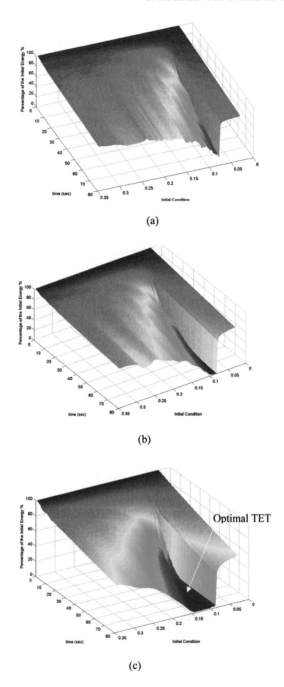

Fig. 3.75 Energy dissipation in system (3.148) when damped IOs are excited for mass assymetry $\varepsilon = 0.03$: (a) $\varepsilon\lambda = 0.015$, (b) $\varepsilon\lambda = 0.003$, (c) $\varepsilon\lambda = 0.006$.

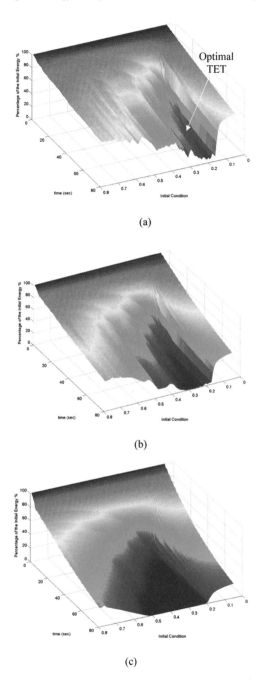

Fig. 3.76 Energy dissipation in system (3.148) when damped IOs are excited for mass assymetry $\varepsilon = 0.1$: (a) $\varepsilon\lambda = 0.005$, (b) $\varepsilon\lambda = 0.01$, (c) $\varepsilon\lambda = 0.02$.

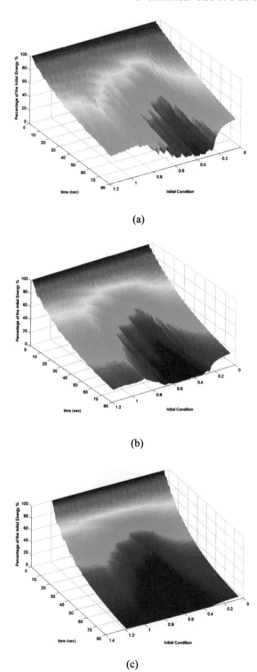

Fig. 3.77 Energy dissipation in system (3.148) when damped IOs are excited for mass assymetry $\varepsilon = 0.2$: (a) $\varepsilon\lambda = 0.01$, (b) $\varepsilon\lambda = 0.02$, (c) $\varepsilon\lambda = 0.04$.

transient resonance capture (TRC) with any highly energetic linear mode irrespective of its frequency, provided, of course, that this mode has no node at the point of attachment of the NES. Then, the NES extracts energy from each specific mode, before escaping from TRC and engaging in transient resonance the next one. In the passive system considered, what controls the order with which modes participate in these RCCs is the initial state of the system, the external excitation (being narrowband or broadband), and the actual rate of energy dissipation due to damping (since the instantaneous energy level of the NES passively 'tunes' its instantaneous frequency). These concepts are discussed in the following sections and demonstrated for the case of a two-DOF linear LO with a SDOF NES attachment. Then, we consider a semi-infinite chain of LOs with a single NES attached to its end, as a first attempt to extend the concept of passive TET to linear waveguides with local essentially nonlinear attachments.

3.5.1 Two-DOF Linear Oscillator with a SDOF NES

The system considered is depicted in Figure 3.78, and consists of a two-DOF damped LO (designated as the primary system) coupled to a SDOF NES. The equations of motion are given by:

$$m_1\ddot{x}_1 + c_1\dot{x}_1 + k_1 x_1 + k_{12}(x_1 - x_2) = 0$$
$$m_2\ddot{x}_2 + c_2\dot{x}_2 + c_v(\dot{x}_2 - \dot{v}) + k_2 x_2 + k_{12}(x_2 - x_1) + C(x_2 - v)^3 = 0$$
$$\varepsilon\ddot{v} + c_v(\dot{v} - \dot{x}_2) + C(v - x_2)^3 = 0 \tag{3.183}$$

The variables $x_1(t)$ and $x_2(t)$ refer to the displacements of the oscillators of the (primary) linear system, whereas $v(t)$ refers to the displacement of the NES. As in the previous sections, a lightweight NES is considered by requiring that $\varepsilon \ll m_1, m_2$, with $0 < \varepsilon \ll 1$ being a small parameter characterizing the strong mass asymmetry of the system.

As in the analysis for the two-DOF system considered in the previous sections, first we discuss the dynamics of the underlying Hamiltonian system obtained by setting all damping terms equal to zero; then we analyze the nonlinear transitions in the weakly damped system and relate these transitions to the Hamiltonian dynamics.

3.5.1.1 Frequency-Energy Plot (FEP) of the Underlying Hamiltonian System

It is not necessary to perform an exhaustive calculation of the periodic orbits of underlying Hamiltonian system of (3.183), since the dynamics governing TET can be studied by considering the following two subsets of orbits in the Hamiltonian FEP: (i) the backbone branches of periodic orbits under conditions of 1:1:1 internal resonance, and (ii) the manifolds of impulsive orbits (IOs). Note that since in this

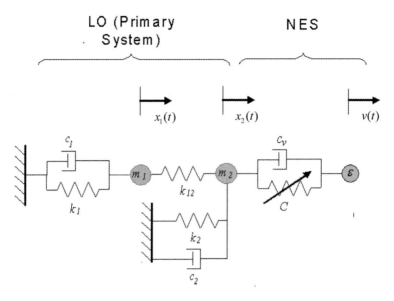

Fig. 3.78 The three-DOF system consisting of a two-DOF primary LO with an essentially nonlinear, lightweight NES.

case the primary linear system possesses two degrees of freedom there exist multiple backbone sub-branches (depending on the relative phases between the three oscillators of the system during 1:1:1 internal resonance), and multiple manifolds of IOs in the FEP.

An analytic approximation of the backbone branches of the Hamiltonian system can be derived by applying the complexification-averaging (CX-A) technique. To this end, the following complex variables are introduced,

$$\psi_1 = \dot{x}_1 + j\omega x_1, \quad \psi_2 = \dot{x}_2 + j\omega x_2, \quad \psi_3 = \dot{v} + j\omega v \quad (3.184)$$

where ω is the common dominant frequency of oscillation during 1:1:1 internal resonance. Following the CX-A procedure as discussed in the previous sections (i.e., averaging over the fast frequency ω, expressing the resulting complex modulations in polar form, and imposing stationarity conditions for the resulting real amplitudes and phases) the following analytical approximation for the NNMs on the backbone branches of the Hamiltonian system is obtained,

$$x_1(t) \approx A \sin \omega t, \quad x_2(t) \approx B \sin \omega t, \quad v(t) \approx D \sin \omega t \quad (3.185)$$

where

$$A = \left[\frac{4\varepsilon\omega^2 c_2}{3C(c_2 - c_1)^3}\right]^{1/2}, \quad D = c_2 A, \quad B = c_1 A,$$

$$c_1 = \left(k_1 + k_{12} - \omega^2 m_1\right)/k_{12}, \quad c_2 = \left[-k_{12} - c_1(\omega^2 m_2 - k_2 - k_{12})\right]/\varepsilon\omega^2$$

The backbone branches can be constructed by varying the frequency ω and calculating the corresponding total energies of the NNMs. Figure 3.79 depicts the backbone branches, denoted by $S111$, of the system with parameters $m_1 = m_2 = k_1 = k_2 = k_{12} = C = 1$ and $\varepsilon = 0.05$. NNMs depicted as projections of the three-dimensional configuration space (v, x_1, x_2) of the system are inset. When the projections of the NNMs are close to horizontal (vertical) lines, the motion is localized to the NES (primary system).

Four characteristic frequencies, f_{1L}, f_{2L}, f_{1H} and f_{2H} are defined in this FEP. At high energy levels and finite frequencies, the essential nonlinearity behaves as a rigid link, and the system is reduced to the following system of two linear coupled oscillators:

$$m_1 \ddot{x}_1 + k_1 x_1 + k_{12}(x_1 - x_2) = 0$$
$$(\varepsilon + m_2) \ddot{x}_2 + k_2 x_2 + k_{12}(x_2 - x_1) = 0 \quad (3.186)$$

For the above parameters the natural frequencies of this system are given by $f_{1H} = 0.9876$ rad/s and $f_{2H} = 1.7116$ rad/s. At low energy levels, the stiffness of the essential nonlinearity tends to zero, and the system is again reduced to the primary two-DOF LO, the natural frequencies of which are given by $f_{1L} = 1$ rad/s and $f_{2L} = \sqrt{3}$ rad/s.

From Figure 3.79, we note that the two frequencies f_{1L} and f_{2L} divide the FEP into three distinct regions. The first region defined by $\omega \geq f_{2L}$, consists of the backbone sub-branch $S111+-+$, where the $(+)$ and $(-)$ signs characterize the relative phases between the three masses of the system, and indicate whether the extremum of the amplitude of the corresponding oscillator during the synchronous 1:1:1 periodic motion (NNM) is positive or negative, respectively. On this sub-branch, the primary LO vibrates in an out-of-phase fashion, and the motion becomes increasingly localized to the LO or the NES as $\omega \to f_{2L}$ or $\omega \to \infty$, respectively. The second region defined by $f_{1L} \leq \omega \leq f_{2H}$, consists of two distict sub-branches, namely $S111+--$ and $S111++-$. These branches coalesce at point $S111+0-$ (depicted as the grey dot in Figure 3.79), where the initial velocity of the mass m_2 is zero. On $S111+--$ the LO vibrates in an out-of-phase fashion, and the motion localizes to the NES as the frequency leaves the neighborhood of f_{2H}. On $S111++-$ the LO oscillates in in-phase fashion, and the vibration localizes to the LO as $\omega \to f_{1L}$. The third region corresponding to $\omega \leq f_{1H}$, consists of the sub-branch $S111+++$, where the LO vibrates in in-phase fashion, and the motion localizes to the NES as the frequency tends away from f_{1H}.

Due to the energy dependence of the NNMs along the sub-branches of $S111$, interesting and strong energy exchanges may occur between the primary LO and the NES when weak damping is introduced in the system. Indeed, the weakly damped system possesses damped NNM manifolds which can be considered as analytic continuations for weak damping of the NNMs of the Hamiltonian system. Since, these manifolds are invariant for the dynamical flow, when a damped response is initiated

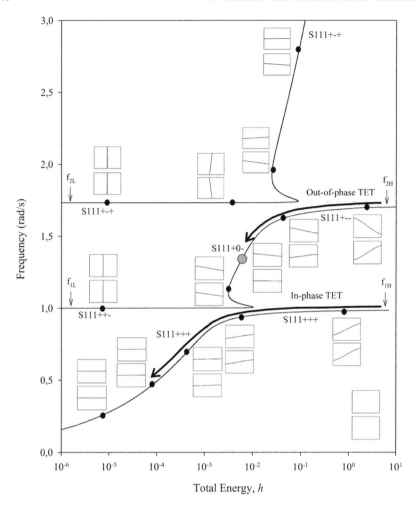

Fig. 3.79 Analytic approximation of the backbone branch of (3.183): NNMs depicted as projections of the three-dimensional configuration space of the system are superposed; the horizontal and vertical axes in these plots are the responses of the nonlinear and primary systems, respectively [top plot (v, x_1), bottom plot (v, x_2)].

on a damped NNM manifold, it stays on it for the entire duration of the decaying oscillation.

Two specific sub-branches, namely $S111+--$ and $S111+++$, play an important role for the realization of fundamental TET in system (3.183). Due to the dependence of the frequency of the damped oscillation on the instantaneous energy, irreversible channeling of vibration energy from the LO to the NES takes place as the damped continuations of the NNMs $S111+--$ and $S111+++$ are traced from high to low frequencies (since the shapes of the corresponding NNMs localize from the LO to the NES as frequency decreases – see Figure 3.79). Hence, both *in-phase* and *out-of-phase fundamental TET* can be realized in this system, corresponding to in-phase or out-of-phase motions of the oscillators of the primary LO, respectively; this shows the adaptivity of the NES to different initial conditions and represents a generalization of the concept of fundamental TET discussed in Section 3.4.2.1 for the two-DOF system. A detailed stability analysis of $S111+--$ and $S111+++$ was not performed, but the following numerical simulations and experimental results show that these are stable oscillations, at least for the parameter values considered in this work.

The backbone of the FEP of the Hamiltonian system can also be computed numerically. Assuming that a NNM is realized for the initial velocity vector $[\dot{x}_1(0) \ \dot{x}_2(0) \ \dot{v}(0)]$ and zero initial displacements, this vector together with the period of the motion, T, are computed by satisfying the following periodicity condition:

$$\begin{bmatrix} x_1(T) \ x_2(T) \ v(T) \ \dot{x}_1(T) \ \dot{x}_2(T) \ \dot{v}(T) \end{bmatrix}^T \\ - \begin{bmatrix} 0 \ 0 \ 0 \ \dot{x}_1(0) \ \dot{x}_2(0) \ \dot{v}(0) \end{bmatrix}^T = \begin{bmatrix} 0 \ 0 \ 0 \ 0 \ 0 \ 0 \end{bmatrix}^T \quad (3.187)$$

The numerical computation was carried out in Matlab® using optimization techniques. For a given value of the period T the objective function to be minimized is the norm of the left-hand side of equation (3.187), and the optimization variables are the non-zero initial conditions. By varying the period, the backbone branch represented in Figure 3.80 is obtained; a small subset of subharmonic tongues (see Sections 3.3.2.2 and 3.3.2.3) has also been identified using this algorithm. We note the close agreement between the backbones computed numerically and analytically (compare Figures 3.79 and 3.80).

Another important feature of the FEP concerns the manifolds of IOs. The essential role of IOs for TET has been discussed extensively in Section 3.4; the periodic IOs of system (3.183) correspond to the special initial conditions, $\dot{x}_1(0) \neq 0$, $\dot{x}_2(0) \neq 0$ and $x_1(0) = x_2(0) = v(0) = \dot{v}(0) = 0$ (or, equivalently, to two impulses applied to the LO with the system initially at rest). Contrary to the two-DOF examined in Sections 3.3.3 and 3.3.4, two distinct families of IOs are realized in the three-DOF under consideration: *in-phase IOs* correspond to two in-phase impulses of identical magnitudes applied to the two masses of the LO at $t = 0$, corresponding to initial conditions, $\dot{x}_1(0) = \dot{x}_2(0) \neq 0$; *out-of-phase IOs* correspond to two out-of-phase impulses of equal magnitude applied to the two masses of the LO, and initial conditions $\dot{x}_1(0) = -\dot{x}_2(0) \neq 0$. Contrary to the two-DOF system examined previously, *no IOs can be realized in the three-DOF system by applying a single impulse to either one of the masses of the LO*. Similarly, however, to the two-DOF system, the excitation of stable IOs localized to the NES, leads to rapid and significant energy transfer from the LO to the NES during a cycle of the oscillation of the three-DOF system; when damping is introduced this leads to effective, fast scale TET from the LO to the NES. It follows that for system (3.183) there exist two distinct IO manifolds, consisting of periodic and quasi-periodic in-phase and out-of-phase IOs, respectively. The computations depicted in Figure 3.80 were re-

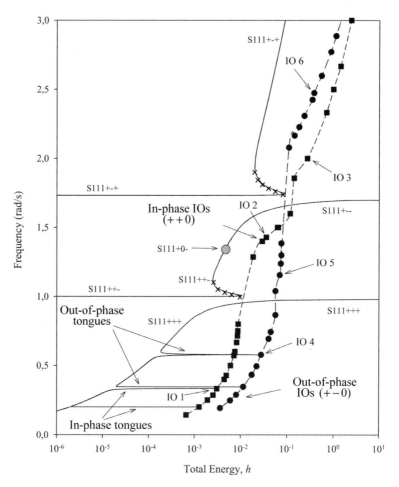

Fig. 3.80 Numerical computation of the FEP (backbone branches and periodic IOs) of system (3.183) for $m_1 = m_2 = k_1 = k_2 = k_{12} = C = 1$ and $\varepsilon = 0.05$; black dots and squares denote out-of-phase and in-phase IOs, respectively; unstable NNMs are denoted by ($\times$); IOs 1–6 refer to Figures 3.81 and 3.82.

stricted to periodic IOs corresponding to low-order internal resonances between the LO and the NES.

As mentioned above, no periodic orbits corresponding to impulsive excitation of only one of the masses of the primary system were detected. However, we conjecture that for this type of impulsive excitation *quasi-periodic impulsive orbits* could still exist, and are such that the NES resonates with a mode of the primary system only above a certain energy threshold. Moreover, it was observed that strong nonlinear interaction of the NES with the in-phase mode of the LO is triggered at lower energy levels compared to the out-of-phase mode.

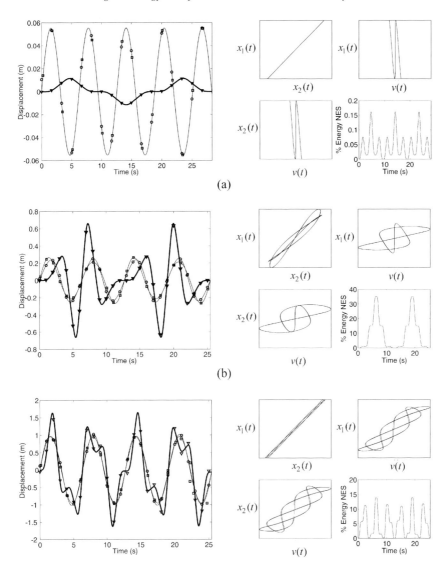

Fig. 3.81 Representative in-phase IOs: (a) IO 1, (b) IO 2, (c) IO 3 (see Figure 3.80); left column: time series; - -■— $x_1(t)$, - -O- - $x_2(t)$, - -∇- - $v(t)$; right column: two-dimensional projections of IOs and instantaneous percentage of total energy carried by the NES during a cycle of the IO.

The in-phase manifold of IOs consists of in-phase impulsive orbits $(++0)$ located on in-phase subharmonic tongues, with the masses of the primary linear system oscillating an in-phase fashion. This manifold is depicted as a smooth curve in the FEP. Representative in-phase IOs labeled as IO 1, IO 2 and IO 3 in Figure 3.80 are illustrated in Figure 3.81. When the phase differences between the masses of

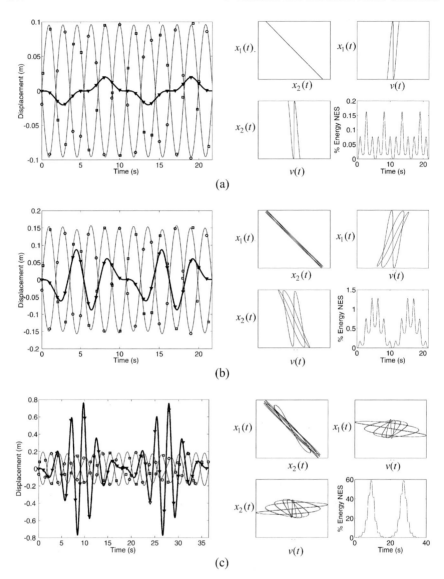

Fig. 3.82 Representative out-of-phase IOs: (a) IO 4, (b) IO 5, (c) IO 6 (see Figure 3.80); left column: time series; - -■- - $x_1(t)$, - -O- - $x_2(t)$, - -∇- - $v(t)$; right column: two-dimensional projections of IOs and instantaneous percentage of total energy carried by the NES during a cycle of the IO.

the system are trivial, the motion of the NNM in the configuration space (x_1, x_2, v) takes the form of a simple curve; in the case of non-trivial phase differences the motion corresponds to a Lissajous curve.

For IO 1, the oscillations of the two masses of the linear primary system are almost identical and nearly monochromatic; the corresponding oscillation of the NES

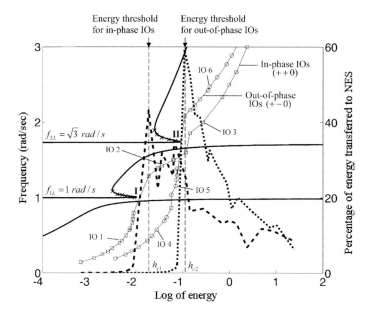

Fig. 3.83 Maximum percentage of energy transferred from the LO to the NES during a cycle of the IO (dashed line: in-phase IOs; dotted line: out-of-phase IOs); the FEP of Figure 3.80 is superimposed to this plot, unstable NNMs are denoted by (×).

has two dominant harmonic components, one equal to the dominant frequency of oscillation of the primary system, and the other equal to one-third of that frequency. Hence, a 1:1:3 internal resonance (IR) between the two masses of the primary system and the NES is realized. The nonlinear beat resulting due to this internal resonance is clearly deduced in the plots of Figure 3.81. For IO 1, the energy exchange between the LO and the NES is insignificant, as the maximum percentage of total energy transferred from the LO to the NES during a cycle is just 0.17%. For IOs 2 and 3, however, which correspond to 3:3:2 and 5:5:2 internal resonances, respectively, energy transfer from the LO to the NES during a cycle of the nonlinear beat is much stronger, reaching levels of 35% and 15%, respectively (the notation $p:p:q$ internal resonance implies that the frequencies of oscillation of the first mass of the LO, the second mass of the LO and the NES are in ratios equal to $p:p:q$).

The out-of-phase manifold of IOs consists of out-of-phase impulsive orbits (+−0) located on out-of-phase subharmonic tongues. This manifold is also represented by a smooth curve in the FEP. Representative out-of-phase IOs (labeled as IO 4–6 in Figure 3.80 and corresponding to 1:1:3, 2:2:3 and 6:6:5 internal resonances, respectively) are shown in Figure 3.82.

In Figure 3.83 we present a study of maximum energy transferred from the LO to the NES during a cycle of the nonlinear beat resulting from excitation of in-phase or out-of-phase IOs. Superimposed to the plot of maximum energy transferred is the FEP of Figure 3.80, indicating the backbone branch and the two manifolds of

IOs. What is evident from this plot is that there exist two critical energy thresholds, one for each of the in-phase and out-of-phase IOs, above which the IOs transfer a significant amount of energy from the LO to the NES during a cycle of the nonlinear beat; moreover, the energy threshold for out-of-phase IOs, h_{c2}, is higher than the corresponding one for in-phase IOs, h_{c1}. For instance, for the out-of-phase IO 4 located below the energy threshold h_{c2}, the maximum energy transferred to the NES during a cycle of the nonlinear beat is approximately 0.15% of the energy of the LO, whereas for the out-of-phase IO 6 located above that threshold the corresponding percentage of energy transferred is nearly 60%.

It is interesting to note that the in-phase and out-of-phase thresholds h_{c1} and h_{c2} are located close to the corresponding energies where the saddle node bifurcations, I for in-phase NNMs and II for out-of-phase NNMs, take place; these bifurcations generate unstable branches of in-phase and out-of-phase NNMs as shown in Figure 3.83. From the discussion of Sections 3.3.5 and 3.4.2.5 we recall that in the two-DOF system a similar bifurcation exists in the corresponding FEP. In that system the homoclinic loops of the unstable NNM generated from the saddle-node bifurcation affect drastically the topologies of nearby IOs, since IOs lying inside the homoclinic loops are localized in phase space and the corresponding motions of the system are predominantly localized to the LO; on the contrary, IOs lying outside the homoclinic loops and being close to the 1:1:1 resonance manifold of the Hamiltonian dynamics make large excursions in phase space and correspond to strong nonlinear beats where significant energy is being exchanged between the linear and nonlinear oscillators. It appears that similar dynamics take place in the three-DOF considered here: the strong energy exchanges for in-phase or out-of-phase IOs in the neighborhoods of the saddle node NNM bifurcations I or II (actually, IOs having energies slightly higher that the energies of these saddle-node bifurcations), are affected by their proximities to homoclinic loops of unstable in-phase or out-of-phase NNMs, respectively, and to the corresponding 1:1:1 resonance manifolds at frequencies f_{1L} and f_{2L}, respectively. Based on the discussion and results of Section 3.4.2.5 we may deduce that the excitation of the damped analogs of these IOs lead to *optimal in-phase and out-of-phase fundamental TET* in the weakly damped three-DOF system.

Another similarity to the dynamics of the two-DOF system is that, the two manifolds of in-phase and out-of-phase IOs of the three-DOF system play important roles regarding fundamental and subharmonic TET in the weakly damped system. However, a distinct feature of the dynamics of the weakly damped three-DOF system is the occurrence of *resonance capture cascades* (RCCs). This is a new feature of TET dynamics, whereby the NES passively extracts energy from *both* modes of the primary LO, as it engages sequentially in transient nonlinear resonance with both of them. This is discussed in the next section.

3.5.1.2 Dynamics of the Damped System: Resonance Capture Cascades

We now consider the dynamics of the weakly damped system (3.183). As in the case of the two-DOF system considered in Section 3.4, the underlying Hamiltonian dynamics determine, in essence, the weakly damped transitions and the energy exchanges between the LO and the NES.

The first series of numerical simulations verifies that both in-phase and out-of-phase fundamental TET can occur in the weakly damped system, corresponding to in-phase or out-of-phase relative motions of the two masses of the LO. The simulations were carried out for the following specific system:

$$\ddot{x}_1 + 0.005\dot{x}_1 + x_1 + (x_1 - x_2) = 0$$
$$\ddot{x}_2 + 0.005\dot{x}_2 + 0.002(\dot{x}_2 - \dot{v}) + x_2 + (x_2 - x_1) + (x_2 - v)^3 = 0$$
$$0.05\ddot{v} + 0.002(\dot{v} - \dot{x}_2) + (v - x_2)^3 = 0 \qquad (3.188)$$

so that the small parameter of the problem is given by $\varepsilon = 0.05$; moreover, the assumption of weak damping is satisfied.

The motion is first initiated on a NNM on the backbone branch $S111+++$, and the resulting motion involves in-phase oscillations of all three masses of the system with the same apparent frequency, as shown in Figures 3.84a, b. The temporal evolution of the instantaneous frequencies of the responses can be followed by superimposing their wavelet transform (WT) spectra to the FEP (as performed in Section 3.4 for the two-DOF system). In Figure 3.84c, the WT spectrum of the relative response $v(t) - x_2(t)$ is superposed to the backbone of the FEP (represented by a solid line). As mentioned in previous sections this representation is purely schematic since it superposes a *damped* WT spectrum to the *undamped* FEP; nevertheless, this representation helps us deduce the essential influence of the underlying Hamiltonian dynamics on the weakly damped transitions, and is only used for purely descriptive purposes.

The plot of Figure 3.84c clearly illustrates that as the total energy in the system decreases due to viscous dissipation, the response closely follows the backbone branch $S111+++$; in actuality, the response takes place on the damped NNM invariant manifold which results as perturbation of $S111+++$ when weak damping is added to the system. The dynamical flow is captured in the neighborhood of a 1:1:1 resonance manifold leading to prolonged 1:1:1 TRC. Figure 3.84d depicts the trajectories of the phase difference $\Delta_1(t) \equiv \phi_v(t) - \phi_{x_1}(t)$ between $v(t)$ and $x_1(t)$, and the phase difference $\Delta_2(t) \equiv \phi_v(t) - \phi_{x_2}(t)$ between $v(t)$ and $x_2(t)$; these phase variables are computed directly from the transient responses $v(t)$, $x_1(t)$ and $x_2(t)$ by applying the Hilbert transform. A non-time-like behavior of the two phase differences is noted, which provides further evidence of the occurrence of 1:1:1 TRC. Figure 3.84e confirms that in-phase fundamental TET, i.e., passive and irreversible (on the average) energy transfer from the LO to the NES, takes place.

In the second simulation the motion is initiated on $S111+--$. In the initial stage of the motion ($0 < t < 100$ s) out-of-phase fundamental TET is realized, with

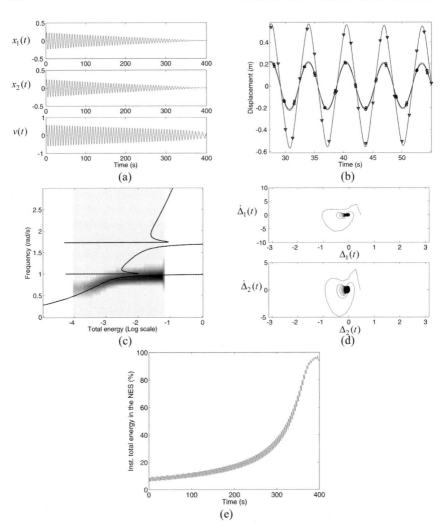

Fig. 3.84 In-phase fundamental TET on the damped NNM invariant manifold $S111+++$: (a) transient responses; (b) close-up of the time series, - -■- - $x_1(t)$, - -●- - $x_2(t)$, - -▼- - $v(t)$; (c) WT spectrum of $v(t) - x_2(t)$ superposed to the FEP; (d) trajectories of phase differences; (e) percentage of instantaneous total energy in the NES.

the two masses of the LO oscillating in an out-of-phase fashion (see Figures 3.85a, b). During this initial regime of the motion, the envelopes of all responses decrease monotonically, but the envelope of the NES seems to decrease more slowly than those of the masses of the linear primary system; TET to the NES occurs during this stage of the motion (see Figure 3.85e). Around $t = 80$ s, the displacement of the second mass of the primary system, $x_2(t)$, becomes very small, and a transition from the out-of-phase damped NNM $S111+--$ to the in-phase damped NNM

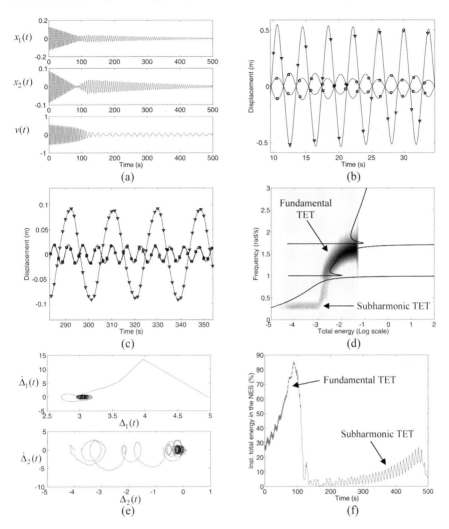

Fig. 3.85 Initial out-of-phase fundamental TET on the damped NNM invariant manifold $S111+--$, followed by 1:3:3 subharmonic TET: (a) transient responses; (b, c) close-ups of the time series during fundamental and subharmonic TET, $--\blacksquare-- \ x_1(t)$, $--\bullet-- \ x_2(t)$, $--\blacktriangledown-- \ v(t)$; (d) WT spectrum of $v(t) - x_2(t)$ superposed to the FEP; (e) trajectories of phase differences; (f) percentage of instantaneous total energy in the NES.

$S111++-$ occurs. When the end of $S111++-$ is traced by the damped dynamics (close to the point of saddle-node bifurcation that eliminates the stable/unstable pair of NNMs in the FEP), escape from 1:1:1 TRC occurs, which results in time-like behavior of the phase differences in Figure 3.85e. The plots of Figures 3.85c, d, f show that this is soon followed by 1:3:3 subharmonic TRC leading to subharmonic TET as the damped motion traces the damped analogue of the in-phase tongue $S113$. Considering the notation used for the subharmonic tongues, we generalize the notation introduced for the subharmonic tongues of the two-DOF, in Section 3.3.1.2: a subharmonic tongue $Sppq$ contains periodic motions with two dominant frequencies, namely ω and $p\omega/q$. We conclude that the NES extracts vibration energy from the LO through two distinct TET mechanisms, that is, initial out-of-phase fundamental TET, followed by subharmonic TET.

We now proceed to verify the existence of energy thresholds above which excitations of IOs can trigger in-phase or out-of-phase fundamental TET. In the results depicted in Figure 3.86, the damped motion is initiated by exciting the in-phase IOs 1 and 2, located below and above the energy threshold h_{c1} of Figure 3.83, respectively; we recall that the in-phase IOs are generated by applying two in-phase identical impulses to the two masses of the LO at $t = 0$. By noting the resulting responses we conclude that the dynamics is markedly different in the two cases. Indeed, when IO 1 is excited, the NES does not extract a significant amount of energy from the LO, as the damped motion is nearly linear and remains localized predominantly to the LO; this is due to the fact that the damped dynamics traces the weakly nonlinear (linearized) branch $S111++-$ with decreasing energy (see Figure 3.86c).

When IO 2 is excited, however, qualitatively different dynamics takes place, since in the initial stage of the response strong nonlinear beats take place leading to TET; during this phase significant energy is dissipated by the damper of the NES. With decreasing energy (and frequency) of the NES escape from the regime of nonlinear beats occurs, and the dynamics makes a transition to the damped NNM $S111+++$, at which point significant in-phase fundamental TET is realized. Overall, multi-frequency TET from the LO to the NES takes place in this case, underscoring the adaptivity of the NES to initial conditions; indeed, depending on the specific initial conditions of the system, the NES passively 'tunes itself' and transiently resonates with different modes of the primary system, absorbing and dissipating vibration energy from the LO.

Likewise, if the damped oscillation is initiated by exciting an out-of-phase IO (i.e., by applying two out-of-phase but equal in magnitude impulses to the two masses of the LO at $t = 0$) located below the energy threshold h_{c2} (IO 4 in Figure 3.83), the response traces the linearized damped NNM $S111+-+$, on which the motion localizes predominantly to the LO throughout. However, if an out-of-phase IO above the energy threshold is excited (IO 6 in Figure 3.83), after an initial regime of nonlinear beats the damped motion makes a transition to the damped NNM $S111+--$ with decreasing energy, and out-of-phase fundamental TET takes place.

Another case of practical importance is when a single impulse is applied to one of the masses of the LO (the primary system). We recall that for single applied impulses to the LO no periodic IOs were detected, but instead both the in-phase and out-of-phase modes of the LO participate in the damped response; hence, a multi-modal response is anticipated in this case, which opens the possibility of interesting multi-frequency nonlinear transitions and energy exchanges in the system.

In the following simulations we consider a slightly modified system (3.183), in the sense that no grounded stiffness for mass m_2 exists (i.e., $k_2 = 0$), and an addi-

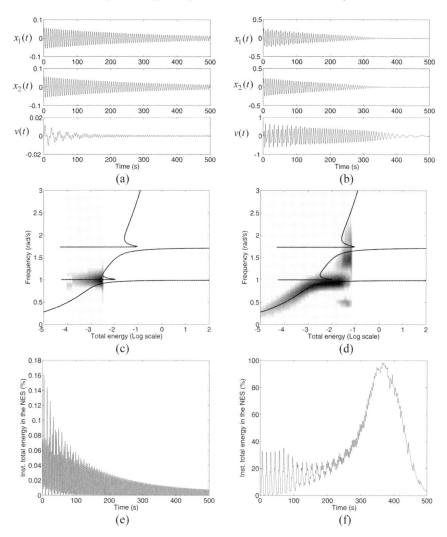

Fig. 3.86 Excitation of IO 1 (a, c, e) and IO 2 (b, d, f): (a) absence of TET on the linearized branch $S111++-$; (b) initial TET through nonlinear beats, followed by inphase TET on $S111+++$; (c, d) WT spectra of $v(t) - x_2(t)$ superposed to the FEP; (e, f) percentages of instantaneous total energy in the NES.

tional dashpot of constant c_{12} is placed between the two masses m_1 and m_2 of the LO. The numerical values of the system parameters were selected to be identical to the ones of an experimental fixture (discussed in Section 3.5.1.3), and are listed in Table 3.4. These parameters were identified using experimental modal analysis and the restoring-force technique (see Section 3.5.1.3). An impulsive force in the form of a half-sine pulse of duration 0.01 s is applied to mass m_1 of the LO; the

Table 3.4 System parameters of the experimental three-DOF system (Figure 3.96).

Parameter	Value
m_1	0.6285 kg
m_2	1.213 kg
ε	0.161 kg
k_1	420 N/m
k_2	0 N/m
k_{12}	427 N/m
C	4.97×10^6 N/m^3
c_1	0.05 to 0.1 Ns/m
c_2	0.5 to 0.9 Ns/m
c_{12}	0.2 to 0.5 Ns/m
c_v	0.3 to 0.35 Ns/m

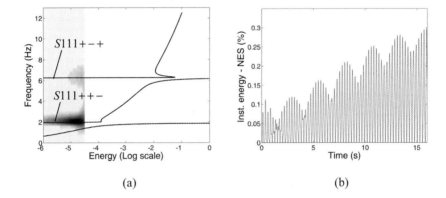

(a) (b)

Fig. 3.87 Damped response for single half-sine force at mass 1 m, with peak 1 N and duration 0.01 s: (a) WT spectrum of $v(t) - x_2(t)$ superposed to the FEP, (b) percentage of instantaneous total energy in the NES.

peak amplitude of the applied impulse was selected in the range 1–40 N to highlight the qualitatively different damped transitions and energy exchanges taking place at different energy levels.

In Figure 3.87 we depict the damped responses for excitation of mass m_1 with a half-sine force with peak equal to 1 N. Although both linear modes participate (at least initially) in the response, the contribution of the in-phase linear mode is dominant and more persistent (see Figure 3.87a). It is clear that the weakly nonlinear damped NNM $S111++-$ is mainly excited in this case, so the response remains localized to the LO and not more than 0.3% of the instantaneous total energy is transferred to the NES at any given time. As a result, negligible TET takes place in this case. Note that there is also a small contribution from the higher weakly nonlinear damped NNM $S111+-+$ but this does not affect significantly TET in this case.

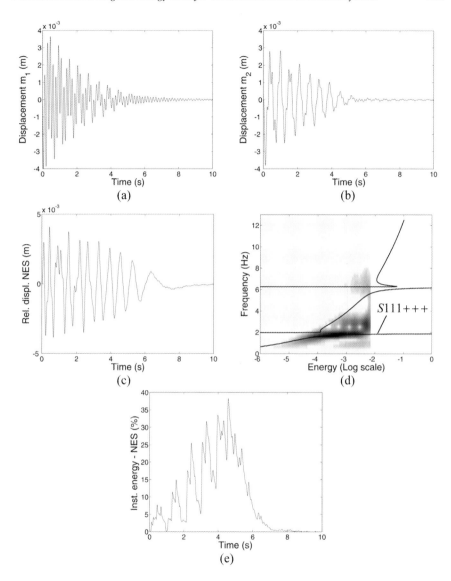

Fig. 3.88 Damped response for single half-sine force at mass m_1, with peak 15 N and duration of 0.01 s: (a–c) transient responses, (d) WT spectrum of $v(t) - x_2(t)$ x t - superposed to the FEP, (e) percentage of instantaneous total energy in the NES.

By increasing the forcing peak to 15 N (see Figure 3.88), the initial energy of the system exceeds the critical threshold for in-phase TET (see the FEP of Figure 3.80). The branch $S111+++$ is excited in this case, and the instantaneous total energy in the NES remains below 40% of the total energy of the system at any given instant of the motion. After $t = 5.5$ s, the participation of the in-phase mode in the system

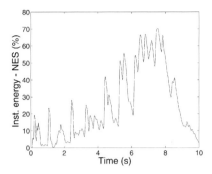

Fig. 3.89 Damped response for single half-sine force at mass m_1, with peak 27 N and duration 0.01 s: percentage of instantaneous total energy in the NES.

response is negligible, a sign that a significant portion of the energy contained in this mode has been transferred to and dissipated by the NES. Higher-frequency components are present in the relative displacement across the nonlinear spring $v(t) - x_2(t)$ (see Figure 3.88c), but these are mainly non-dominant harmonics of the damped response and do not correspond to a nonlinear resonance interaction of the NES with the out-of-phase linear mode. Hence, the energy initially imparted to the out-of-phase linear mode remains in that mode, and is dissipated by the dampers of the LO; this explains the relatively weak TET evidenced for this force level.

The damped dynamics remains qualitatively unchanged until the force peak reaches 27 N, where the percentage of instantaneous energy transferred to the NES reaches levels of up to 70% (Figure 3.89).

A qualitatively different picture of the damped dynamics, however, occurs when the force peak increases to 28 N (see Figure 3.90). This is due to the fact that for this level of impulsive force the initial energy of the system exceeds the threshold for occurrence of out-of-phase TET in the system (see Figure 3.80). From the numerical results of Figure 3.90 it is clear that in this case the damped dynamics possess two distinct regimes. In the initial regime of the motion ($0 < t < 2$ s) the NES engages in 1:1:1 TRC with the high-frequency out-of-phase linear mode (see Figure 3.90d) as it traces the NNM branches $S111+-+$ and $S111+--$. During this initial stage of the dynamics there occurs strong out-of-phase TET at a fast time scale, so that at $t \approx 1$ s, the NES carries 89% of the instantaneous total energy, and the participation of the out-of-phase linear mode in the damped response drastically decreases with time as it looses energy to the NES. We conclude that in the initial stage of the motion the NES extracts energy from the out-of-phase linear mode and locally dissipates it without 'spreading it back' to the LO. In terms of the previously introduced notation out-of-phase fundamental TET takes place during this initial stage of the damped dynamics, and the motion resembles that depicted in Figure 3.85. In that case, however, at the later stage of the motion the response underwent a transition to a low-frequency subharmonic tongue, whereas in the present case a different damped transition follows after the initial excitation of $S111+--$.

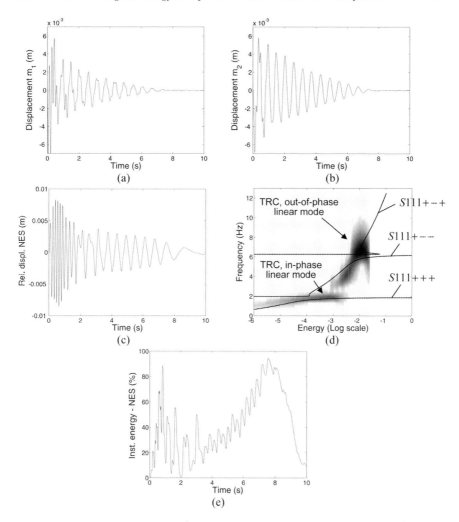

Fig. 3.90 Damped response for single half-sine force at mass m_1, with peak 28 N and duration 0.01 s : (a–c) transient responses, (d) a resonance capture cascade (RCC) in the WT spectrum of $v(t) - x_2(t)$ superposed to the FEP, (e) percentage of instantaneous total energy in the NES.

Indeed, for $t > 2$ s there occurs a damped transition to the damped NNM $S111{+}{+}{+}$, as the NES escapes TRC with the out-of-phase linear mode and engages in TRC with the in-phase linear mode of the LO; as a result, starting from $t = 2$ s the NES starts extracting energy from the in-phase linear mode and, from $t = 3.5$ s strong in-phase TET to the NES occurs, with the instantaneous total energy in the NES reaching levels of 90% of total instantaneous energy of the system.

This is an example of occurrence of a *resonance capture cascade (RCC)*, i.e., of a *sequential transient resonance interaction of the NES with both modes of the*

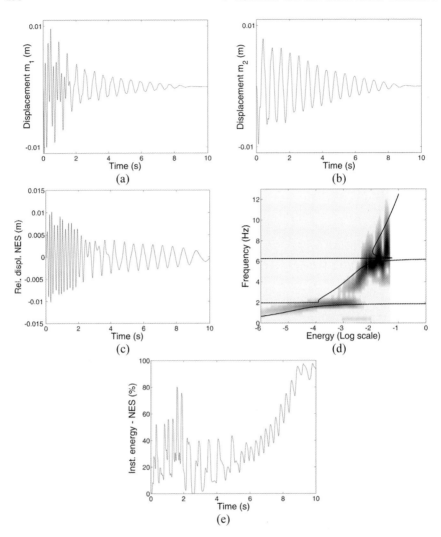

Fig. 3.91 Damped response for single half-sine force at mass m_1, with peak 40 N and duration 0.01 s: (a–c) transient responses, (d) a resonance capture cascade WT spectrum of $v(t) - x_2(t)$ superposed to the FEP, (e) percentage of instantaneous total energy in the NES.

primary system. The NES first extracts and dissipates almost the entire energy of the out-of-phase linear mode, before engaging in resonance and extracting energy from the in-phase mode linear mode. What triggers the RCC is the dependence of the instantaneous frequency of the NES on its energy, and, more importantly, the lack of a preferential resonance frequency of the NES due to its essential stiffness nonlinearity. It follows that depending on its instantaneous energy, the NES is capable of resonantly interacting with both linear modes, extracting energy from the

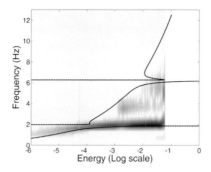

Fig. 3.92 Damped response for single half-sine force at mass m_1, with peak 6 N and duration 0.25 s: WT spectrum of $v(t) - x_2(t)$ superposed to the FEP.

higher-frequency mode before engaging the lower frequency one. What is especially notable is that *the process of RCC is adaptive and purely passive*, as the NES 'tunes itself' with the most highly energetic linear modes irrespective of their frequencies, before making a transition to modes with lower energies. RCC gives rise to multi-frequency TET from the LO to the NES, which becomes increasingly more broadband as the number of linear modes participating in the RCC increases (see Figure 3.90c). We emphasize *the capacity of the NES to engage in TRC with modes of the primary system at arbitrary frequency ranges* (provided, of course, that these modes do not posses nodes close to the point of attachment of the NES), *as this underlines the broadband feature of nonlinear TET; this is qualitative different from the narrowband action of the classical linear vibration absorber, and is a feature of the NES that renders it especially suitable for practical applications*. Moreover, the phenomenon of RCC is a distinct feature of MDOF LOs with attached NESs, as it cannot be realized in the two-DOF system examined in previous sections.

Figure 3.91 proves that RCCs are robust and persist for higher peak force amplitudes. For the increased impulsive level of 40 N considered in that Figure, the initial TRC of the NES dynamics with the out-of-phase linear mode (resulting in suppression of the out-of-phase linear mode during the first few cycles of the damped response), and the subsequent damped transition to TRC with the in-phase linear mode are even more evident.

Finally, in Figure 3.92 we compare the damped transitions of the previous case (force peak of 40 N and duration 0.01 s) to the ones occurring for an impulsive force of longer duration (0.25 s) but smaller peak (6 N) so that the total initial energy imparted to the system by the impulse remains constant. We note that due to the increased peak duration, the participation in the damped response of the out-of-phase linear mode drastically decreases (compare Figures 3.91 and 3.90d), so that in-phase TET occurs from the beginning of the motion and no RCC occurs. This case is similar to the case presented in Figure 3.84, where direct excitation of the backbone branch $S111+++$ was considered (the only difference being the stronger higher harmonics that occur in the present case).

From the previous results we conclude that the duration and amplitude of the applied half-sine pulse have an important influence on the damped dynamics and TET in the system. Depending on these parameters (but also on damping), different branches of the FEP may be excited or traced during the damped nonlinear transitions, affecting the strength of TET.

To provide an additional example of the complex, multi-frequency transitions that can take place in coupled oscillators with essentially nonlinear local attachments, we consider the following alternative three-DOF system (Kerschen et al., 2006a):

$$\ddot{x}_2 + \omega_0^2 x_2 + \lambda_2 \dot{x}_2 + d(x_2 - x_1) = 0$$

$$\ddot{x}_1 + \omega_0^2 x_1 + \lambda_1 \dot{x}_1 + \lambda_3(\dot{x}_1 - \dot{v}) + d(x_1 - x_2) + C(x_1 - v)^3 = 0$$

$$\varepsilon \ddot{v} + \lambda_3(\dot{v} - \dot{x}_1) + C(v - x_1)^3 = 0 \tag{3.189}$$

with parameters $\omega_0^2 = 136.9$, $\lambda_1 = \lambda_2 = 0.155$, $\lambda_3 = 0.544$, $d = 1.2 \times 10^3$, $\varepsilon = 1.8$, and $C = 1.63 \times 10^7$, corresponding the linearized natural frequencies $\omega_1 \equiv 2\pi f_1 = 11.68$ rad/s and $\omega_2 \equiv 2\pi f_2 = 50.14$ rad/s. In Figure 3.93a we present the relative response $v(t) - x_1(t)$ of the system for initial displacements $x_1(0) = 0.01$, $x_2(0) = v(0) = -0.01$ and zero initial velocities. The multi-frequency content of the transient response is evident, and is quantified in Figure 3.93b, where the instantaneous frequency of the time series is computed by applying the numerical Hilbert transform (Huang et al., 1998).

As energy decreases due to damping dissipation, an interesting RCC takes place, involving as many as eight TRCs. The complexity of the RCC is evidenced by the fact that of these eight TRCs only two (labeled IV and VII in Figure 3.93b) involve the linearized in-phase and out-of-phase modes of the linear oscillator, while the remaining ones correspond to essentially nonlinear interactions of the NES with a number of low- and high-frequency nonlinear modes of the system (which apparently have no analogues in the linearized dynamics). During each TRC there occur energy exchanges between the NES and the the nonlinear mode involved in the resonance capture, after which escape from TRC occurs and the NES engages in transient resonance with the next mode of the series. Clearly, the main 'tuning' parameter that controls this purely passive RCC is the instantaneous energy of the system and its rate of decrease due to damping dissipation. In essence, *the NES acts as a passive, broadband boundary controller*, absorbing, confining and eliminating vibration energy from the linear oscillator.

In the two additional applications that follow, we demonstrate the occurrence of RCCs in coupled MDOF oscillators with essentially nonlinear attachments. In the first application we consider the six-DOF system

$$\ddot{x}_1 + 0.014\dot{x}_1 + 2x_1 - x_2 = 0$$

$$\ddot{x}_2 + 0.014\dot{x}_2 + 2x_2 - x_1 - x_3 = 0$$

$$\ddot{x}_3 + 0.014\dot{x}_3 + 2x_3 - x_2 - x_4 = 0$$

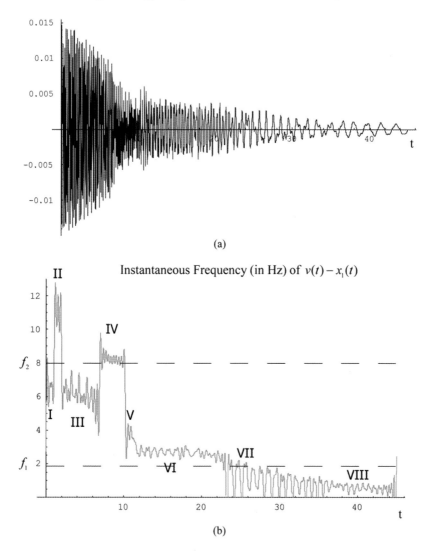

Fig. 3.93 Resonance capture cascade (RCC) in the damped transient dynamics of system (3.189): (a) relative response $v(t) - x_1(t)$, (b) instantaneous frequency of $v(t) - x_1(t)$ computed by the Hilbert transform (eight TRCs indicated).

$$\ddot{x}_4 + 0.014\dot{x}_4 + 2x_4 - x_3 - x_5 = 0$$
$$\ddot{x}_5 + 0.0141\dot{x}_5 - 0.0001\dot{v} + 2x_5 - x_4 + (x_5 - v)^3 = 0$$
$$0.05\ddot{v} + 0.0001(\dot{v} - \dot{x}_5) + (v - x_5)^3 = 0 \qquad (3.190)$$

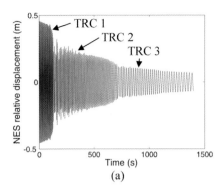 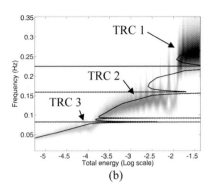

Fig. 3.94 RCC in the damped dynamics of the six-DOF system (3.190) following direct excitation of the fourth linear mode: (a) relative response $v(t) - x_5(t)$, (b) WT spectrum of $v(t) - x_5(t)$ superposed to the FEP.

with initial excitation of only the fourth mode of the linear primary system. In Figure 3.94 we depict the relative response $v(t) - x_5(t)$, along with its WT spectrum superimposed to the FEP of the underlying Hamiltonian system of (3.190); for clarity, only the first four linear modes are depicted in the FEP.

We note that a RCC occurs in this case, leading to multi-frequency TET from the primary system to the NES. After an initial TRC of the NES dynamics with the fourth linear mode (labeled TRC 1 in Figure 3.94), a damped transition occurs after which the NES engages in TRC with the second linear mode (TRC 2). At a later stage of the dynamics a second damped transition occurs leading to final TRC of the NES dynamics with the first linear mode (TRC 3). This application illustrates clearly the usefulness of the utilization of combined WT spectra and FEPs as tools for interpreting useful nonlinear transitions.

In the next application we consider an $(N + 1)$-DOF linear chain of coupled oscillators (the primary system) with a grounded NES (Configuration I – see Section 3.1) attached to its end (Vakakis et al., 2003). Each linear oscillator of the chain possesses unit mass and grounding stiffness ω_0^2, and is coupled to its neighboring oscillators by linear stiffnesses of characteristic d. The primary system possesses $(N+1)$ mass-normalized eigenvectors $\phi^{(i)} = [\phi_0^{(i)} \ldots \phi_N^{(i)}]^T$ and (N+1) corresponding distinct eigenfrequencies ω_i, $i = 0, 1, \ldots, N$. The responses of the oscillators of the primary system are denoted by $x_0(t), \ldots, x_N(t)$, where $x_0(t)$ is the response of the point of attachment to the NES. These responses are then expressed in modal series:

$$x_i(t) = \sum_{k=0}^{N} \phi_i^{(k)} a_k(t), \quad i = 0, 1, \ldots, N$$

We express the equations of motion of the system using modal coordinates for the primary system

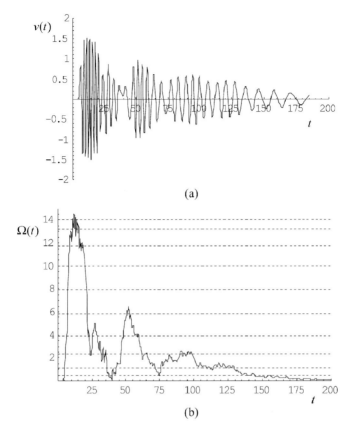

Fig. 3.95 RCC in the damped dynamics of the 11-DOF system (3.191) with $N = 9$: (a) multi-frequency response of the NES, (b) instantaneous frequency $\Omega(t)$ of the NES versus time (the 10 linear modes of the chain are denoted by dashed lines).

$$\ddot{v}(t) + Cv^3(t) + \varepsilon\lambda\dot{v}(t) + \varepsilon\left[v(t) - \sum_{k=0}^{N}\phi_0^{(k)}a_k(t)\right] = 0$$

$$\ddot{a}_m(t) + \omega_m^2 a_m(t) + \varepsilon\lambda\dot{a}_m(t) + \varepsilon\left[\sum_{k=0}^{N}\phi_0^{(k)}\phi_0^{(m)}a_k(t) - \phi_0^{(m)}v(t)\right] = 0$$

(3.191)

with $m = 0, 1, \ldots, N$.

For the numerical simulation we considered a chain of ten linear oscillators ($N = 9$) with parameters $\omega_0^2 = 0.4$, $d = 3.5$, $C = 5.0$, $\lambda = 0.5$, $\varepsilon = 0.1$ and initial conditions $v(0) = \dot{v}(0) = 0$, $x_m(0) = 0$, $m = 0, 1, \ldots, 9$ and $\dot{x}_m(0) = 0$, , $m = 0, 1, \ldots, 8$, $\dot{x}_9(0) = 70$. This corresponds to an impulsive excitation being applied at $t = 0$ to the oscillator of the chain most distant from the NES. In Figure 3.95 we present the transient response of the attachment $v(t)$, together with its instantaneous frequency of oscillation $\Omega(t)$ versus time. Note the strong RCC taking place in the damped dynamics involving as many as six of the linearized modes of the chain, including both modes located at the boundaries of the frequency spectrum of the chain. This application further demonstrates the capacity of the NES for broadband TET from the primary system.

The final series of numerical simulations demonstrates the superior performance of an essentially nonlinear attachment (NES) as passive absorber of shock energy of a linear MDOF system of coupled oscillators, when compared to the classical linear absorber (or tuned mass damper – TMD). To this end, we consider the following eleven-DOF system with a strongly nonlinear end attachment (Ma et al., 2008):

$$\varepsilon \ddot{v} + \varepsilon \lambda (\dot{v} - \dot{x}_0) + C(v - x_0)^3 = 0$$
$$\ddot{x}_0 + \varepsilon \lambda \dot{x}_0 + \omega_0^2 x_0 - \varepsilon \lambda (\dot{v} - \dot{x}_0) - C(v - x_0)^3 + d(x_0 - x_1) = 0$$
$$\ddot{x}_j + \varepsilon \lambda \dot{x}_j + \omega_0^2 x_j + d(2x_j - x_{j-1} - x_{j+1}) = 0, \quad j = 1, \ldots, 8$$
$$\ddot{x}_9 + \varepsilon \lambda \dot{x}_9 + \omega_0^2 x_9 + d(x_9 - x_8) = 0 \tag{3.192}$$

In this example we consider an ungrounded lightweight NES (of Configuration II – see Section 3.1) by assuming that $0 < \varepsilon \ll 1$. We assume that the system is initially at rest, and an impulse of magnitude X is applied at $t = 0$ to the left boundary of the linear chain, corresponding to initial conditions, $v(0) = \dot{v}(0) = 0$; $x_p(0) = 0$, $p = 0, \ldots, 9$; $\dot{x}_9(0+) = X$; and $\dot{x}_k(0) = 0, k = 0, \ldots, 8$.

To study TET efficiency, i.e., the capacity of the NES to passively absorb and locally dissipate impulsive energy from the linear chain, we employ the instantaneous and asymptotic energy dissipation measures (EDMs) defined by relations (3.4), suitably modified for system (3.192):

$$E_{\text{NES}}(t) = \frac{\lambda_2 \int_0^t [\dot{v}(\tau) - \dot{x}_0(\tau)]^2 d\tau}{(X^2/2)} \times 100, \quad E_{\text{NES},t \gg 1} = \lim_{t \gg 1} E_{\text{NES}}(t)$$

In Figure 3.96a we present the plot of the EDM $E_{\text{NES},t \gg 1}$ as function of the stiffness characteristic C of the NES, for impulse strength $X = 4.3$, system parameters $\omega_0^2 = 1.0$, $d = 2.0$, and two values of damping, namely, $\varepsilon \lambda = 0.0125$, and 0.025 (Ma et al., 2008). For comparison, we also depict the corresponding EDMs for a chain with a linear TMD attached at its end, with identical parameter values. Clearly, the TMD proves to be effective only in a narrow band of small stiffness values, i.e., in the neighborhood of resonance with the chain. On the contrary, *the NES proves to be more effective than the TMD, since it is capable of passively absorbing a significant portion of the impulsive energy of the chain over a wide range of values of* C; this is due to the capacity of the NES to engage in resonance capture and passively absorb energy from *any* of the modes of the chain, irrespective of their actual natural frequencies. We note that as much as 37% of input energy is passively

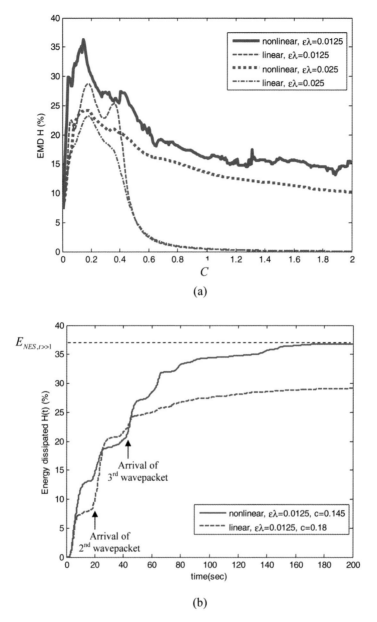

Fig. 3.96 Comparison of the TET for the case of linear (TMD) and strongly nonlinear NES) attachments: (a) EDM $E_{\text{NES},t\gg 1}$ for varying stiffness C, and two damping values; (b) EDM $E_{\text{NES}}(t)$ for specific values of C and fixed damping $\varepsilon\lambda = 0.0125$.

absorbed and eventually dissipated by the NES, and that, even away from the region of optimal TET, the NES is capable of significant TET. In Figure 3.96b we depict the instantaneous EDM $E_{\text{NES}}(t)$ for the case of optimal TET and $\varepsilon\lambda = 0.0125$, and compare it to the corresponding EDM for the linear TMD with optimal parameters. It is interesting to note that the sequence of early-time arrivals to the attachments of reflected wavepackets from the boundaries of the chain are associated with sudden increases of the rates of energy dissipation.

In Ma et al. (2008) the capacity for TET of the system (3.192) is related to the shapes and energies of the underlying proper orthogonal modes (POMs) of the transient dynamics (Cusumano et al., 1994; Georgiou et al., 1999; Azeez and Vakakis, 2001; Ma and Vakakis; 1999). It is shown that enhanced TET is related to excitation of dominant highly energetic POMs that localize to the NES. This observation is then used for constructing accurate low-dimensional reduced-order models for the TET dynamics.

3.5.1.3 Experimental Demonstration of Multi-Frequency TET

Although experimental TET results will be presented in detail in Chapter 8, in this section we provide some preliminary experimental evidence in support of the previous theoretical findings. The experimental measurements reported here were performed using the fixture depicted in Figure 3.97, composed of a two-DOF linear oscillator (the primary system) coupled to an essentially nonlinear ungrounded SDOF attachment (an NES of Configuration II). The primary system consists of two cars made of aluminum angle stock which are supported on a straight air track (that reduces friction forces during the oscillation). The NES consists of a shaft supported by two linear bearings; steel plates on the shaft clamp two steel wires configured with practically no pretension, realizing the essential cubic stiffness nonlinearity C (see Section 2.6 for a discussion on the practical realization of essential cubic stiffness nonlinearity, and also Chapter 8). The wires are connected to the primary system through clamps at their outer ends.

A short half-sine force pulse representative of a broadband input is applied to the left car (of mass m_1) of the primary system (see Figure 3.97), and the damped responses of the three oscillators are measured using accelerometers. Estimates of velocities and displacements are obtained by numerically integrating the measured acceleration time series, and the resulting signals are high-pass filtered to remove spurious components introduced by the integration procedure.

The parameters of the experimental fixture were measured before the experimental tests. Prior to system identification, the cars of the primary system and the NES were weighed as $m_1 = 0.6285$ kg, $m_2 = 1.213$ and $\varepsilon = 0.161$ kg, respectively, which implies a low mass ratio equal to 8.7%. Experimental modal analysis was then carried out to measure the stiffness and damping parameters of the integrated three-DOF experimental system. First, the primary system was disconnected from the NES, and experimental modal analysis was performed using the stochastic subspace identification method (Van Overschee and De Moor, 1996) to provide the

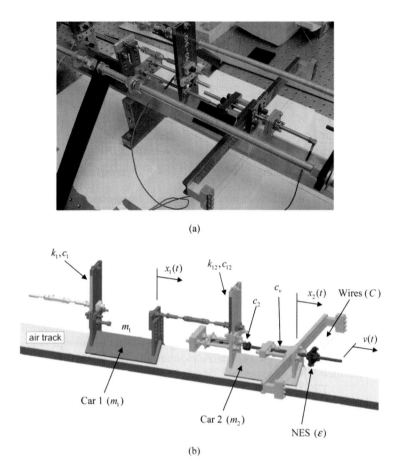

Fig. 3.97 Experimental fixture: (a) NES, (b) schematic of the two-DOF primary system and the SDOF NES.

two natural frequencies estimates of 1.95 Hz and 6.25 Hz, respectively. Because the masses of the primary system were known, the stiffness and damping parameters k_1, k_{12}, c_1, c_2, and c_{12} could be deduced from this experimental modal analysis, and are listed in Table 3.4.

In the second step of modal analysis the primary system was clamped, an impulsive force was applied to the NES using a modal hammer, and the NES acceleration and applied force were measured. The restoring force surface method (Masri and Caughey, 1979) was then used to estimate the coefficient of the essential nonlinearity C and the damping coefficient c_v of the NES. For further details about the procedure, the reader is referred to Chapter 8. The identified system parameters of the experimental fixture are listed in Table 3.4. Damping estimation is a difficult problem in this fixture due to the presence of several ball joints and bearings, and of

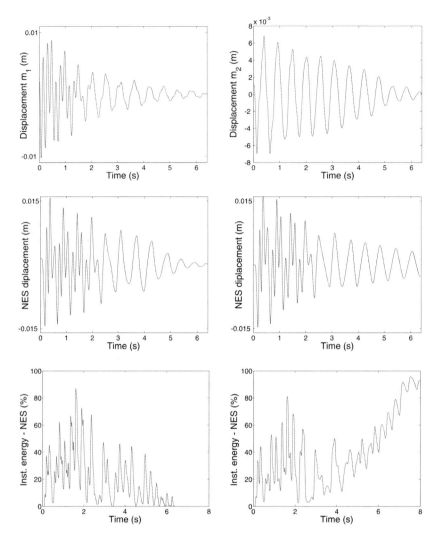

Fig. 3.99 Experimental RCC, Case IV: (a–c) measured responses; (d) theoretically predicted NES response; (e, f) measured and theoretically predicted percentage of instantaneous total energy in the NES.

dicted by the numerical simulations, the out-of-phase damped NNM $S111+--$ is excited in the initial stage of the motion (leading to out-of-phase fundamental TET), followed by a transition of the dynamics to TRC with the in-phase damped NNM $S111+++$ (and in-phase fundamental TET). For the system with increased mass ratio (Figure 3.100b) a similar, albeit weaker RCC is observed in the experimental measurements.

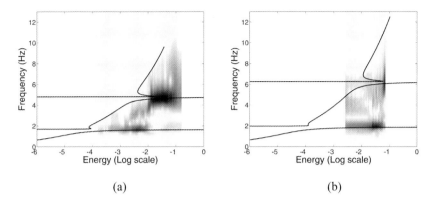

Fig. 3.100 RCCs in the WT spectra of experimental relative responses $v(t) - x_2(t)$ superimposed to the FEP: (a) 6%; (b) 8.7% mass ratio, peak duration of 0.15 s.

3.5.2 Semi-Infinite Chain of Linear Oscillators with an End SDOF NES

In this final section we study the dynamics of a semi-infinite linear chain of coupled oscillators with an essentially nonlinear attachment (NES) at its boundary. Considering first the undamped system we analyze families of localized nonlinear standing waves situated inside the lower or upper attenuation zones of the dynamics of the linear chain, with energy being predominantly confined to the NES. In addition, we estimate the energy radiated from the NES back to the chain, when the NES is excited under non-resonant conditions by wavepackets with dominant frequencies inside the propagation zone of the dynamics of the chain. We show that *in this system TET from the semi-infinite chain to the NES is possible even in the absence of damping*. The TET dynamics, however, is qualitatively different in this case: instead of TET through TRCs as in the case of finite-DOF weakly damped oscillators considered previously, *TET in the undamped infinite-DOF system relies on the excitation of in-phase standing waves localized to the NES*. Passive TET from the semi-infinite linear chain to the NES is confirmed numerically. The analysis of the undamped system follows closely the work by Manevitch et al. (2003).

Then we analyze the weakly damped semi-infinite linear chain with a weakly damped essentially nonlinear oscillator attached (Vakakis, 2001). Using a reduction approach, we reduce the dynamics to a complex integro-differential equation and then analyze TET using the complexification-averaging approach (CX-A). We show that TET in the weakly damped system is generated by TRCs as in the case of finite-dimensional discrete oscillators discussed in previous sections.

3.5.2.1 Dynamics of the Chain-NES Interaction

The dynamics of linear or nonlinear periodic chains with local attachments (or 'defects') is a research area with many interesting applications, such as in the areas of optical and magneto-optical waveguide periodic arrays, semiconductor superlattices, layered composite media, micro- or nano-lattices used as thermal barriers, in photonic band-gap materials (photonic crystals), and bio-molecular engines (see, for example, the works by Chen and Mills, 1987; Eggleton et al., 1996; Akozbek and John, 1998). Gendelman and Manevitch (2000) examined the dynamics of a semi-infinite string with a strongly nonlinear oscillator attached to its end, and studied energy transfer from the string to the attached oscillator through impeding short rectangular pulses. They found that excitation of vibrations in the oscillator was possible through this nonlinear interaction. Lazarov and Jensen (2007) studied the influence of stiffness nonlinearities on the filtering properties (i.e., the low-frequency bands) of infinite linear chains with attached nonlinear oscillators; they found that the position of low-frequency bands in these systems depended on the form of the nonlinearity and the level of energy of the motion. Goodman et al. (2004) analyzed the dynamic interaction of a nonlinear Schrödinger soliton with a local defect and proposed a mechanism of resonance energy transfer from the impeding soliton to a nonlinear standing wave localized at the defect. Additional works (Kivshar et al., 1990; Forinash et al., 1994; Goodman et al., 2002a, b) examined nonlinear interactions of standing or traveling waves in infinite nonlinear media with local defects.

The system under consideration is a semi-infinite chain of coupled linear oscillators, whose free end is weakly coupled to an essentially nonlinear attachment. We wish to study the possibility of passive TET from the chain to the nonlinear attachment, which then acts, in essence, as an NES. Each oscillator of the chain is grounded and possesses only next-neighbor interactions. Assuming no damping in the system, the set of equations governing the dynamics is given by

$$\ddot{x}_k + c^2(2x_k - x_{k-1} - x_{k+1}) + \omega_0^2 x_k = 0, \quad k < 0$$
$$\ddot{x}_0 + c^2(x_0 - x_{-1}) + \varepsilon(x_0 - v) + \omega_0^2 x_0 = 0$$
$$\ddot{v} + 8av^3 - \varepsilon(x_0 - v) = 0 \quad (3.193)$$

where x_k denotes the response of the k-th oscillator of the linear chain, v the response of the NES, c^2 the linear coupling stiffness between adjacent oscillators, and ω_0^2 the linear grounding stiffness of each oscillator. The dimensionless perturbation parameter $0 < \varepsilon \ll 1$ scales the weak coupling between the linear chain and the NES, and the parameter a denotes the strength of the essential (nonlinearizable) stiffness nonlinearity of the attachment. Note that in this case we consider a grounded form of NES (of the type presented in Figure 3.1 – Configuration I), and the mass of the NES is not assumed to be small (as in previous sections). Instead, in the following analysis the small parameter characterizes the weak coupling between the semi-infinite chain and the NES.

Before discussing the chain-NES dynamic interaction, we examine briefly the dynamics of the infinite chain with no boundaries and no nonlinear attachment. The dispersion relation of the infinite linear chain is composed of two *attenuation zones* (AZs) and a single *propagation zone* (PZ) in the frequency domain (Brillouin, 1953; Mead, 1975). In the AZs the chain supports two families of standing waves with exponentially decaying envelopes, which represent near field solutions of the infinite chain. The lower AZ is in the frequency range $\omega \in [0, \omega_0)$, whereas the upper AZ extends up to arbitrarily large frequencies, $\omega \in (\sqrt{\omega_0^2 + 4c^2}, \infty)$. In the PZ, $\omega \in (\omega_0, \sqrt{\omega_0^2 + 4c^2})$, the infinite chain supports two families of traveling waves that propagate unattenuated in opposite directions of the chain. It is well known that *energy through the chain can only propagate by means of traveling waves*, i.e., only with frequencies inside the PZ. The bounding frequencies $\omega_{b1} = \omega_0$ and $\omega_{b2} = \sqrt{\omega_0^2 + 4c^2}$ that separate the two AZs from the PZ correspond to in-phase and out-of-phase normal mode oscillations (i.e., synchronous non-decaying standing waves) of the infinite chain (Mead, 1975).

Now suppose that the integrated semi-infinite chain-NES system is initially at rest, and at $t = 0$ an impulse $F\delta(t)$ is applied to an oscillator of the chain. Then, the motion of the system at $t = 0+$ is a conservative free oscillation and the energy transfer through the chain and to the nonlinear oscillator may be approximately analyzed with the help of linear theory. The first basic problem of the chain-NES dynamics is to establish the type of excitation of the NES by the chain. Clearly, the excitation of the nonlinear oscillator is caused by an initial right-going traveling wave propagating through the chain; depending on the form of this wave, the initial chain-NES dynamic interaction may occur under resonance or non-resonance conditions.

Resonance interaction is most probable if the time and distance needed for the wave to travel through the chain and impede to the NES suffice for the formation of a wave packet with primary frequency $\Omega > \omega_0$. Then the excitation of the NES occurs approximately under condition of 1:1 resonance. In that case, the chain may be approximately simulated as a single particle acting on the nonlinear oscillator with prescribed force, i.e., possessing certain amplitude and frequency, and applied during a known time interval. All these parameters may be obtained simply by solving the linear problem for the chain. On the other hand, *non-resonant interaction* between the chain and the NES corresponds to the situation when the wave packet disturbance in the chain does not have sufficient time and space to form into a cohesive wave form, and, as a result the force that excites the nonlinear oscillator is non-harmonic.

The next basic problem of the chain-NES dynamics focuses on the radiation (backscattering) of energy from the NES back to the chain after the initial wave has impeded to it. This process is the most interesting from an analytical point of view, and as shown below, can be divided into two essentially different parts: (a) the transient radiation of excess energy from the NES back to the chain through traveling or near-field waves; and (b) the formation at the NES of a localized standing wave mode. In the following analysis we discuss these issues separately.

We first consider radiation (backscattering) of energy from the NES to the semi-infinite chain through *traveling waves*, i.e., waves with frequencies in the PZ. Specifically, we consider the state of the system after the main impulsive excitation of the chain commences. The NES is excited with a wave packet with predominant frequency $\Omega > \omega_0$ (i.e., in the PZ on the linear chain), as these are the only waves that can travel from the source of the excitation through the chain and impede to the NES; moreover, this frequency most probably belongs to the zone of moderate wavenumbers corresponding to the maximum of group velocity. Therefore, under conditions of 1:1 resonance the energy of the NES is radiated back to the chain in the diapason of moderate wavenumbers, and for qualitative purposes the energy radiation may be studied in the continuum approximation. These assumptions regarding the radiation process will be proved and validated *a posteriori* by the derived results.

In this case we propose the following ordering of the variables of system (3.193), $v = O(1)$, $x_k = O(\varepsilon)$, $k = 0, -1, -2, \ldots$. Hence, to a first approximation we consider the following continuum approximation of equations (3.193):

$$\frac{\partial^2 x(s,t)}{\partial t^2} - c^2 r_0^2 \frac{\partial^2 x(s,t)}{\partial s^2} + \omega_0^2 x(s,t) \approx 0, \quad s \leq 0$$

$$c^2 r_0 \frac{\partial x(0,t)}{\partial s} \approx \varepsilon v(t)$$

$$v''(t) + 8\alpha v^3(t) \approx 0 \tag{3.194}$$

where r_0 is the distance between oscillators, so that the k-th oscillator corresponds to the position $s = kr_0$, $k = 0, -1, -2, \ldots$ of the one-dimensional continuum, and primes denote differentiation with respect to s. In deriving (3.194) we replaced the infinite set of variables $x_k(t)$, $k \leq 0$ by the continuous variable $x(s,t)$, $s \leq 0$, and the semi-infinite set of ordinary differential equations of (3.193) by a single partial differential equation [the first of relations (3.194)].

The last equation in (3.194) describes (to a first approximation) a vibration of the nonlinear oscillator with constant amplitude and frequency. In fact, this is only an approximation since in actuality the amplitude and frequency of the nonlinear attachment varies slowly with time due to energy loss by energy radiation to the chain. However, it will be shown that this radiation effect is of order $O(\varepsilon^2)$, and, therefore, the variations of the amplitude and the frequency of the nonlinear oscillator are nearly adiabatic up to $O(\varepsilon^2)$.

The flow of energy through the chain in the continuum limit may be estimated by recalling that the energy stored in the spatial interval $a < s < b$ of the chain is computed as

$$E_{ab}(t) = \frac{1}{2} \int_a^b \left[\left(\frac{\partial x}{\partial t}\right)^2 + c^2 r_0^2 \left(\frac{\partial x}{\partial s}\right)^2 + \omega_0^2 x^2 \right] ds \tag{3.195}$$

so the flow of energy in the chain is approximated as

$$\frac{dE_{ab}}{dt} = \int_a^b (\dot{x}\ddot{x} + c^2 r_0^2 x' \dot{x}' + \omega_0^2 x \dot{x}') \, ds$$

$$= \int_a^b \left[\dot{x}\ddot{x} + c^2 r_0^2 x' \dot{x}' + (c^2 r_0^2 x'' - \ddot{x}) \dot{x} \right] ds$$

$$= c^2 r_0^2 \int_a^b (x' \dot{x})' \, ds = c^2 r_0^2 \left(x' \dot{x} \big|_{x=b} - x' \dot{x} \big|_{x=a} \right) \quad (3.196)$$

Therefore, the rate of total energy radiated from the nonlinear oscillator back to the chain is estimated by setting $a = -\infty$, $b = 0$ in (3.196), and taking into account that (due to causality) the chain is motionless in the far field $s \to -\infty$:

$$\frac{dE_{\text{chain}}}{dt} = c^2 r_0^2 \, x'(0) \, \dot{x}(0) = -\frac{dE_{\text{NES}}}{dt} \quad (3.197)$$

The rate of energy loss of the NES due to radiation is the negative of the corresponding energy gain by the chain, which is a consequence of the lack of damping dissipation in system (3.193). We note that the energy contained in the NES depends only on its instantaneous frequency of oscillation, and this fact is crucial in our discussion. Indeed, considering the dominant harmonic component at frequency ω of the (approximately) periodic response $v(t)$ of the NES, the outgoing radiated harmonic traveling wave in the chain may be expressed as

$$x_\omega(s, t) \approx A_\omega e^{j(\omega t + \beta s)}, \quad \beta = (cr_0)^{-1} (\omega^2 - \omega_0^2)^{1/2}, \quad \omega \geq \omega_0,$$

where $j = (-1)^{1/2}$. Due to the fact that this is a traveling wave emanating from the NES due to energy backscattering, it propagates in the direction of decreasing negative s, i.e., away from the NES and towards the far field $s \to -\infty$. The amplitude of this wave may be computed from the second of equations (3.194),

$$A_\omega \approx \frac{-j \varepsilon Z_\omega}{c \sqrt{\omega^2 - \omega_0^2}} \quad (3.198)$$

where Z_ω is the amplitude of the harmonic of $v(t)$. Substituting this result into (3.197), and averaging over the period $T = 2\pi/\omega$, we derive the following approximate expression for the rate of energy radiation at frequency ω in the PZ of the linear chain:

$$\frac{dE_{\text{chain}}}{dt} \approx \frac{\varepsilon^2 \omega |Z_\omega|^2}{2c \sqrt{\omega^2 - \omega_0^2}}, \quad \omega \geq \omega_0 \quad (3.199)$$

Hence, energy radiation is indeed of $O(\varepsilon^2)$ which validates our previous assertions and assumptions. In actuality, the energy of the oscillator decreases slowly due to energy radiation back to the chain, and so does its instantaneous frequency of oscillation until it approaches the neighborhood of the lower bounding frequency $\omega_{b1} = \omega_0$. Clearly, expression (3.199) is not valid in the neighborhood of this bounding frequency, since the assumed scaling of v and x_k does not hold there; this means that as ω slowly decreases towards ω_{b1} the traveling wave *ansatz* becomes invalid since the dynamics of the system approach the qualitatively different state of 1:1 resonance, which should be considered separately.

The previous scenario is supported by the findings reported in Vakakis (2001) (and in Section 3.5.2.3), where numerical simulations of the dynamic interaction of a damped NES with a damped linear chain of coupled oscillators are presented. It is numerically shown (and analytically proven), that after some initial irregular transients (corresponding to the energy radiation phase described previously), 1:1 TRC between the in-phase normal mode of the chain (at frequency ω_0) and the NES takes place. During this TRC strong energy exchanges between the two systems occur.

The results (3.195–3.199) concerning monochromatic energy radiation from the nonlinear oscillator to the chain can be extended to the case of transient energy radiation. To show this, we Laplace-transform the first two linear equations of (3.194), assuming that the chain is initially at rest, and imposing the far field condition $\lim_{s \to -\infty} x(s, t) = 0$. This leads to the following expression for the Laplace transform $U(0, p) = \mathcal{L}[u(0, t)]$, where p is the Laplace transform variable,

$$U(0, p) = \frac{\varepsilon V(p)}{\varepsilon + c \left(p^2 + \omega_0^2\right)^{1/2}}$$

$$= \varepsilon V(p) \left[\frac{1}{c \left(p^2 + \omega_0^2\right)^{1/2}} - \frac{\varepsilon}{c^2 \left(p^2 + \omega_0^2\right)} + O(\varepsilon^2) \right] \quad (3.200)$$

where $V(p) = \mathcal{L}[v(t)]$. Inverse Laplace-transforming the above expression and substituting the result into the last of equations (3.194) we obtain the following nonlinear integro-differential equation governing the transient energy radiation from the NES back to the chain:

$$\ddot{v}(t) + 8\alpha v^3(t) =$$
$$- \varepsilon \left[v(t) - \frac{\varepsilon}{c} \int_0^t v(\tau) J_0 \left[\omega_0(t - \tau) \right] d\tau \right.$$
$$\left. + \frac{\varepsilon^2}{c^2 \omega_0} \int_0^t v(\tau) \sin \omega_0(t - \tau) d\tau + O(\varepsilon^3) \right] \quad (3.201)$$

In agreement with the previous simplifying analysis, the integral terms on the right-hand side that govern energy radiation to the chain are of $O(\varepsilon^2)$. In Section 3.5.2.3 we discuss in detail the solution of this integro-differential equation.

From the above discussion we conclude that after a wavepacket impedes to the NES, its energy is slowly radiated back to the chain in an $O(\varepsilon^2)$ nonlinear dynamic interaction, until the dynamics approaches a regime of 1:1 resonance close to the lower bounding frequency $\omega_{b1} = \omega_0$. The dynamics of this resonance interaction is studied in the next section.

3.5.2.2 Nonlinear Resonance Interactions and TET

We now focus in nonlinear resonance interactions occurring between the NES and the semi-infinite chain in the neighborhood of the lower bounding frequency of the PZ of the infinite chain. Later we will extend the analysis to resonance interactions occurring the the neighborhood of the upper bounding frequency.

We commence right from the beginning that the problem of resonance in the system under consideration is by no means trivial, as we deal with a problem possessing an infinite number of DOFs and a local essential (strong) nonlinearity; moreover, the transient nature of the examined dynamical interactions complicates even further the analysis. Since no common ways exist to proceed with this problem, we need to apply some simplifying propositions that will enable us to analytically approximate in a self-consistent way the dynamic phenomena under investigation. We follow closely the analysis by Manevitch et al. (2003). However, as in the previous section, the validity of the assumptions made has to be checked *a posteriori* when the analytical results are derived.

First, we assume that 1:1 resonance between the semi-infinite chain and the NES occurs at a frequency smaller than the lower bounding frequency $\omega_{b1} = \omega_0$, i.e., inside the lower AZ of the dispersion relation of the linear chain. It follows that the amplitudes of the responses of the oscillators of the chain decay exponentially with increasing distance from the NES. This basic simplifying assumption will be checked (and validated) through numerical simulations later. An additional simplification is achieved by supposing that the shape of this exponential amplitude decay is fairly approximated by a single exponent which is consistent with the dispersion relation of the linear chain,

$$x_j \approx x_0 e^{\kappa j}, \quad j \leq 0, \quad \omega_0^2 - \Omega^2 \approx 2c^2(\cosh \kappa - 1) \qquad (3.202)$$

where Ω denotes the *fast frequency* of oscillation of the oscillators of the chain [as explained below in relation (3.205)], and κ the frequency-dependent rate of exponential decay. It follows that, in contrast to the analysis of the previous section, we now seek *standing-wave solutions* localized to the NES. The assumption (3.202) introduces an approximation in the analysis, since it omits nonlinear effects in the decay rate which are present in the system; for an asymptotic study of near field solutions in nonlinear layered media we refer to Vakakis and King (1995).

Substituting (3.202) into (3.193) we reduce approximately the dynamics to a system of two coupled ordinary differential equations:

$$\ddot{x}_0 + x_0[c^2(1 - e^{-\kappa}) + \omega_0^2] + \varepsilon(x_0 - v) = 0$$
$$\ddot{v} + 8\alpha v^3 - \varepsilon(x_0 - v) = 0 \qquad (3.203)$$

This indicates that the problem of studying the resonance interaction of the NES with the semi-infinite chain can be reduced approximately to the simpler problem of resonance interaction between the NES and the nearest to it oscillator of the chain. Clearly, the biggest advantage gained by the above reduction is that the study of the resonance interaction may be performed by applying the CX-A method introduced in previous sections.

To this end, we introduce the complex variables $\psi_1 = \dot{x}_0 + j\omega x_0$ and $\psi_2 = \dot{v} + j\omega v$ with $\omega = \sqrt{c^2(1 - e^{-\kappa}) + \omega_0^2 + \varepsilon}$, which reduces (3.203) to the following set of first-order complex modulation equations:

$$\dot{\psi}_1 - (j\omega/2)(\psi_1 - \psi_1^*) - (j\omega/2)(\psi_1 - \psi_1^*) + (j\varepsilon/2\omega)(\psi_2 - \psi_2^*) = 0$$
$$\dot{\psi}_2 - (j\omega/2)(\psi_2 - \psi_2^*) + (ja/\omega^3)(\psi_2 - \psi_2^*)^3$$
$$+ (j\varepsilon/2\omega)(-\psi_2 + \psi_2^* + \psi_1 - \psi_1^*) = 0 \qquad (3.204)$$

where asterisk denotes complex conjugate. We now introduce the following approximate slow-fast partition of the dynamics, implying that *in the studied resonance interactions there exists a single dominant fast frequency* Ω:

$$\psi_k = \varphi_k(t) e^{j\Omega t}, \quad k = 1, 2 \qquad (3.205)$$

The fast frequency Ω is assumed to be in the neighborhood of $\omega_{b1} = \omega_0$, and the complex amplitudes $\phi_k(t)$ to be slowly-varying; this implies that $\dot{\phi}_k(t) = O(\varepsilon)$ or smaller. Substituting (3.205) into (3.204) and averaging over the fast frequency we obtain the following set of modulation equations governing the evolutions of the slow-varying complex amplitudes,

$$\dot{\varphi}_1 - j\mu_1\varphi_1 + (j\varepsilon/2\omega)\varphi_2 = 0$$
$$\dot{\varphi}_2 + j\mu_2\varphi_2 - (3ja/\omega^3)|\varphi_2|^2 \varphi_2 + (j\varepsilon/2\omega)\varphi_1 = 0 \qquad (3.206)$$

where $\mu_1 = \omega - \Omega$ and $\mu_2 = \Omega - (\omega/2) + (\varepsilon/2\omega)$.

There are two different ways to proceed with the analysis of (3.206), both of which are equivalent. In the first approach we express the complex variables in in polar form, $\phi_k(t) = a_k(t) e^{j\beta_k(t)}$, $k = 1, 2$, substitute into (3.206) and set the real and imaginary parts separately equal to zero. Then the following system of real modulation equations results:

$$\left. \begin{array}{l} \dot{a}_1 - (\varepsilon/2\omega) a_2 \sin(\beta_2 - \beta_1) = 0 \\ \dot{a}_2 + (\varepsilon/2\omega) a_1 \sin(\beta_2 - \beta_1) = 0 \end{array} \right\} \Rightarrow a_1^2 + a_2^2 = \rho^2 \qquad (3.207a)$$

$$a_1\dot{\beta}_1 - \mu_1 a_1 + (\varepsilon/2\omega) a_2 \cos(\beta_2 - \beta_1) = 0$$

$$a_2\dot\beta_2 + \mu_2 a_2 - (3a\,a_2^3/\omega^3) + (\varepsilon/2\omega)\,a_1\cos(\beta_2 - \beta_1) = 0 \quad (3.207b)$$

Provided that $a_1 a_2 \neq 0$, we define the new phase difference variable $\theta = \beta_2 - \beta_1$ and combine equations (3.207b) to get:

$$\dot\theta + \mu_2 + \mu_1 - \frac{3\alpha a_2^2}{\omega^3} + \frac{\varepsilon}{2\omega}\left(\frac{a_1}{a_2} - \frac{a_2}{a_1}\right) = 0 \quad (3.207c)$$

Equations (3.207a, c) form an autonomous set of nonlinear evolution equations. The integral relation between the two amplitudes in (3.207a) is an energy-like expression indicating conservation of the total energy of the undamped system during the motion. Indeed, we note that for the type of localized standing waves considered here the total energy of the integrated semi-infinite chain-NES is finite and conserved.

The stationary solutions of (3.207a, c) correspond to (approximately) time-periodic localized standing waves of the integrated system. These are computed by solving the following set of nonlinear algebraic equations:

$$a_1^2 + a_2^2 = \rho^2,\; \theta = 0$$

$$\frac{\omega}{2} + \frac{\varepsilon}{2\omega} - \frac{3\alpha a_2^2}{\omega^3} + \frac{\varepsilon}{2\omega}\left(\frac{a_1}{a_2} - \frac{a_2}{a_1}\right) = 0 \quad (3.208)$$

where we recall that $\omega = \sqrt{c^2(1 - e^{-\kappa}) + \omega_0^2 + \varepsilon}$ and that the exponential decay factor κ is expressed in terms of the fast frequency Ω by the second of relations (3.202), i.e., the linear dispersion relation of the infinite chain. Combining all these results we derive the following expression relating the fast frequency of oscillation Ω to the decay factor κ (through the frequency ω):

$$\omega = \{\omega_0^2 + \varepsilon + (1/2)(\Omega^2 - \omega_0^2) + (1/2)[(\omega_0^2 - \Omega^2)^{1/2}(\omega_0^2 - \Omega^2 + 4c^2)^{1/2}]\}^{1/2} \quad (3.209)$$

Since we are interested in localized standing waves with frequencies close to the lower bounding frequency $\omega_{b1} = \omega_0$ but inside the lower AZ, we introduce at this point a frequency detuning parameter $\delta\omega$ defined by the relation:

$$\Omega^2 = \omega_0^2 - \varepsilon^2 \delta\omega^2$$

This leads to the following algebraic relations governing the amplitudes and decay factors of the nonlinear standing wave motions:

$$\omega = \omega_0 + [\varepsilon(1 + c\delta\omega)/2\omega_0] + O(\varepsilon^2), \quad a_1^2 + a_2^2 = \rho^2$$

$$\frac{1}{2}[\omega_0 + (\varepsilon/2\omega_0)(1 + c\delta\omega)] + (\varepsilon/2\omega_0) - 3a a_2^2[\omega_0^{-3} - (3\varepsilon/2\omega_0^5)(1 + c\delta\omega)]$$

$$+ (\varepsilon/2\omega_0)\left(\frac{a_1}{a_2} - \frac{a_2}{a_1}\right) + O(\varepsilon^2) = 0 \quad (3.210)$$

The frequency of the slow modulation corresponding to the stationary solution is obtained by considering the phase relations (3.207b) and taking into account that $\theta = 0 \Rightarrow \beta_1 = \beta_2$:

$$\dot{\beta}_1 = \dot{\beta}_2 = O(\varepsilon)$$
$$= \omega - \Omega - (\varepsilon/2\omega)(a_2/a_1)$$
$$= (\varepsilon/2\omega)\left[1 + c\delta\omega - (a_2/a_1)\right] + O(\varepsilon^2) \quad (3.211)$$

This result is consistent with our assumption of slowly-varying phases.

For fixed energy ρ and detuning frequency $\delta\omega$ the set (3.210) is solved numerically for the amplitudes a_1 and a_2. Then the corresponding phases are computed by means of (3.211). The localized standing wave solutions with frequency close to the lower bounding frequency of the chain are then approximated as follows:

$$x_0(t) \approx (a_1/\omega)\sin[\Omega t + \beta_1(t)],$$
$$\dot{x}_0(t) \approx a_1 \cos[\Omega t + \beta_1(t)], \quad x_p(t) \approx x_0(t) e^{\kappa p}, \quad p \leq 0$$
$$v(t) \approx (a_2/\omega)\sin[\Omega t + \beta_2(t)],$$
$$\dot{v}(t) \approx a_2 \cos[\Omega t + \beta_2(t)] \quad (3.212)$$

This is a synchronous oscillation with constant amplitude, fast frequency $\Omega = (\omega_0^2 - \varepsilon^2 \delta\omega^2)^{1/2}$, and effective frequency $\omega_{\text{effective}} = \Omega + \dot{\beta}_1 = \omega_0 + (\varepsilon/2\omega_0)[1 + c\delta\omega - (a_2/a_1)] + O(\varepsilon^2)$. In order to comply with the assumptions of the analysis these quantities should satisfy the relations, $|\dot{\beta}_1| = |\dot{\beta}_2| \ll \Omega$ (as this separates the slow and fast dynamics), and $\omega_{\text{effective}} = \Omega + \dot{\beta}_1 < \omega_0$ (since this satisfies the condition that the frequency of the standing waves lies inside the lower AZ of the chain).

In Figure 3.101 we depict the energy dependence of $\omega_{\text{effective}}$ for parameters $c^2 = 1$, $\omega_0^2 = 0.4$, $\varepsilon = 0.1$, $\alpha = 5/8$ and varying frequency detuning $\delta\omega$. These solutions correspond to $a_1 > 0$ and $a_2 > 0$, i.e., to in-phase motions between the NES and the adjacent oscillator of the chain, localized to the NES. Hence, the 1:1 resonance interaction between the NES and the chain close to the lower bounding frequency $\omega_{b1} = \omega_0$ gives rise to a continuous family of localized, slowly modulated standing waves that lie inside the lower AZ of the chain; the decay rates of these waves increase as the frequency detuning $\delta\omega$ is increased, further inside the lower AZ.

We now discuss a second approach for analyzing the averaged set of complex slow modulations (3.206) that takes in account the integrability features of this set of equations. We start by noting that the set (3.206) is completely integrable, since it possesses the following two first integrals of motion:

$$\rho^2 = |\varphi_1|^2 + |\varphi_2|^2$$
$$H = -j\mu_1|\varphi_1|^2 - j\mu_2|\varphi_2|^2 - (3j\alpha/2\omega^3)|\varphi_2|^4 + j\lambda(\varphi_1^*\varphi_2 + \varphi_2^*\varphi_1) \quad (3.213)$$

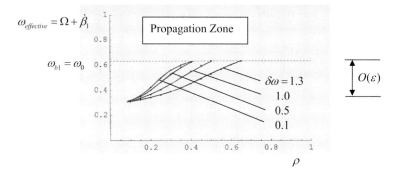

Fig. 3.101 Energy dependence of the effective frequencies of in-phase localized standing waves inside the lower AZ, for varying frequency detuning $\delta\omega$.

where $\lambda = \varepsilon/2\omega$, $\mu_1 = \omega - \Omega$ and $\mu_2 = \Omega - (\omega/2) + (\varepsilon/2\omega)$. We recall that ω is the reference frequency related to the rate of exponential decay κ, whereas Ω is the (single) dominant fast frequency of the 1:1 resonance interaction.

Taking into account the first integral of motion, we express the slowly-varying complex amplitudes as $\varphi_1 = \rho \cos\phi \, e^{j\delta_1}$ and $\varphi_2 = \rho \sin\phi \, e^{j\delta_2}$, where φ and δ_k, $k = 1, 2$ are time-dependent angle variables. Employing the second integral of motion the set modulation equations (3.206) is transformed as follows:

$$\dot{\varphi} = \lambda \sin\delta$$
$$\dot{\delta}_1 = \mu_1 - \lambda \cos\delta \tan\varphi$$
$$\dot{\delta} = (\mu_1 + \mu_2) - (3\alpha\rho^2/\omega^3)\sin^2\varphi - \lambda\cos\delta\,(\tan\varphi - \cot\varphi)$$
$$C = (\mu_1 + \mu_2)\sin^2\varphi - \left(3\alpha\rho^2/\omega^3\right)\sin^4\varphi + 2\lambda\sin\varphi\cos\varphi\cos\delta \quad (3.214)$$

where C is a first integral of the motion, and $\delta = \delta_1 - \delta_2$. Setting $Z \equiv \sin^2\varphi$ we can solve exactly this first-order slow flow approximation. Indeed, the following analytic solution of (3.214) can be derived:

$$\dot{Z} = \{4\lambda^2 Z(1-Z) - [C - Z(\mu_1 - \mu_2) + (3\alpha\rho^2/2\omega^3)Z^2]^2\}^{1/2}$$
$$\Rightarrow \int \{4\lambda^2 Z(1-Z) - [C - Z(\mu_1 - \mu_2) + (3\alpha\rho^2/2\omega^3)Z^2]^2\}^{-1/2} dZ = t + S$$
$$(3.215)$$

where S is a constant of integration, and the integral in (3.215) can be explicitly expressed in terms of elliptic integrals.

Returning to the slow flow (3.214), it is of interest to study the case when the effective frequency of oscillation $\omega_{\text{effective}}$ is exactly equal to the prescribed fast frequency of the resonance, i.e., $\omega_{\text{effective}} = \Omega$. In that case there is no slow frequency modulation of the fast oscillation (since then it holds that the slow-phases are stationary, i.e., $\dot{\beta}_1 = \dot{\beta}_2 = 0$), and the integrated chain-NES system executes purely time-periodic oscillations at the fast frequency Ω. This special (pure fast frequency) solution is computed by solving the following (extended) set of stationary equations:

$$0 = \sin\delta \Rightarrow \delta = 0, \pi \Rightarrow \mu_1 \mp \lambda \tan\varphi = 0$$

$$(\mu_1 + \mu_2) - (3\alpha\rho^2/2\omega^3)\sin^2\phi \pm \lambda(\tan\phi - \cot\phi) = 0 \quad (3.216)$$

which leads to the following amplitude-frequency relation for this special, purely fast-frequency solution:

$$\rho = \rho(\Omega) = \left\{ \frac{\Omega^3}{3\alpha} \left[\frac{\lambda^2(\Omega)}{\mu_1(\Omega)} - \mu_2(\Omega) \right] \left[\frac{\mu_1^2(\Omega) + \lambda^2(\Omega)}{\mu_1^2(\Omega)} \right] \right\}^{1/2} \quad (3.217)$$

This solution exists only in the finite interval $\omega_{min} < \omega < \omega_0$ inside the lower AZ, where ω_{min} is the solution of the equation $\mu_1(\Omega)\mu_2(\Omega) - \lambda^2(\Omega) = 0$. A typical plot depicting this solution is presented in Figure 3.102a. Note the breakdown of the analytical approximation in the neighborhood of the lower bounding frequency $\omega_{b1} = \omega_0 = 1$. The corresponding physical energy of the oscillation is given by

$$E(\Omega) = \frac{\rho^2(\Omega)}{2(\lambda^2 + \mu_1^2)} \left[\frac{\lambda^2}{1 - \exp(-2\kappa)} + \mu_1^2 \right] \quad (3.218)$$

where κ is the exponential decay rate of the localized standing wave. The corresponding plot is presented at Figure 3.102b. Note the abrupt energy increase as the PZ is approached, a feature consistent with the fact that inside the PZ the standing wave solution is transformed to a traveling wave propagating in the semi-infinite chain and corresponding to unbounded energy.

In summary, we proved the existence of a family of nonlinear standing wave solutions localized to the NES, and possessing effective frequencies situated inside the lower AZ of the dispersion relation of the linear chain. Physically, during these motions the chain executes synchronous in-phase oscillations, which are also in-phase with the NES responses. In the following analysis we prove the existence of a similar family of localized standing waves with effective frequencies situated in the upper AZ of the linear chain, corresponding to out-of-phase oscillations of adjacent pairs of oscillators.

Hence, we consider localized standing waves of (3.193) with frequencies in the range $\left[\omega_{b2} = \sqrt{\omega_0^2 + 4c^2}, +\infty \right)$. Following the procedure outlined previously, we introduce the following assumption of exponential decay for the amplitudes of the oscillators of the chain,

$$x_k = (-1)^k x_0 e^{\nu k}, k \leq 0, \quad \omega_0^2 - \Omega^2 = 2c^2(\cosh\kappa - 1), \quad \kappa = p\pi + \nu, p \in Z \quad (3.219)$$

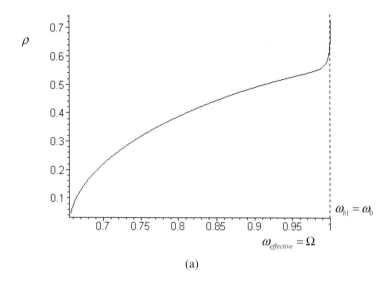

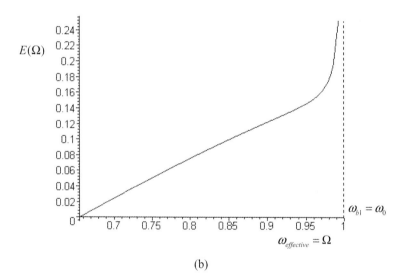

Fig. 3.102 Purely fast localized standing waves for parameters $\omega_0 = 1$, $a = 5/8$, $c = 1$ and $\varepsilon = 0.1$: (a) dependence of the energy-like variable ρ on the effective frequency $\omega_{\text{effectiveective}} = \Omega$, (b) dependence of energy on frequency.

where Ω denotes again the (common) fast frequency of the linear oscillators and the NES, and out-of-phase motions are assumed. Substituting the *ansatz* (3.219) into (3.193) we reduce the problem of computing localized standing waves inside the upper AZ to the following system of coupled oscillators:

$$\ddot{x}_0 + x_0[c^2(1 - e^{-v}) + \omega_0^2 + \varepsilon] - \varepsilon v = 0$$
$$\ddot{v} + 8av^3 - \varepsilon(x_0 - v) = 0 \qquad (3.220)$$

Introducing the reference frequency $\omega = [c^2(1 - e^{-v}) + \omega_0^2 + \varepsilon]^{1/2}$ the analysis of system (3.220) follows the steps outlined above for the reduced system (3.203), but for an important modification. This is dictated by the fact that, in contrast to the family of in-phase localized standing waves considered previously, out-of-phase standing waves can exist only above a certain energy threshold since their frequencies must exceed the upper bound ω_{b_2}. Moreover since the oscillation is expected to be strongly localized to the NES it is logical to impose the additional requirement that $|a_2| \gg |a_1|$. This amounts to rescaling the amplitudes of the slow flow in terms of the small parameter of the problem according to, $a_1 = \varepsilon b_1$, $a_2 = b_2$ and $b_1, b_2 = O(1)$.

Taking into account these assumptions, and performing a similar analysis to that adopted for the in-phase localized standing waves, we derive the following stationary solutions corresponding to time-periodic, out-of-phase, localized standing waves of the chain-NES system:

$$b_2^2 = \rho^2 + O(\varepsilon^2), \quad \theta \equiv \beta_1 - \beta_2 = 0,$$
$$b_1 = \rho[\omega^2 - (6\rho^2/\omega^2) + \varepsilon]^{-1} + O(\varepsilon^2)$$
$$\dot{\beta}_1 = \dot{\beta}_2 = \omega - \Omega - (b_2/2\omega b_1)$$
$$\Omega^2 = \omega_0^2 + 4c^2 + \varepsilon^2 \delta\omega$$
$$\omega = (\omega_0^2 + 2c^2)^{1/2} + (\varepsilon/2)(\omega_0^2 + 2c^2)^{-1/2}[1 - 2^{-1/2}c\,\delta\omega] + O(\varepsilon^2)$$
$$(3.221)$$

This set of stationary conditions is similar to the set (3.208, 3.211) for in-phase, localized standing waves. Moreover, the solution (3.221) is valid only when the conditions $|\dot{\beta}_1| \ll \Omega$ and $\Omega + \dot{\beta}_1 > (\omega_0^2 + 4c^2)^{1/2}$ hold.

In Figure 3.103 we depict the dependence of the effective frequency $\omega_{\text{effective}} = \Omega + \dot{\beta}_1$ with respect to the energy-like quantity $\rho = (a_1^2 + a_2^2)^{1/2}$ for the family of out-phase localized standing waves. These computations were performed for $c^2 = 1$, $\omega_0^2 = 0.4$, $\varepsilon = 0.1$, $\alpha = 5/8$ and varying frequency detuning parameter $\delta\omega$. We note that close to the upper bounding frequency there exists an approximately linear dependence of the effective frequency on energy. The localized solution corresponds to $a_1 < 0$, $a_2 > 0$, $a_2 \gg |a_1|$, i.e., to out-of-phase oscillations between the NES and the nearest to it linear oscillator of the chain.

An analytical estimate for the energy threshold for the family of out-of-phase localized standing waves is now derived. To this end, we express the energy-like quantity ρ as $\rho = \rho^{(0)} + \varepsilon r$, where $\rho^{(0)}$ is a constant and r is the variation of the energy in the neighborhood of the upper bounding frequency ω_{b1}. Substituting this expression into the third of equations (3.221) provides a way for determining the constant $\rho^{(0)}$; indeed, $\rho^{(0)}$ is chosen so to eliminate the $O(1)$ term from the

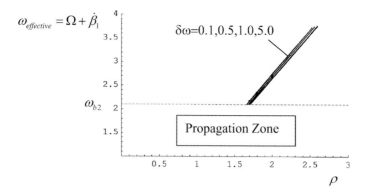

Fig. 3.103 Energy dependence of the effective frequencies of out-of-phase localized standing waves inside the upper AZ, for varying frequency detuning $\delta\omega$.

frequency of the slow variation thus rendering the analytical solution consistent with the assumptions made. This leads to the following estimate:

$$\rho^{(0)} = [(\omega_0^2 + 2c^2)^{3/2}/3\alpha][(\omega_0^2 + 4c^2)^{1/2} - (1/2)(\omega_0^2 + 2c^2)^{1/2}] \quad (3.222)$$

As a result, the frequency of the slow modulation becomes an $O(\varepsilon)$ quantity, given by the following expression:

$$\dot{\beta}_1 = \dot{\beta}_2 = \varepsilon \left\{ \Delta - \frac{1}{2\rho^{(0)}} \left[\frac{r}{(\omega_0^2 + 2c^2)^{1/2}} - \frac{\Delta\rho^{(0)}}{(\omega_0^2 + 2c^2)} \right] \right.$$

$$\times \left[(\omega_0^2 + 2c^2) - \frac{6\alpha\rho^{(0)2}}{(\omega_0^2 + 2c^2)} \right]$$

$$\left. + \frac{B}{2\rho^{(0)}(\omega_0^2 + 2c^2)^{1/2}} \left[(\omega_0^2 + 2c^2) - \frac{6\alpha\rho^{(0)2}}{(\omega_0^2 + 2c^2)} \right]^2 \right\} + O(\varepsilon^2)$$

(3.223)

where

$$\Delta = \Delta(\delta\omega) = (1/2)(\omega_0^2 + 2c^2)^{-1/2}(1 - 2^{-1/2}c\delta\omega)$$

and

$$B = B(r, \delta\omega) = -\rho^{(0)} \left[1 + 2\Delta(\omega_0^2 + 2c^2)^{1/2} - \frac{12\alpha\rho^{(0)}r}{(\omega_0^2 + 2c^2)} + \frac{12\alpha\rho^{(0)2}\Delta}{(\omega_0^2 + 2c^2)^{3/2}} \right]$$

$$\times \left[(\omega_0^2 + 2c^2) - \frac{6\alpha\rho^{(0)2}}{(\omega_0^2 + 2c^2)} \right]^{-2} + r \left[(\omega_0^2 + 2c^2) - \frac{6\alpha\rho^{(0)2}}{(\omega_0^2 + 2c^2)} \right]^{-1}$$

Finally, the amplitudes of oscillation of the problem are approximated as follows:

$$a_1 = \varepsilon \rho^{(0)} \left[\left(\omega_0^2 + 2c^2 \right) - \frac{6\alpha \rho^{(0)2}}{\left(\omega_0^2 + 2c^2 \right)} \right]^{-1} + \varepsilon^2 B(r, \delta\omega) + O(\varepsilon^3)$$

$$a_2 = \rho^{(0)} + \varepsilon r + O(\varepsilon^2) \qquad (3.224)$$

This solution indicates that close to the upper bounding frequency the effective frequency $\omega_{\text{effective}} = \Omega + \dot{\beta}_1$ of the out-of-phase localized standing waves vary linearly with increasing energy, a result which is consistent with the numerical result of Figure 3.102. The energy threshold for the existence of this family is given by $\rho_{\text{cr}}(\delta\omega) = \rho^{(0)} + \varepsilon\, r_{\text{cr}}(\delta\omega) + O(\varepsilon^2)$, and is approximated by the requirement that on the threshold it must be satisfied that $\omega_{\text{effective}} = \omega_{b2}$, or, $\Omega + \dot{\beta}_1 = \omega_{b2} \Rightarrow \dot{\beta}_1 = 0 + O(\varepsilon^2)$. This leads to the following algebraic expression for determining $r_{\text{cr}}(\delta\omega)$:

$$\Delta(\delta\omega) - \frac{1}{2\rho^{(0)}} \left[\frac{r_{\text{cr}}}{\left(\omega_0^2 + 2c^2 \right)^{1/2}} - \frac{\Delta(\delta\omega)\rho^{(0)}}{\left(\omega_0^2 + 2c^2 \right)} \right] \left[\left(\omega_0^2 + 2c^2 \right) - \frac{6\alpha\rho^{(0)2}}{\left(\omega_0^2 + 2c^2 \right)} \right]$$

$$+ \frac{B(r_{\text{cr}}, \delta\omega)}{2\rho^{(0)} \left(\omega_0^2 + 2c^2 \right)^{1/2}} \left[\left(\omega_0^2 + 2c^2 \right) - \frac{6\alpha\rho^{(0)2}}{\left(\omega_0^2 + 2c^2 \right)} \right]^2 = 0 \qquad (3.225)$$

This completes the analytical study of the out-of-phase localized standing waves in the system (3.193). In the remainder of this section we perform a series of numerical simulations in order to highlight the role that the computed families of localized standing waves play on TET from the chain to the NES. We note that, in contrast to our previous studies of TET in weakly damped finite-DOF coupled oscillators, *TET in the present problem takes place even in the absence of damping*. This is due to the fact that the energy radiation from the NES to the far-field of the semi-infinite chain [i.e., as $s \to -\infty$ in the continuum approximation (3.194)] has an equivalent effect to damping dissipation in finite-DOF discrete oscillators, and, hence, induces the necessary frequency variation of the NES response required for TET.

We performed a series of numerical simulations with a chain composed of 200 oscillators with an essentially nonlinear oscillator (the NES) attached to its right end. In the first series of simulations the initial conditions of all oscillators are set equal to zero, except for $\dot{x}_{-3}(0) = X \neq 0$; in essence, this simulates an initial impulse of magnitude X applied to the fourth oscillator from the NES. The total instantaneous energy of the system was monitored to verify energy conservation and ensure accuracy of the numerical simulations. In addition, care was taken to select the time window of the simulations small enough to avoid the interference due to reflected waves from the left free end of the chain in the measurements.

For a small enough impulse neither in-phase nor out-of-phase localized standing waves (modes) are excited (see Figure 3.104). For a sufficiently strong impulsive magnitude, however, excitation of the in-phase localized standing wave occurs. This

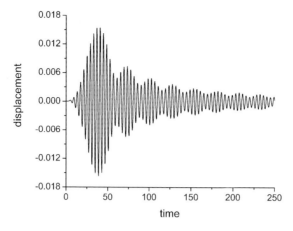

Fig. 3.104 Numerical simulation of the chain-NES interaction for weak impulse excitation of the fourth oscillator of the chain: Response of the NES.

is shown the the numerical simulations depicted in Figure 3.105 for impulsive magnitude $X = 50$, and system parameters $\omega_0 = 1.5$, $c^2 = 2.0$, $8\alpha = 0.5$ and $\varepsilon = 0.3$. In Figure 3.105a we depict the transient responses of the NES and its neighboring oscillator of the chain, whereas in Figure 3.105b we depict the temporal evolution of the instantaneous frequency $\omega_{NL}(t)$ of the NES. In the plot of Figure 3.105a we note an initial regime of strong dynamic interaction between the chain and the NES, after which the system settles into a time-periodic localized standing wave motion, with energy predominantly confined to the NES. This time-periodic solution is the theoretically predicted localized in-phase standing wave inside the lower AZ of the infinite chain. This is confirmed by the fact that its frequency (i.e., the asymptotic value reached by $\omega_{NL}(t)$ in Figure 3.105b) is equal to $1.497 < \omega_{b1} = \omega_0$; by the near-exponential decay of the amplitudes of the oscillators (see Figure 3.105c – the small discrepancies noted for distant oscillators is due to the fact that they have not reached a complete steady state motion at the time of the measurement); and by the near in-phase oscillations of the chain and the NES. In Figure 3.105d we depict the instantaneous fraction of initial energy contained in the leading 26 oscillators of the chain and the NES; as time increases this energy reaches an asymptotic value that represents the fraction of total initial energy transferred to the localized standing wave.

Hence, *passive TET from the undamped semi-infinite chain to the undamped NES occurs through the excitation of the in-phase standing wave localized to the NES.* This is qualitatively different compared to the mechanisms of TET for finite-DOF, weakly damped oscillators, which relied either on fundamental and subharmonic TRCs or on the excitation of nonlinear beats. In the absence of damping in the infinite-dimensional system, radiation to the far field provides an energy dissipation mechanism similar to damping, which drives the dynamics to the domain of attraction of the localized in-phase standing wave, and, hence, generates TET.

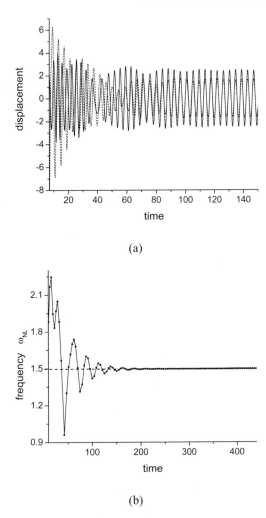

Fig. 3.105 Numerical simulation of the chain-NES interaction for strong impulse excitation: (a) — $v(t)$, - - - $x_0(t)$; (b) instantaneous frequency $\omega_{NL}(t)$ of the NES.

In the second series of numerical simulations we study the excitation of the localized out-of-phase standing wave inside the upper AZ of the linear chain. In our simulations *we could not establish the occurrence of TET in the impulsively excited chain through excitation of the out-of-phase family of localized standing waves*, i.e., we could not reproduce the scenario for TET discussed above, which relied on the excitation of the in-phase family of standing waves. As an alternative, we wish to numerically demonstrate the existence of the out-of-phase family of localized waves. To this end, we initiate the system by exponentially decaying out-of-phase initial conditions for the 25 leading oscillators of the chain, and observe an initial regime of chain – NES dynamic interaction, after which a time-periodic localized

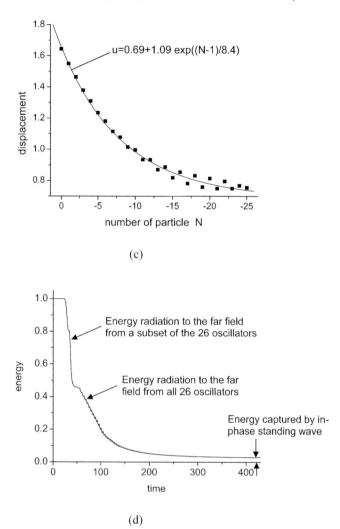

Fig. 3.105 Numerical simulation of the chain-NES interaction for strong impulse excitation: (c) near exponential decay of the amplitudes of the oscillators when the localized in-phase standing wave is excited; (d) evolution of instantaneous normalized energy of the leading 26 oscillators and the NES.

out-of-phase standing wave is formed, with energy predominantly confined to the NES. In Figure 3.106a we depict the corresponding evolution of the instantaneous frequency $\omega_{NL}(t)$ of the NES, which eventually enters into the higher AZ, above the upper bounding frequency $\omega_{b2} = 3.2015$. In Figure 3.106b we depict the instantaneous normalized energy of the leading 18 oscillators of the chain and the NES, representing the portion of the total energy 'trapped' in the localized standing wave.

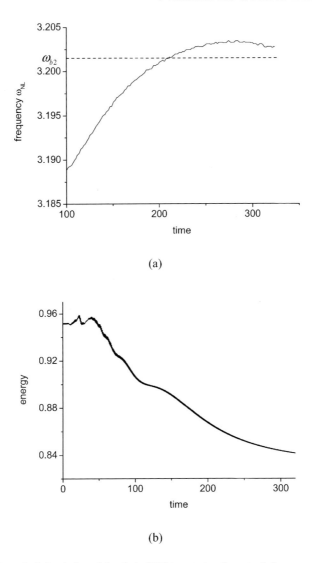

Fig. 3.106 Numerical simulation of the chain-NES interaction for out-of-phase exponentially decaying initial excitation of the leading 25 oscillators: (a) instantaneous frequency $\omega_{NL}(t)$ of the NES; (b) evolution of instantaneous normalized energy of the leading 18 oscillators and the NES.

The main conclusion drawn from the analytical and numerical results of this section is that passive TET can occur in the undamped semi-infinite chain of linear oscillators with a weakly coupled, essentially nonlinear end attachment; that is, impulsive energy from the chain can be transferred irreversibly to the nonlinear oscillator (which acts as an NES) under conditions of nonlinear 1:1 resonance. The only scenario for TET established by the numerical simulations is through the excitation of families of in-phase standing waves (nonlinear modes) situated inside the lower AZ of the linear chain, and localized to the NES.

Based on the previous theoretical and numerical results we can formulate the following scenario for passive TET from the semi-infinite chain to the nonlinear oscillator. An initial impulsive excitation of the chain causes energy to propagate towards the NES (and also away from the NES to the far field) through traveling wavepackets with predominant frequencies inside the PZ $\omega \in (\omega_{b1}, \omega_{b2})$ of the chain (actually, the only way to transfer energy through the linear chain is by exciting traveling waves). After these traveling wavepackets impede to the nonlinear oscillator they excite it initially with frequencies inside the PZ of the chain, under non-resonant conditions (this is confirmed by the numerical result of Figure 3.105b). These initial non-resonant interactions cause initial near-adiabatic radiation of energy from the nonlinear oscillator back to the chain, a process that reduces its instantaneous frequency; indeed the radiation of energy from the nonlinear oscillator back to the chain has the same effect as energy dissipation due to damping in finite-DOF discrete coupled oscillators. After sufficient radiation of energy, the instantaneous frequency of the nonlinear oscillator reaches from above the lower bounding frequency $\omega_{b1} = \omega_0$ of the chain, where conditions for 1:1 resonance between the chain and the nonlinear oscillator are established. This eventually leads to excitation of an in-phase localized standing wave (mode) of the integrated chain-attachment system. Once this localized mode is excited, energy is 'trapped' in the nonlinear oscillator, and no further energy radiation back to the chain is possible afterwards, since the motion takes place on an invariant nonlinear normal mode manifold localized to the nonlinear oscillator. As a result, there occurs confinement of energy to the NES and passive TET.

An interesting feature of this TET scenario is that it is realized in the *absence of damping*. This contrasts to our studies of finite-DOF systems of coupled oscillators, where TET occurred only in the presence of damping dissipation, through TRCs in neighborhoods of the corresponding resonant manifolds. In the infinite-DOF undamped system considered in this section the far field acts as an effective energy dissipater, 'absorbing' irreversibly energy in the form of traveling waves propagating away from the nonlinear oscillator. Hence, in the scenario outlined above for the undamped infinite-DOF system TET is realized through the eventual excitation of a standing wave localized to the NES rather than through TRCs. Due to the invariance property of the family of localized standing waves, once such a standing wave is excited the motion remains confined to the NES and no energy radiation to the semi-infinite chain is possible afterwards. In the next section we formulate an alternative analytical methodology for studying TET in the corresponding weakly damped system, and examine the mechanisms for TET in that case.

3.5.2.3 Integro-Differential Formulation

To study TET in the weakly damped, semi-infinite chain with the nonlinear end attachment we adopt a different methodology by reducing the dynamics to a single

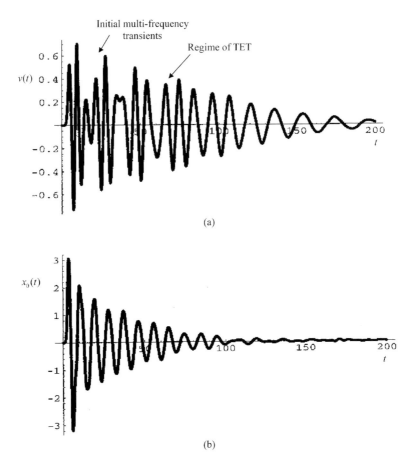

Fig. 3.108 Case of TET from the chain to the NES: (a) response of the NES, (b) response of the neighboring to the NES linear oscillator.

no approximation) to a single integro-differential equation. To perform this task we make use of the analytical results of Lee (1972) and Wang and Lee (1973) who, in essence, derived the Green's functions of the free and forced damped chain of linear oscillators in explicit form.

To this end, the response of the k-th oscillator of the chain (3.226) can be symbolically expressed as follows:

$$x_k(t) = X[G_{k-p}(t) + G_{k+p-3}(t)] + \varepsilon[v(t) - x_0(t)] * [G_k(t) + G_{k+1}(t)], \quad k \geq 0 \tag{3.227}$$

where $(*)$ denotes the convolution operation. The kernel $G_m(t) = G_{-m}(t)$ is defined as

$$G_m(t) = e^{-\varepsilon\lambda t/2} \int_0^t J_0\left[(\omega_0^2 - \varepsilon^2\lambda^2/4)^{1/2}(t^2 - \tau^2)^{1/2}\right] J_{2m}(2c\tau)d\tau$$

$$\equiv e^{-\varepsilon\lambda t/2} H_m(t) \tag{3.228}$$

where $J_{2m}(\bullet)$ denotes the Bessel function of the first kind of order $2m$. Using (3.227) we express the response $x_0(t)$ of the linear oscillator adjacent to the NES in the following integro-differential form:

$$x_0(t) = X[G_p(t) + G_{p+1}(t)] + \varepsilon\left\{v(t) - X[G_p(t) + G_{p+1}(t)]\right\}$$
$$* [G_0(t) + G_1(t)] + O(\varepsilon^2) \tag{3.229}$$

with $p \geq 0$.

Substituting (3.229) into the last of equations (3.226) we obtain the following reduced dynamical system, in the form of a single integro-differential equation governing the motion of the NES:

$$\ddot{v} + \varepsilon\lambda\dot{v} + Cv^3 + \varepsilon v =$$
$$\varepsilon X[G_p(t) + G_{p+1}(t)] + \varepsilon^2\left\{v(t) - X[G_p(t) + G_{p+1}(t)]\right\}$$
$$* [G_0(t) + G_1(t)] + O(\varepsilon^3) \tag{3.230}$$

This equation is supplemented by the initial conditions $v(0) = \dot{v}(0) = 0$.

It follows that the problem of studying the dynamics of TET in system (3.226) is reduced to the equivalent problem of studying the dynamics of the integro-differential equation (3.230) with zero initial conditions. Clearly, direct application of the CX-A technique developed in the previous section is not possible at this point, due to the apparent lack of a single 'fast' frequency in the non-homogeneous term on the right-hand side of (3.230). Hence, before proceeding with the analysis of this equation it is necessary to examine carefully the frequency content of the non-homogeneous term; if this term can be approximated by a slowly modulated fast monochromatic oscillation, it will render the integro-differential equation (3.230) amenable to direct CX-A analysis.

Since the quantity

$$G_{m-1}(t) + G_m(t) \equiv e^{-\varepsilon\lambda t/2}[H_{m-1}(t) + H_m(t)]$$

appears repeatedly in (3.230) we start our analysis by studying the spectral content of this quantity. As shown by Wang and Lee (1973), $H_m(t)$ can be expressed in the following alternative form (which highlights its spectral content):

$$H_m(t) = \pi^{-1}\int_0^\pi \frac{\cos m\theta}{2j\omega(\theta)}\left[e^{j\omega(\theta)t} - e^{-j\omega(\theta)t}\right]dt,$$
$$\omega(\theta) = \left[\omega_0^2 + 4c^2\sin^2(\theta/2)\right]^{1/2} \tag{3.231}$$

By (3.231) $H_m(t)$ is expressed as a superposition of a continuum of harmonics with frequencies in the range $[\omega_0, (\omega_0^2 + 4c^2)^{1/2}]$, which is coincident with the PZ of the infinite undamped linear chain, where time-harmonic traveling waves can propagate unattenuated upstream or downstream through the chain. Outside this frequency range (in the two AZs) the chain acts as a filter, exponentially attenuating harmonic signals and producing merely near-field solutions. Hence, a first conclusion is that *the reduced system (3.230) highlights the fact that the NES is forced by a continuum of impeding harmonics in the range of the PZ of the linear chain.*

We now asymptotically analyse (3.232) in order to show that *after some initial multi-frequency transients, $H_m(t)$ performs oscillations dominated by the single 'fast' frequency* $\omega_{b1} = \omega_0$, which is the lower bounding frequency of the dispersion relation of the chain; this finding will pave the way for applying the CX-A methodology to the reduced system. Considering the time dependence of the integral (3.231) we note that for $t \gg 1$ the harmonic terms in the integrand perform fast oscillations; it follows that for sufficiently long times we can apply the method of stationary phase (Bleistein and Handelsman, 1986) to asymptotically approximate $H_m(t) + H_{m-1}(t)$ as follows:

$$[H_m(t) + H_{m-1}(t)]_{(t \gg 1)} = \frac{e^{j(\omega_0 t + \pi/4)}}{2j} \left(\frac{2}{\pi c^2 \omega_0 t}\right)^{1/2}$$

$$- \frac{e^{j(\omega_0 t + 3\pi/4)}}{32 j \pi^{1/2} \omega_0} \left(\frac{4\omega_0}{c^2 t}\right)^{3/2} \left\{\frac{c^2}{\omega_0^2} + 2[m^2 + (m+1)^2] - 1\right\}$$

$$+ O(t^{-5/2}) + \text{cc}, \quad t \gg 1 \quad (3.232)$$

where 'cc' denotes complex conjugate and $m = 0, 1, 2, \ldots$. We note that at sufficiently long times (i.e., after the multi-frequency early transients have died out) the quantity $H_m(t) + H_{m-1}(t)$ settles approximately to a fast oscillation with frequency ω_0 modulated by an algebraically decaying 'slow' envelope. Similar algebraic time decay rates for anharmonic chains were derived by Sen et al. (1996).

A short time analytic approximation for $H_m(t) + H_{m-1}(t)$ is derived by Taylor-expanding the exponentials in (3.232) close to $t = 0$, and performing successive integrations with respect to θ of the resulting coefficients of powers of t,

$$[H_m(t) + H_{m-1}(t)]_{(t \ll 1)} \approx \pi^{-1} \sum_{i=1,3,5,\ldots} [I_i(m) + I_i(m-1)] \frac{t^i}{i!}, \quad t \ll 1$$

$$(3.233)$$

where

$$I_i(m) = \int_0^\pi \omega^{(i-1)}(\theta) \cos m\theta \, d\theta, \quad i = 1, 3, 5, \ldots$$

An interesting observation is that for fixed i the quantity $I_i(m)$ becomes zero for $m \geq (i+1)/2$. It follows that as the order m increases we must consider higher orders of t in the early time expansion (3.233) to obtain accurate approximations. This observation is consistent with the existence of exceedingly larger initial 'silent' regions in $G_p(t) + G_{p+1}(t)$ in the non-homogeneous term of the reduced integrodifferential equation with increasing p (i.e., as the impulse is shifted further downstream away from the NES).

The point of matching of the short- and long-time approximations can be computed by imposing an appropriate criterion, for example, by minimizing in time the error quantity

$$Er(t) = \{[g_{(t \ll 1)}(t) - g_{(t \gg 1)}(t)]^2 + [\dot{g}_{(t \ll 1)}(t) - \dot{g}_{(t \gg 1)}(t)]^2\}^{1/2}$$

$$g(t) \equiv [H_m(t) + H_{m-1}(t)]$$

This quantitative criterion provides the time interval $[0, t^*]$ of validity of the Taylor-series based approximation, and the beginning of the range of validity of the long-term asymptotic approximation (3.232). The error at the point of matching, $Er(t^*)$, can be made arbitrarily small by including a sufficient number of terms in the two approximations. Similar matching techniques of short- and long-time local solutions have been introduced in previous works (for example, Salenger et al., 1999) to construct global analytical approximations of strongly nonlinear responses of coupled oscillators.

In Figure 3.109a we depict a comparison of the short and long time approximations with the (exact) numerical simulation for the quantity $[H_p(t) + H_{p+1}(t)]$ for the chain whose responses are shown in Figure 3.108. Since the impulse is applied in the fourth particle of the system we have that $p = 2$; the short time approximation was derived up to $O(t^3)$, whereas the long time asymptotic approximation up to $O(t^{-3/2})$. In the same figure we depict the error $Er(t)$ versus time from where the instant of transition t^* is determined. Better approximations can be obtained by improving the accuracy of the long time asymptotic approximation.

The previous discussion proves that, in the TET regime and after certain initial multi-frequency transients the non-homogeneous term of the reduced equation (3.230) possesses a dominant harmonic with fast frequency ω_0. This finding enables us to apply the CX-A method to analyze TET from the chain to the NES. The solution of the reduced system is developed in two steps. For $t \in [0, t^*)$ the short-term solution of the reduced system is expressed in Taylor series whose coefficients are computed by matching respective powers of t on the left- and right-hand sides. For $t \geq t^*$ we express the quantities $[G_p(t) + G_{p+1}(t)]$ and $[G_0(t) + G_1(t)]$ on the right-hand side of (3.230) using the long-time asymptotic approximation (3.232). We then apply the CX-A method by partitioning the dynamics into fast and slow-components using as initial condition the state of the system at t^* (as computed by the Taylor series expansions of the previous step).

Elaborating further on the second step, to approximate $v(t)$ we introduce the complex variable $\psi(t) = \dot{v}(t) + j\omega_0 v(t)$, and express $\psi(t)$ in polar form, $\psi(t) = \varphi(t)e^{j\omega_0 t}$, where $\phi(t)$ represents the slowly varying modulation of the fast oscillation $e^{j\omega_0 t}$. Moreover, it is of help to introduce the complex amplitude $\sigma(t)$ defined by $\phi(t) = \sigma(t)e^{-\varepsilon \lambda t/2}$. Finally, we use the following compact notation for the long-time asymptotic solution (3.232),

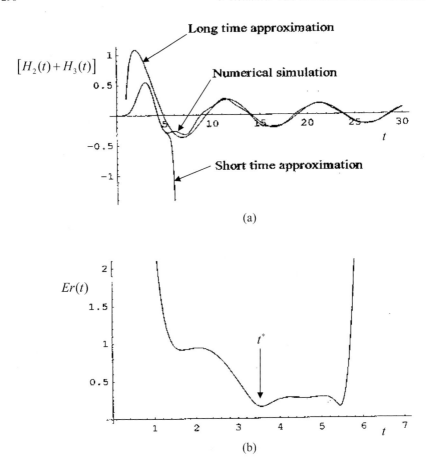

Fig. 3.109 Matching the local approximations (3.232) and (3.233): (a) short and long time approximations for $[H_p(t) + H_{p+1}(t)]$, $p = 2$, compared to the exact numerical simulation for the response depicted in Figure 3.107; (b) error function $Er(t)$ for the same system determining the transition point t^*.

$$[H_{m-1}(t) + H_m(t)]_{(t \gg 1)} \equiv h(t; m) \, e^{j\omega_0 t} + O(t^{-5/2}) + \text{cc} \qquad (3.234)$$

where

$$h(t; m) = \frac{e^{j\pi/4}}{2j} \left(\frac{2}{\pi c^2 \omega_0 t} \right)^{1/2} - \frac{e^{j3\pi/4}}{32\pi^{1/2} \omega_0 j} \left(\frac{4\omega_0}{c^2 t} \right)^{3/2}$$
$$\times \left\{ \frac{c^2}{\omega_0^2} + 2[m^2 + (m+1)^2] - 1 \right\}$$

Introducing the new variables, $\tilde{t} = t - t^*$, $\sigma(t) = \sigma(\tilde{t} + t^*) \equiv \tilde{\sigma}(\tilde{t})$, $h(t; p) = h(\tilde{t} + t^*; p) \equiv \tilde{h}(\tilde{t}; p)$, $\tilde{t} \geq 0$, and omitting the tildes from the resulting expressions we derive the following slow flow approximation governing the dynamics of the complex modulation

$$\dot{\sigma} + \frac{j(\omega_0^2 - \varepsilon)}{2\omega_0}\sigma - \frac{3jCe^{-\varepsilon\lambda t^*}e^{-\varepsilon\lambda t}}{8\omega_0^3}|\sigma|^2\sigma$$

$$= \varepsilon X h(t; p+1) + \frac{\varepsilon^2}{2j\omega_0}\int_0^t \sigma(\tau) h(t - \tau; p+1)\,d\tau$$

$$- \varepsilon^2 X \int_0^t h(\tau; p+1)\,h(t - \tau; 1)\,d\tau + O(\varepsilon^3) \tag{3.235}$$

The initial condition $\sigma(0)$ is determined by computing the Taylor series solution at the transition point $t = t^* \Rightarrow \tilde{t} = 0$. We note that due to the approximations involved, the solution of (3.235) is expected to be valid only up to times of $O(1/\varepsilon^2)$.

Hence, the problem of studying TET in the weakly damped system (3.226) is reduced approximately to the analyis of the dynamics of the complex modulation equation (3.235). This analysis is similar to the ones performed in previous sections for studying the slow flow of the two-DOF system, and is not carried out further. We state, however, that the reduction of the dynamics to (3.235) indicates that TET in the weakly damped system (3.225) is due to TRC of the NES dynamics in the neighborhood of a 1:1 resonance manifold at frequency ω_0; in that sense, TET in the weakly damped system can be regarded as qualitatively different from the TET mechanism in the corresponding undamped system which was due to excitation of an in-phase family of standing waves localized to the NES. Viewed in a different context, however, the TET dynamics in the damped and undamped systems possess a similarity. Indeed the spectral study of the non-homogeneous term of the reduced system carried out in this section confirms the TET scenario of the previous section, namely, that TET from the semi-infinite chain to the NES occurs when the frequency of the NES approches from above the lower bound of the PZ of the chain. Similar results were obtained in Dumcum (2007) where the analysis was extended to semi-infinite linear chains with lightweight ungrounded NESs (of Configuration II – see Section 3.1).

A final note concerns the initial multi-frequency transients that occur after a traveling wavepacket propagating in the semi-infinite chain impedes on the NES (see Figure 3.108). In this regime the NES interacts with traveling waves possessing frequencies inside the PZ of the chain, and radiates energy to the far field of the chain. Traveling waves, however, can be regarded as the continuum limit of the closely packed resonances of a chain composed of a large (but finite) number of coupled oscillators, as this number tends to infinity. Viewed in that context, the initial multi-frequency transients resulting from the dynamic interaction of the NES with im-

Vakakis, A.F., Inducing passive nonlinear energy sinks in vibrating systems, *J. Vib. Acoust.* **123**, 324–332, 2001.

Vakakis, A.F., Gendelman, O.V., Energy pumping in nonlinear mechanical oscillators: Part II – Resonance capture, *J. Appl. Mech.* **68**, 42–48, 2001.

Vakakis, A.F., King, M.E., Nonlinear wave transmission in a mono-coupled elastic periodic system, *J. Acoust. Soc. Am.* **98**(3), 1534–1546, 1995.

Vakakis, A.F., Rand, R.H., Non-linear dynamics of a system of coupled oscillators with essential stiffness non-linearities, *Int. J. Nonlinear Mech.* **39**, 1079–1091, 2004.

Vakakis, A.F., Manevitch, L.I., Mikhlin, Y.V., Pilipchuk, V.N., Zevin, A.A., *Normal Modes and Localization in Nonlinear Systems*, Wiley Interscience, New York, 1996.

Vakakis, A.F., Manevitch, L.I., Gendelman, O., Bergman, L., Dynamics of linear discrete systems connected to local essentially nonlinear attachments, *J. Sound Vib.* **264**, 559–577, 2003.

Van Overschee, P., De Moor B., *Subspace Identification for Linear Systems: Theory, Implementation, Applications*, Kluwer Academic Publishers, Boston, 1996.

Veerman, P., Holmes, P., The existence of arbitrarily many distinct periodic orbits in a two-DOF Hamiltonian System, *Physica D* **14**, 177–192, 1985.

Veerman, P., Holmes, P., Resonance bands in a two-DOF Hamiltonian system, *Physica D* **20**, 413–422, 1986.

Verhulst, F., *Methods and Applications of Singular Perturbations*, Springer Verlag, Berlin/New York, 2005.

Wang, Y.Y., Lee, K.H., Propagation of a disturbance in a chain of interacting harmonic oscillators, *Am. J. Physics* **41**, 51–54, 1973.

Wiggins S., *Introduction to Applied Nonlinear Dynamical Systems and Chaos*, Springer-Verlag, Berlin/New York, 1990.

Chapter 4
Targeted Energy Transfer in Discrete Linear Oscillators with Multi-DOF NESs

4.1 Multi-Degree-of-Freedom (MDOF) NESs

In the previous chapter we considered targeted energy transfer (TET) from linear discrete primary systems to single-degree-of-freedom (SDOF) essentially nonlinear attachments (or nonlinear energy sinks – NESs). In this chapter we extend our discussion of nonlinear TET to multi-DOF essentially nonlinear NESs. The reason for doing so is twofold. First, we aim to show that through the use of MDOF NESs it is possible to passively extract vibration energy *simultaneously* from multiple linear modes of primary systems. This feature normally does not appear in the case of SDOF NESs, since as shown in Chapter 3, in such attachments multi-frequency TET (involving resonance interactions of the NESs with multiple linear modes) can only occur through resonance capture cascades (RCCs); i.e., through sequential transient resonance captures (TRCs) involving only one linear mode at a time. Second, we wish to show that by using MDOF NESs we can improve the efficiency and robustness of TET, *even at small energy levels*. This represents a qualitatively new feature in TET dynamics, since as we discussed in Chapter 3, strong TET from primary discrete systems to SDOF NESs can be realized only when the energy exceeds a well-defined critical threshold (e.g., see Figure 3.4).

The general study of the nonlinear dynamical interactions of linear primary systems with MDOF essentially nonlinear NESs is a formidable problem from an analytical point of view, due to the high-order degeneracies of the governing dynamics that lead to high-co-dimension bifurcations (Guckenheimer and Holmes, 1983; Wiggins, 1990). However, we will show in this chapter that if the aim of the analysis is narrowed to focus on TET dynamics, asymptotic analysis can still be applied to study analytically certain aspects of the problem. The following exposition draws results from the thesis by Tsakirtzis (2006). Additional works on MDOF NESs were performed by Gourdon et al. (2005, 2007) and Gourdon and Lamarque (2005), whereas Musienko et al. (2006) studied nonlinear energy transfers from a linear oscillator to a system of two attached SDOF NESs. Ma et al. (2008) studied TET from a chain of particles to a two-DOF essentially nonlinear attachment at its end by ap-

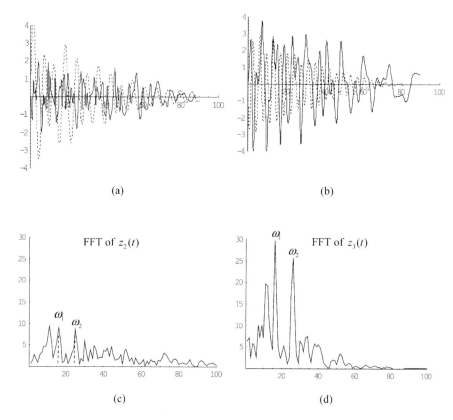

Fig. 4.4 Transient responses of system (4.2): (a) $x_1(t)$ —, $z_2(t)$ - - - - - -, (b) $x_2(t)$ —, $z_3(t)$ - - -, (c) FFT of $z_2(t)$, (d) FFT of $z_3(t)$.

behavior, though its interaction with the out-of-phase linear mode takes place after some initial time delay. This mode also absorbs energy in a multi-frequency fashion, and resonates with both linear modes of the primary system; the presence of the lower NNM is again noted in the in-phase nonlinear modal response.

The results presented in this section provide a numerical demonstration that, indeed, MDOF NESs can act as passive energy absorbers of vibration energy over wide frequency ranges. This is due to the occurrence of simultaneous TRCs at different frequency ranges, resulting from resonance interactions of multiple NNMs of the NES with multiple linear modes of the primary system. Motivated by this preliminary numerical evidence, we now proceed to a more systematic numerical study of targeted energy transfer (TET) phenomena from linear oscillators to attached MDOF NESs.

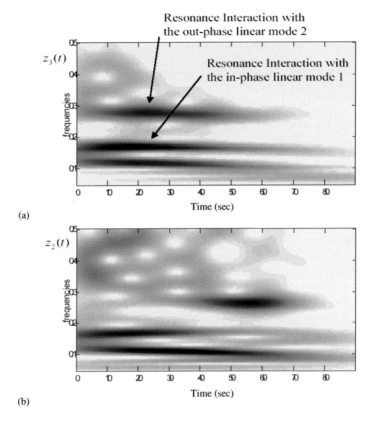

Fig. 4.5 Wavelet transform spectra of the transient responses of the MDOF NES (frequencies in Hz): (a) $z_3(t)$, (b) $z_2(t)$.

4.1.2 Numerical Evidence of TET in MDOF NESs

In this section we will study systematically the efficiency of passive TET in the system depicted in Figure 4.2. The study follows closely Tsakirtzis (2006) and Tsakirtzis et al. (2007). This system consists of a two-DOF primary linear oscillator connected through a weak linear stiffness of constant ε (which is the small parameter of the problem, i.e., $0 < \varepsilon \ll 1$) to a three-DOF NES with essential stiffness nonlinearities. Each mass of the primary system is normalized to unity, and the stiffnesses of the NES possess pure cubic characteristics with constants C_1 and C_2. Each mass of the nonlinear attachment is equal to μ, and both linear and nonlinear subsystems possess linear viscous dampers with small constants $\varepsilon\lambda$. Assuming that impulsive excitations $F_1(t)$ and $F_2(t)$ are applied to the primary system and that no direct forcing excites the nonlinear attachment, the equations of motion are given by

$$\ddot{u}_1 + (\omega_0^2 + \alpha)u_1 - \alpha u_2 + \varepsilon\lambda_1 \dot{u}_1 = F_1(t)$$

$$\ddot{u}_2 + (\omega_0^2 + \alpha + \varepsilon)u_2 - \alpha u_1 - \varepsilon v_1 + \varepsilon\lambda_1 \dot{u}_2 = F_2(t)$$

$$\mu\ddot{v}_1 + C_1(v_1 - v_2)^3 + \varepsilon(v_1 - u_2) + \varepsilon\lambda_2(\dot{v}_1 - \dot{v}_2) = 0$$

$$\mu\ddot{v}_2 + C_1(v_2 - v_1)^3 + C_2(v_2 - v_3)^3 + \varepsilon\lambda_2(2\dot{v}_2 - \dot{v}_1 - \dot{v}_3) = 0$$

$$\mu\ddot{v}_3 + C_2(v_3 - v_2)^3 + \varepsilon\lambda_2(\dot{v}_3 - \dot{v}_2) = 0 \tag{4.4}$$

As mentioned in the previous section, in the limit $\varepsilon \to 0$ system (4.4) is decomposed into two uncoupled oscillators: a two-DOF linear primary system with natural frequencies $\omega_1 = \sqrt{\omega_0^2 + 2\alpha}$ and $\omega_2 = \omega_0 < \omega_1$ corresponding to the out-of-phase and in-phase linear modes, respectively; and a three-DOF NES with a rigid body mode, and two flexible nonlinear normal modes – NNMs (Tsakirtzis et al., 2005). Our first aim is to study the dynamics of system (4.4), and, in particular, the efficiency (strength) of TET from the forced primary system to the NES. In this section the study of the damped dynamics is performed through direct numerical simulations of the equations of motion and post-processing of the transient results. We do this in order to establish the ranges of parameters for which efficient targeted energy transfer from the primary system to the NES takes place. In later sections we will study TET in (4.4) using analytic techniques.

An extensive series of numerical simulations is performed over different regions of the parameter space of the system, in order to establish the system parameters for which optimal passive TET from the primary system to the NES occurs. Moreover, by varying the linear coupling stiffness α of the primary system, we study the influence of the spacing of the two eigenfrequencies ω_1, and ω_2 on TET. The numerical simulations are carried out by assigning different sets of initial conditions of the primary system, with the NES always being initially at rest. To assess the strength of passive TET from the primary system to the NES, the following energy dissipation measure (EDM) is numerically computed:

$$E(t) = \frac{\varepsilon\lambda_2}{E_{\text{in}}} \int_0^t \left[(\dot{v}_1(\tau) - \dot{v}_2(\tau))^2 + (\dot{v}_2(\tau) - \dot{v}_3(\tau))^2\right] d\tau \tag{4.5}$$

where E_{in} is the input energy provided to the system by the initial conditions. This non-dimensional EDM represents the instantaneous portion of input energy dissipated by the NES up to time instant t; it follows that by means of (4.5) we can obtain a qualitative measure of the effectiveness of the MDOF NES to passively absorb and locally dissipate vibration energy from the primary system. Clearly, due to the fact the system examined is purely passive (with energy being continuously lost due to damping dissipation) the instantaneous EDM should reach a definite asymptotic limit which is symbolically denoted as

$$E_{\text{NES}} = \lim_{t \gg 1} E(t) \tag{4.6}$$

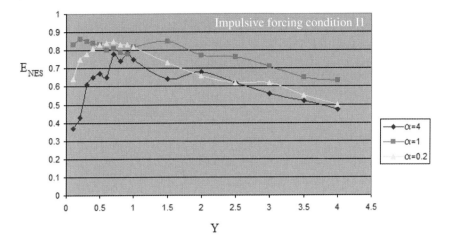

Fig. 4.6 EDM for varying values of single impulse Y (impulsive forcing condition I1) and coupling stiffness a of the primary system.

This asymptotic EDM represents the portion of input energy that is *eventually* dissipated by the NES. In the following exposition, the asymptotic evaluation (4.6) is used as a measure of the efficiency of TET from the primary system to the MDOF NES. We note, however, that the EDMs (4.5) and (4.6) can not describe the time scale of TET, i.e., how rapidly energy gets transferred and dissipated by the NES; clearly, in certain applications the time scale of energy transfer is an important factor for assessing NES efficiency but this issue will not be pursued further in this section (however, it will be revisited in later sections and chapters). It suffices to state that the use of NESs with non-smooth nonlinearities drastically decreases the time scale of energy dissipation (Georgiadis et al., 2005); in addition, as shown in Section 3.4.2.5 the excitation of impulsive orbits affects the time-scale of TET dynamics.

As shown below, for weak coupling between the primary system and the NES, efficient passive TET from the primary system to the NES can be achieved for small values of the mass parameter μ and nonlinear characteristic C_2 of the NES with all other parameters being quantities of $O(1)$. This combination of system parameters leads to large relative displacements between the particles of the NES, which, in turn, leads to large energy dissipation by the dampers of the NES. Hence, a basic conclusion drawn from the numerical study is that lightweight MDOF NESs with weak nonlinear stiffnesses C_2 are effective energy absorbers and dissipators; this is an interesting conclusion from the practical point of view, since it renders such lightweight NESs applicable for a diverse set of engineering applications.

The numerical simulations were performed for the following system parameters:

$$\varepsilon = 0.2, \quad \alpha = 1.0, \quad C_1 = 4.0, \quad C_2 = 0.05, \quad \varepsilon\lambda_1 = \varepsilon\lambda_2 = \varepsilon\lambda = 0.01,$$
$$\mu \to \varepsilon^2\mu = 0.08, \quad \omega_0^2 = 1.0$$

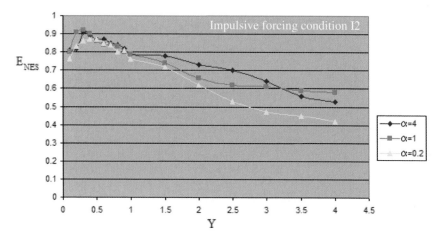

Fig. 4.7 EDM for varying values of the in-phase impulses Y (impulsive forcing condition I2) and coupling stiffness a of the primary system.

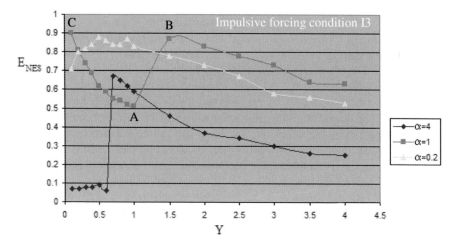

Fig. 4.8 EDM for varying values of the out-of-phase impulses Y (impulsive forcing condition I3), and coupling stiffness a of the primary system; symbols A and B at the plot corresponding to $\alpha = 1$ refer to the results depicted in Figures 4.11 and 4.12, respectively.

and three types of impulsive forcing conditions – IFCs (or, equivalently, initial conditions – velocities) for the primary system: (i) *single IFC* designated by I1, corresponds to $F_1(t) = Y\delta(t)$ (or, equivalently, $\dot{u}_1(0) = Y$), and all other initial conditions zero; (ii) *in-phase IFC* designated by I2, with $F_1(t) = F_2(t) = Y\delta(t)$ and all other initial conditions zero; and (iii) *out-of-phase IFC* I3, with $F_1(t) = -F_2(t) = Y\delta(t)$ and all other initial conditions zero.

In Figures 4.6–4.8 we depict the asymptotic EDM E_{NES} (e.g., the portion of input energy eventually dissipated by the NES) as function of the magnitude of the

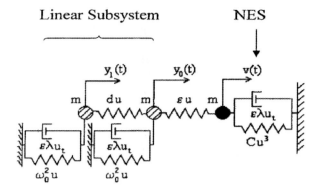

Fig. 4.9 The system with SDOF NES attachment whose dynamics is compared to the system depicted in Figure 4.2.

impulse Y and the linear coupling stiffness of the primary system α, for the above three types of IFCs. In all cases, a significant portion (reaching as high as 86% for IFC I1; 92% for IFC I2; and 90% for IFC I3) of the input energy gets passively absorbed and dissipated by the MDOF NES. This significant passive TET occurs in spite of the fact that the (directly forced) primary linear system and the NES have identical dashpots. Moreover, the energy transfer is *broadband*, since the vibration energy absorption takes place over wide frequency ranges.

Whereas the portion of energy eventually dissipated at the NES depends on the level of energy input and the closeness of the natural frequencies of the primary system (as expected, since the system considered is nonlinear), this dependence is less pronounced compared to the case of the SDOF NES. This is concluded when comparing the performance of the MDOF NES to that of the SDOF NES depicted in Figure 4.9 (Vakakis et al., 2004) – this is performed in the comparative plot of Figure 4.10 for a system with $\alpha = 0.2$, and IFCs I1-I3 – and also by considering the results reported in Chapter 3. The system with SDOF NES whose response is depicted in Figure 4.10 is identical to that of Figure 4.2, but with the MDOF NES being replaced by a single mass of magnitude 3μ grounded by means of an essential cubic stiffness nonlinearity with characteristic $C = 1.0$ and weak viscous damper $\varepsilon\lambda$. So it is clear that a significant improvement of efficiency of TET is achieved by using the multi-DOF NES; in addition, TET for the case of the MDOF NES is more robust to variations of the input force compared to the SDOF case.

Particularly notable is the capacity of the MDOF NES to absorb a significant portion of the input energy *even for low applied impulses*. Such low-energy targeted energy transfer is markedly different from the performance of SDOF NESs, where, as reported in previous works (Vakakis et al., 2004; McFarland et al., 2004) and in Chapter 3 of this work, TET is 'activated' only when the magnitude of input energy exceeds a certain critical threshold. For the case of the MDOF NES such a critical energy threshold can only be detected in the energy plot for $\alpha = 4$ of Figure 4.8, e.g., only in the case when the primary system possesses well separated natural

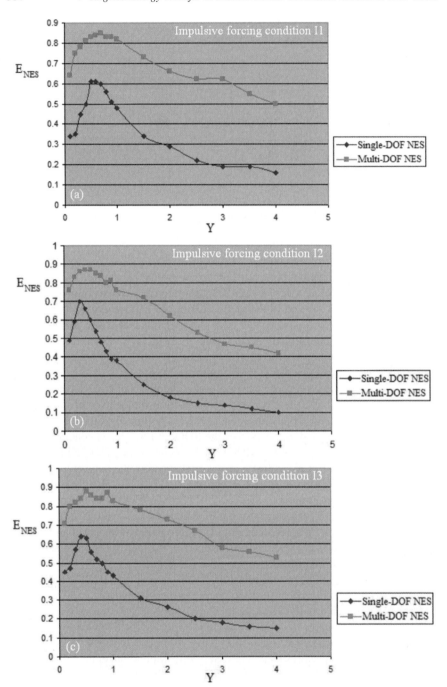

Fig. 4.10 Comparisons of EDMs for primary systems attached to SDOF and MDOF NESs, and impulsive forcing conditions: (a) I1, (b) I2, and (c) I3.

frequencies and is excited by out-of-phase initial conditions. In all other cases (Figures 4.6–4.8) no such critical input energy threshold is identified. This interesting dynamical feature of the MDOF NES will be reconsidered in more detail in a later section; here it suffices to state that the capacity of MDOF NES for low-energy TET is enabled by the rich structure of periodic orbits (NNMs) of the underlying Hamiltonian system, a subset of which localize to the NES with decreasing energy due to damping dissipation.

Of particular interest is the plot of E_{NES} depicted in Figure 4.8 corresponding to $\alpha = 1$ (for the case when the natural frequencies of the uncoupled primary system are equal to $\omega_1 = 1.7321$, $\omega_2 = 1.0$ rad/sec) and out-of-phase impulse excitations. In that plot we note that for sufficiently small impulse magnitudes, the portion of energy dissipated by the MDOF NES develops an initial local minimum before reaching higher values. To gain insight into the dynamics of targeted energy transfer in that region, in Figures 4.11 and 4.12 the numerical spectra of Cauchy wavelet transforms (WTs) of the internal relative NES displacements $[v_2(t) - v_1(t)]$ and $[v_3(t) - v_2(t)]$ at points labeled A and B of Figure 4.8 are depicted. Point A corresponds to the case of relatively weak TET from the primary system to the MDOF NES, whereas, point B to a case where nearly 90% of the input energy gets absorbed and eventually dissipated by the NES. The WT spectra depict the amplitude of the WT as function of frequency (vertical axis) and time (horizontal axis). Heavy shaded areas correspond to regions where the amplitude of the WT is high whereas lightly shaded regions correspond to low amplitudes. Such plots enable one to deduce the temporal evolutions of the dominant frequency components of the signals analyzed.

Comparing the two responses of Figures 4.11 (point A) and 4.12 (point B), it is clear that the enhanced TET noted in the later case is due mainly to the large-amplitude transient relative response $[v_3(t) - v_2(t)]$. Moreover, judging from the corresponding WT spectrum, this time series consists of a 'fast' oscillation with frequency close to ω_1, that is modulated by a large-amplitude 'slow' envelope. Additionally, one notes that this modulated response is not sustained over time, but takes place only in the initial phase of the motion and escapes from this regime of the motion at approximately $t = 50$. Similar behavior is noted for the time series of the other relative response, $[v_2(t) - v_1(t)]$ depicted in Figure 4.12. It is well established (Vakakis et al., 2004; Panagopoulos et al., 2004; McFarland et al., 2004) that this represents a TRC of the NES dynamics on a resonance manifold near the out-of-phase linear mode of the uncoupled primary system, which results in enhanced and irreversible energy transfer from the primary system to the NES. Comparing the responses of Figures 4.12 and 4.11, it is clear that in the later case (where weaker TET occurs) the transient responses are dominated by sustained frequency components indicating excitation of NNMs, rather than occurrence of TRCs. The frequencies of some of the excited NNMs differ from the linearized natural frequencies ω_1 and ω_2, indicating the presence of essentially nonlinear modes in the response, having no linear analogs.

From the above discussion it is clear that the transient dynamics of the dissipative system of Figure 4.2 is rather complex. Moreover, the numerical results depicted in

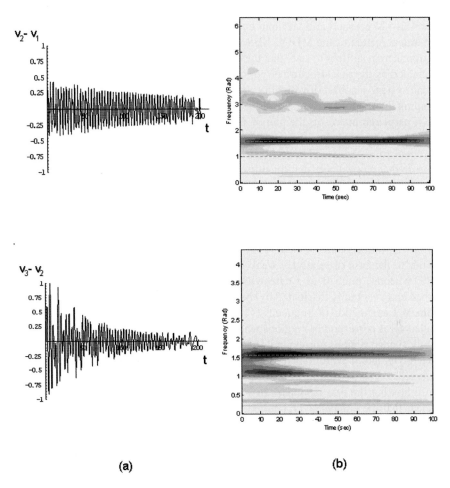

Fig. 4.11 Internal NES relative displacements for out-of-phase impulses (IFC I3) with $Y = 1$ and $\alpha = 1$ (point A in Figure 4.8): (a) Time series, (b) Cauchy wavelet transforms; the linear natural frequencies of the uncoupled ($\varepsilon = 0$) primary system are indicated by dashed lines.

Figures 4.6–4.12 indicate that the MDOF NES leads to enhanced TET compared to the SDOF NES, a conclusion that provides ample motivation for a systematic and detailed study of the corresponding transient dynamics. This is performed in the following sections. We start our study by considering the underlying Hamiltonian system (i.e., the corresponding system with no dissipation), and show that the Hamiltonian dynamics influences drastically the weakly damped responses and, hence, controls TET.

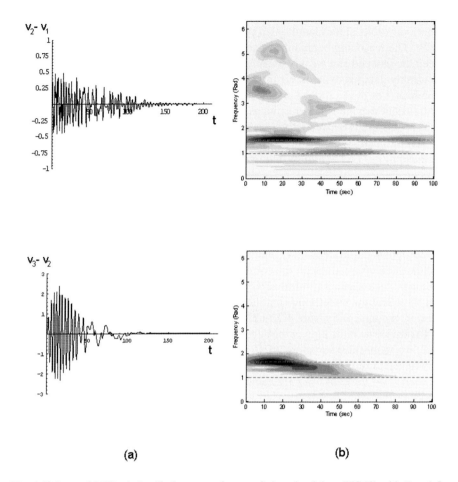

Fig. 4.12 Internal NES relative displacements for out-of-phase impulses (IFC I3) with $Y = 1.5$ and $\alpha = 1$ (point B in Figure 4.8): (a) Time series, (b) Cauchy wavelet transforms; the linear natural frequencies of the uncoupled ($\varepsilon = 0$) primary system are indicated by dashed lines.

4.2 The Dynamics of the Underlying Hamiltonian System

The results reported in the previous sections provide ample motivation to study the dynamics of the system depicted in Figure 4.2. Our aim is to better understand the different regimes of the motion, and the dynamic mechanisms that govern passive TET from the directly excited primary system to the MDOF NES (Tsakirtzis, 2006; Tsakirtzis et al., 2007). A first step towards analyzing the dynamics of system (4.4) is to study the structure of the periodic orbits of the corresponding Hamiltonian system (with no damping terms, $\varepsilon\lambda = 0$). Then, to show that passive TET as well as other type of complicated transient dynamics of the weakly damped system (4.4) can be explained and interpreted in terms of transitions between different branches of periodic orbits of the Hamiltonian system in an appropriate frequency-energy plot (FEP). The reasoning behind this plan has to do with the intricate relationship between the damped and weakly undamped systems, and on the paradoxical fact that the weakly damped dynamics is mainly determined by the underlying Hamiltonian dynamics (Lee et al., 2005; Kerschen et al., 2006). Indeed, the effect of damping in the transient dynamics is parasitic, as it does not generate new dynamics but only invokes transitions between different branches of solutions (NNMs) of the underlying Hamiltonian system. It follows that *although damping is prerequisite for TET, the dynamics of TET is mainly determined by the underlying Hamiltonian structure of the dynamics.*

We will employ both analytical and numerical techniques to show that the undamped (Hamiltonian) system possesses a surprisingly complicated structure of periodic orbits that give rise to complicated phenomena and damped transitions. This result should not be unexpected given the high degeneracy of the linear structure of the dynamical system (4.4), which is expected to lead to complicated, high-codimension bifurcations on the corresponding high-dimensional center manifold. Although such a general bifurcation study is beyond the scope of this work, we will show that the underlying Hamiltonian dynamics influence the weakly damped transient dynamics of Figure 4.2, and, in essence, governs TET.

To provide an indication of the degeneracy of the system with an attached MDOF NES, we reconsider equations (4.4) and set the damping parameters and forcing terms equal to zero. Changing into modal coordinates of the primary (linear) system, $w_1 = u_1 + u_2$, $w_2 = u_1 - u_2$, the equations of motion can be placed in the following form:

$$\ddot{w}_1 + \omega_0^2 w_1 + (\varepsilon/2)(w_1 - w_2) - \varepsilon v_1 = 0$$
$$\ddot{w}_2 + (\omega_0^2 + 2\alpha)w_2 - (\varepsilon/2)(w_1 - w_2) + \varepsilon v_1 = 0$$
$$\mu \ddot{v}_1 + C_1(v_1 - v_2)^3 + \varepsilon[v_1 - (1/2)(w_1 - w_2)] = 0$$
$$\mu \ddot{v}_2 + C_1(v_2 - v_1)^3 + C_2(v_2 - v_3)^3 = 0$$
$$\mu \ddot{v}_3 + C_2(v_3 - v_2)^3 = 0 \qquad (4.7)$$

Placing these equations into state form we obtain:

$$[\dot{w}_1 \quad \ddot{q}_1 \quad \dot{w}_2 \quad \ddot{q}_2 \quad \dot{v}_1 \quad \ddot{q}_3 \quad \dot{v}_2 \quad \ddot{q}_4 \quad \dot{v}_3 \quad \ddot{q}_5]^T =$$

$$\begin{bmatrix} 0 & 1 & & & & & & & & 0 \\ -\omega_0^2-(\varepsilon/2) & 0 & (\varepsilon/2) & 0 & \varepsilon & & & \bullet & & \\ 0 & 0 & 0 & 1 & & & & \bullet & & \\ (\varepsilon/2) & 0 & -(\omega_0^2+2\alpha)-(\varepsilon/2) & 0 & -\varepsilon & & & \bullet & & \\ 0 & 0 & 0 & 0 & 0 & 1 & 0 & 0 & 0 & 0 \\ (\varepsilon/2\mu) & 0 & -(\varepsilon/2\mu) & & 0 & (-\varepsilon/\mu) & 0 & 0 & 0 & 0 \\ & & & \bullet & & & 0 & 1 & & \\ & & & \bullet & & & 0 & 0 & & \\ & & \bullet & & & & & & 0 & 1 \\ 0 & & & & & & & & 0 & 0 \end{bmatrix} \begin{bmatrix} w_1 \\ q_1 \\ w_2 \\ q_2 \\ v_1 \\ q_3 \\ v_2 \\ q_4 \\ v_3 \\ q_5 \end{bmatrix} +$$

$$-(C_1/\mu)\begin{bmatrix} 0 & 0 & 0 & 0 & 0 & (v_1-v_2)^3 & 0 & (v_2-v_1)^3 + \kappa(v_2-v_3)^3 & 0 & \kappa(v_3-v_2)^3 \end{bmatrix}^T$$

(4.8)

with $\kappa = C_2/C_1$. In the limit of zero coupling between the primary system and the MDOF NES, $\varepsilon \to 0$, the combined system degenerates into a system with two pairs of imaginary eigenvalues and three double zero eigenvalues. Clearly, this is a highly degenerate dynamical system, with a ten-dimensional center manifold (which coincides with the entire phase space of the system). Such degenerate dynamical systems possess highly co-dimensional bifurcation structures, which give rise to complicated regular and chaotic dynamics (Guckenheimer and Holmes, 1983; Wiggins, 1990), and their study is beyond the current state-of-the-art. However, by narrowing our aim to the study of TET, it is possible to apply analytical techniques to the study of the dynamics of this highly degenerate system.

Hence, we reconsider the dynamics of the five-DOF essentially nonlinear Hamiltonian system which is derived by removing the damping terms from equations (4.4). There are various numerical algorithms that compute the periodic orbits of this system, and in this work the numerical algorithm described in Tsakirtzis et al. (2005) is followed. To compute the periodic orbits of this system, first it is assumed that a periodic orbit of the Hamiltonian system is realized for the initial velocity vector $[\dot{u}_1(0) \quad \dot{u}_2(0) \quad \dot{v}_1(0) \quad \dot{v}_2(0) \quad \dot{v}_3(0)]$ with zero initial displacements; then, the algorithm computes this initial condition vector together with the period T, for which the following periodicity condition is satisfied:

$$[u_1(T) \, u_2(T) \, v_1(T) \, v_2(T) \, v_3(T) \, \dot{u}_1(T) \, \dot{u}_2(T) \, \dot{v}_1(T) \, \dot{v}_2(T) \, \dot{v}_3(T)]^T$$
$$- [0 \, 0 \, 0 \, 0 \, 0 \, \dot{u}_10 \, \dot{u}_20 \, \dot{v}_10 \, \dot{v}_20 \, \dot{v}_30]^T = 0 \quad (4.9)$$

The algorithm has been implemented in Matlab® using optimization techniques. For a given value of the period T, the objective function to minimize is the norm of the left-hand side of equation (4.9), with the optimization variables being the five non-zero initial velocities. By varying the period, a frequency-energy plot (FEP) can be drawn, depicting the dominant frequency of a periodic motion (NNM) as function of the corresponding (conserved) energy of the Hamiltonian system. When more than one dominant frequencies exist (for example, when two coordinates have different dominant frequency components), the lowest of these dominant frequencies is depicted in the FEP.

In the following sections we consider two configurations of MDOF NESs, principally distinguished by the order of magnitude of their masses. The aim of the study is to assess the influence of the masses of the NESs on TET.

4.2.1 System I: NES with $O(1)$ Mass

The first system configuration considered (referred to from now on as 'System I') consists of a relatively heavy nonlinear attachment, and system parameters:

$$\mu = 1.0, \quad \omega_0^2 = 1.0, \quad \alpha = 1.0, \quad \varepsilon = 0.1, \quad C_1 = 2.0, \quad C_2 = \varepsilon^2 \quad \text{(System I)}$$

The small value of the nonlinear characteristic C2 was dictated by the numerical results of the previous section, where it was found that for small values of C2 enhanced TET from the primary system to the NES was realized. First, we discuss certain features of the dynamics of this system in the frequency-energy plot (FEP).

A first observation related to the system of equations (4.4), is that for no damping and forcing, and *in the limit of small energy and finite frequencies* the dynamics of the system is approximately governed by the following linear subsystem of equations (4.4):

$$\ddot{u}_1 + (\omega_0^2 + \alpha)u_1 - \alpha u_2 = 0$$

$$\ddot{u}_2 + (\omega_0^2 + \alpha + \varepsilon)u_2 - \alpha u_1 - \varepsilon v_1 = 0 \quad \text{(Limit of low energies finite frequencies)}$$

$$\mu \ddot{v}_1 + \varepsilon(v_1 - u_2) = 0 \tag{4.10}$$

In that case the periodic orbits of the full undamped and unforced nonlinear system tend to the three eigenmodes of the linear subsystem (4.10), with corresponding eigenfrequencies, $f_1 = 1.7473$, $f_2 = 1.0265$, and $f_3 = 0.3054$ rad/s.

A second observation is that *in the limit of high energies and finite frequencies* the essentially nonlinear stiffnesses of system (4.4) behave approximately as massless rigid links, resulting in the following alternative approximate linear subsystem:

$$\ddot{u}_1 + (\omega_0^2 + \alpha)u_1 - \alpha u_2 = 0$$

$$\ddot{u}_2 + (\omega_0^2 + \alpha + \varepsilon)u_2 - \alpha u_1 - \varepsilon v_1 = 0 \quad \text{(Limit of high energies finite frequencies)}$$

$$3\mu \ddot{v}_1 + \varepsilon(v_1 - u_2) = 0 \tag{4.11}$$

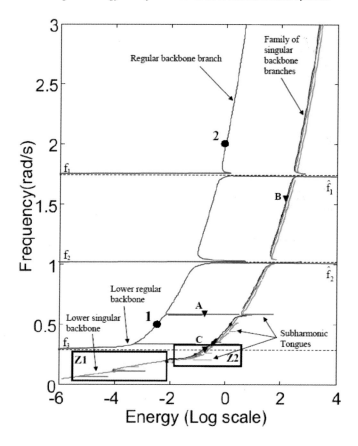

Fig. 4.13 Frequency-energy plot (FEP) of the Hamiltonian dynamics of System I; symbols A, B and C are the initial conditions for the damped transitions depicted in Figures 4.26, 4.27 and 4.28, respectively.

Then the periodic motions of the Hamiltonian system tend asymptotically to the linear eigenfrequencies, $\hat{f}_1$, $\hat{f}_2$ and $\hat{f}_3$ of subsystem (4.11). For System I these frequencies are equal to $\hat{f}_1 = 1.766$, $\hat{f}_2 = 1.0248$, and $\hat{f}_3 = 0.1766$ rad/s.

These observations are important in order to understand the complicated structure of periodic orbits of the Hamiltonian System I in the FEP. This will lead also to clear interpretations of multi-frequency damped transitions, as sudden jumps between distinct branches of solutions in the FEP. The FEP for the periodic orbits of System I is depicted in Figure 4.13, together with two enlarged regions Z1, and Z2 showing in detail certain domains of the plot (see Figures 4.14, 4.15). Indicated also in the plot are the natural frequencies f_i, $\hat{f}_i$ of the limiting linear systems (4.10) and (4.11). Unless in the neighborhood of one of the six natural frequencies f_i, $\hat{f}_i$, $i = 1, 2, 3$, the response of the primary subsystem is small, and the motion is localized to the nonlinear attachment.

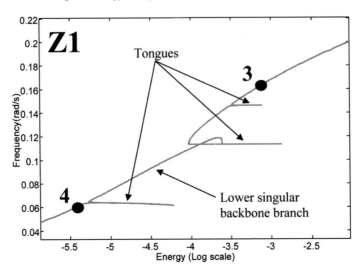

Fig. 4.14 Enlarged region Z1.

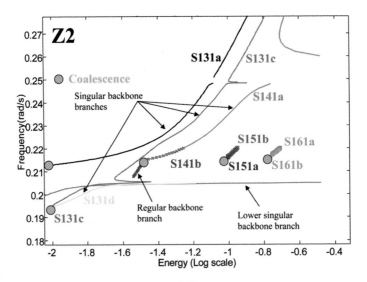

Fig. 4.15 Enlarged region Z2.

Regarding the general features of the FEP, we note that it contains two basic types of branches: *backbone (global) branches* consisting of multi-frequency periodic motions defined over extended frequency and energy ranges; and *local branches* termed *subharmonic tongues* consisting of multi-frequency periodic motions, with frequencies defined only in neighborhoods of certain basic frequencies. Each tongue is defined over a finite energy range, and consists of two subharmonic branches of periodic solutions (NNMs), which at a critical energy value coalesce in a bifurcation that

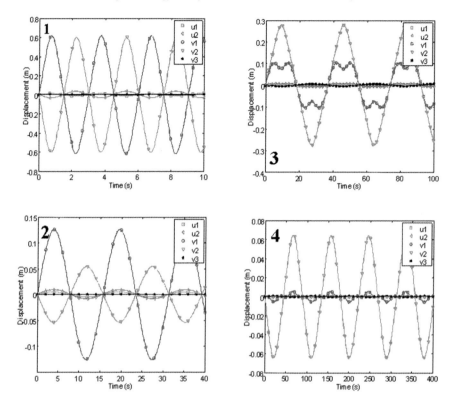

Fig. 4.16 Time series of the periodic motions of System I corresponding to the indices indicated at the FEP of Figures 4.13 and 4.14.

signifies the end of that particular tongue and the elimination of the corresponding subharmonic motion.

Moreover, there exists a *regular backbone branch* where the last mass of the nonlinear attachment (the NES) has nearly zero amplitude (e.g., $v_3 \approx 0$). Periodic motions (NNMs) on this regular backbone are approximately *monochromatic*, that is, all coordinates of System I vibrate approximately in-unison with identical dominant frequencies; NNMs on that regular backbone branch correspond to either in-phase or out-of-phase relative motions of the particles of the system. On this branch, the motion is always localized to the first two masses of the nonlinear attachment, except in the vicinity of the natural frequencies of the low-energy limiting linear subsystem (4.10), and at the extremities of the two lower tongues observed in Figure 4.13; one of these tongues occurs at $f_1/3 = 0.58$ rad/s, and the other at $f_2/3 = 0.34$ rad/s.

A countable infinity of additional subharmonic tongues occurs in the neighborhoods of frequencies that are in rational relationships to the basic frequencies f_1, f_2 and f_3 of subsystem (4.10), but these are not represented in the FEP of Figure 4.13. The time histories depicted in Figures 4.16a, b (points 1 and 2) show that the motion on the regular backbone branch is mainly monochromatic, and that, indeed, the MDOF NES vibrates with the same frequency as the primary system. At point 1 the displacements of the two masses of the primary system oscillate in out-of-phase fashion, whereas at point 2 in in-phase fashion.

Another interesting feature of the FEP of System I is that, besides the regular backbone branch, there exist *additional singular backbone branches* at higher values of the energy (the term 'singular' is justified by the analysis of the next section). Each of the singular backbone branches may also carry tongues of subharmonic periodic motions. For instance, a lower tongue appears around $f_2/3 = 0.34$ rad/s for each of the three singular backbone branches depicted in Figure 4.14. There are basic qualitative differences between the additional singular backbone branches and the main backbone branch: first, the amplitude of oscillation of the last mass of the NES takes finite values at the singular backbone branches; second, for periodic motions on the singular backbone branches the particles of the system oscillate with differing dominant frequency components (this contrasts to the regular backbone branch where all particles oscillate with identical dominant frequency components). Indeed, *the singular backbone branches consist of subharmonic motions that are defined over wide frequency ranges of the FEP*, in contrast to subharmonic motions on the tongues that are localized to frequencies rationally related to f_i and $\hat{f}_i$.

It is interesting that the additional family of backbone curves of System I is not limited to the three singular branches depicted in Figure 4.13. Indeed, as shown later a countable infinity of singular backbone branches exists in the FEP, a result substantiated by numerical evidence. In particular, an extended computation of the periodic orbits of System I performed at a fixed dominant frequency $\omega = 1.5$ rad/s, yielded as many as eleven distinct periodic orbits (NNMs) distinguished by their energy and frequency contents (i.e., they possess different composition of harmonics); however, some of these orbits are unstable. The computed initial conditions of these orbits are listed in Table 4.1, together with their corresponding energies. All these periodic orbits (NNMs) on the singular backbones have two common features: first, the motion of System I is always strongly localized to the nonlinear attachment; second, they all correspond to approximately the same motion of the primary system, since the linear out-of-phase mode is predominantly excited at this particular frequency.

The difference between these periodic solutions becomes clear when the Fast Fourier Transforms (FFTs) of the corresponding time series are considered. Whereas the relative displacement $[v_2(t) - v_3(t)]$ contains always the dominant component at $\omega = 1.5$ rad/s, the dominant harmonic component of $[v_1(t) - v_2(t)]$ varies depending on the specific orbit considered. This enables us to label the singular backbone branches with the notation $S1jp$. The first index refers to the dominant frequency of the primary system (in this case $\omega = 1.5$ rad/s), whereas the second indicates that the dominant frequency of $[v_1(t) - v_2(t)]$ is j times the dominant frequency of the primary system; the third index indicates that the dominant frequency of $[v_2(t) - v_3(t)]$ is p times of that of the primary system. Following this notation, the regular backbone branch of the FEP of Figure 4.13 is labeled as $S111$ (since it is approximately monochromatic, i.e., all particles oscillate with identical domi-

Table 4.1 Initial conditions and energies of the periodic orbits of System I for $\omega = 1.5$ rad/s.

Solutions	Feature	$\dot{u}_1(0)$	$\dot{u}_2(0)$	$\dot{v}_1(0)$	$\dot{v}_2(0)$	$\dot{v}_3(0)$	Energy
1	1:1	0.9388	−0.2456	−7.4273	−6.2384	13.2846	2.1327
2	1:1	0.9649	−0.2484	−8.1159	−5.5059	13.2263	2.1337
3	1:3	0.7854	−0.2296	−2.4476	−10.9023	13.0534	2.1701
4	1:3	0.7855	−0.2269	−3.0519	−10.5010	13.2539	2.1701
5	1:3	0.7851	−0.2240	−3.6277	−10.0961	13.4223	2.1701
6	1:4	0.7333	−0.2291	3.8334	−15.0532	10.9568	2.2576
7	1:4	0.7278	−0.1484	−14.3179	1.5783	12.4052	2.2576
8	1:5	0.7300	−0.1582	−19.7692	7.2405	12.1996	2.4718
9	1:5	0.7105	−0.2134	8.1081	−19.7099	11.3402	2.4649
10	1:6	0.7083	−0.2091	13.9267	−25.9124	11.7220	2.7004
11	1:6	0.7081	−0.1551	−25.9220	14.0430	11.5617	2.7004

nant frequency), and the additional backbones as $S131, S141, S151, \ldots$. Generally speaking, the higher the dominant harmonic of $[v_1(t) - v_2(t)]$ is, the higher is the energy of the corresponding periodic orbit.

Starting from $\omega \approx 0.22$ rad/s coalescences between different backbone branches occur sequentially as shown in Figure 4.15; these are saddle-node (SN) bifurcations. Coalescences occur between two branches with similar motion, labeled by (a) and (b) (for instance, S151a coalesces with S151b). At the coalescence points, the motion is identical to that on the regular backbone branch, meaning that the coalescing branches meet the regular backbone branch at the coalescence points. Hence, with diminishing frequency the different families of singular backbone branches eventually disappear through coalescences, and a single low frequency singular backbone branch eventually emerges, termed lower singular backbone branch. On this branch, the last mass of the NES has very small displacement but the overall motion of System I is still localized to the first two masses of the nonlinear attachment; this is confirmed by the simulations of Figures 4.16c, d corresponding to points 3 and 4 on the lower singular backbone branch.

Summarizing, the most interesting feature of the frequency-energy plot (FEP) of System I is the existence of a countable infinity of closely spaced singular backbone branches that extend over wide ranges of frequencies and energies. This feature of the dynamics is novel, and differs from the FEPs discussed in Chapter 3 corresponding to SDOF NESs. In the following section we consider the same primary system – MDOF NES configuration but with $O(\varepsilon)$ masses, in order to assess the effect on the dynamics of a reduction of the NES masses.

4.2.2 System II: NES with O(ε) Mass

We now reconsider the system depicted in Figure 4.2 with weak nonlinear stiffness C_2 and small NES masses; this system we label as 'System II'. It is shown that by reducing the masses of the NES the complexity of the dynamics increases, and the capacity of the system for TET is significantly enhanced. Hence, the unforced and undamped system (4.4) is considered again with parameters:

$$\varepsilon = 0.2, \quad \alpha = 1.0, \quad C_1 = 4.0, \quad C_2 \to \varepsilon^2 C_2 = 0.05,$$
$$\mu \to \varepsilon^2 \mu = 0.08, \quad \omega_0^2 = 1.0 \quad \text{(System II)}$$

We are interested to study the effect on the dynamics of a reduction of the masses of the nonlinear attachments, and to relate TET to the topological structure of periodic orbits of the FEP of the underlying Hamiltonian system. Moreover, we wish to compare the FEP of this system to that of System I. The underlying Hamiltonian system in this case takes the form:

$$\ddot{u}_1 + (\omega_0^2 + \alpha)u_1 - \alpha u_2 = 0$$
$$\ddot{u}_2 + (\omega_0^2 + \alpha + \varepsilon)u_2 - \alpha u_1 - \varepsilon v_1 = 0$$
$$\varepsilon^2 \mu \ddot{v}_1 + C_1(v_1 - v_2)^3 + \varepsilon(v_1 - u_2) = 0$$
$$\varepsilon^2 \mu \ddot{v}_2 + C_1(v_2 - v_1)^3 + \varepsilon^2 C_2(v_2 - v_3)^3 = 0$$
$$\varepsilon^2 \mu \ddot{v}_3 + \varepsilon^2 C_2(v_3 - v_2)^3 = 0 \quad (4.12)$$

Regarding ε as a perturbation parameter of the problem, system (4.12) is expected to possess complicated dynamics as $\varepsilon \to 0$ since it is essentially (strongly) nonlinear, high-dimensional, and singular [in three of equations (4.12) the highest derivatives are multiplied by the perturbation parameter squared].

The periodic orbits of System II were computed utilizing the numerical algorithm described in the previous section for System I. In Figure 4.17 the periodic orbits of (4.12) are presented in a FEP, and in Figure 4.18 some representative orbits are presented. Since the numerical algorithm could not reliably capture the lowest frequency branch, this was analytically computed (as discussed later) and superimposed to the numerical results. These results provide an indication of the complexity of the dynamics.

As for the case of System I, the FEP contains both a regular backbone and a family of singular backbones. In this case, however, the singular backbone branches are not densely packed as in System I. Moreover, for System II the backbone branches of periodic orbits (NNMs) are defined over wider frequency and energy ranges compared to System I, and no subharmonic tongues were revealed. Hence, it appears that *by reducing the masses of the NES the local subharmonic tongues are eliminated*; that is, there are no subharmonic motions at frequencies rationally related to the natural frequencies f_1, f_2, f_3 of the linear subsystem (4.10) (for System II these frequencies assume the values $f_1 = 1.8529$, $f_2 = 1.5259$, $f_3 = 0.9685$ rad/s). As for the case of System I, in the limit of high energies and moderate frequencies, System II reaches the linear limiting system (4.11), with corresponding limiting natural frequencies given by $\hat{f}_1 = 1.7734$, $\hat{f}_2 = 1.1200$, and $\hat{f}_3 = 0.7960$ rad/s.

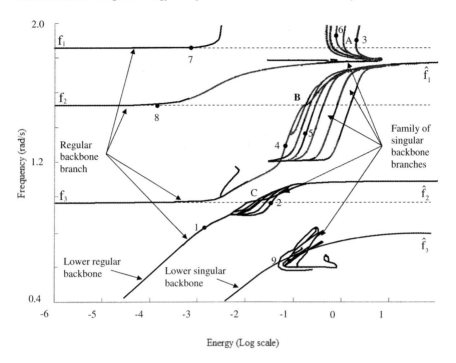

Fig. 4.17 FEP of the periodic orbits of System II; indices refer to the time series depicted in Figure 4.18.

It is interesting to consider the dynamics of System II at points A, B and C of the plot of Figure 4.17, i.e., at points where the regular backbone branch crosses the natural frequencies of the low-energy limiting linear system (4.10). At these points it holds approximately that, $v_1 \approx v_2$ and $v_2 \approx -v_3$, so the system may be approximately decomposed into two subsystems: the subsystem (4.10) (the limiting linear system for low energies and finite frequency), and a strongly nonlinear system composed of the first two masses of the NES with their center of mass being approximately motionless. At points A, B and C the linear subsystem vibrates on one of its linear modes at frequencies f_1, f_2 or f_3, whereas the nonlinear attachment adjusts its energy to oscillate with the same frequency. Hence, the energy of the nonlinear subsystem (together with the energy of the linear subsystem) determines the points of crossing A,B and C of the regular backbone curve with each of the natural frequencies of the linear limiting subsystem (4.10).

An additional remark is that the reduction of the masses of the NES causes a 'spreading out' of the closely spaced members of the family of singular backbones of System I. As a result, multiple subharmonic periodic orbits coexist over wider energy ranges compared to System I (though some of these orbits are unstable and, hence, not physically realizable). The elimination of the subharmonic tongues and the spreading of the family of singular backbone curves imply that in System II

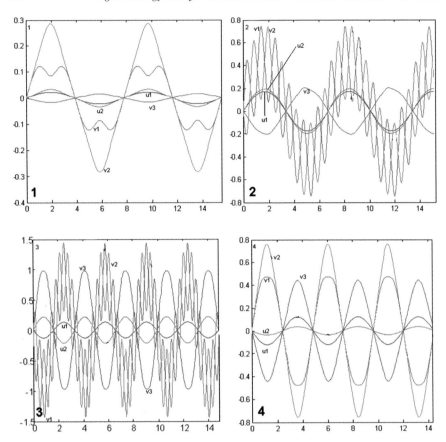

Fig. 4.18 Periodic orbits at specific points (indicated by numbers) in the FEP of Figure 4.17 (System II).

subharmonic motions are realized only on the singular backbone curves (instead of tongues as in System I) that extend over wide regions of the FEP. These features of the FEP will have profound effects on the transient responses of the weakly damped System II, which are examined in the next section. Moreover, it will be shown that System II possesses enhanced TET properties compared to System I.

4.2.3 Asymptotic Analysis of Nonlinear Resonant Orbits

In this section we initiate the analytical study of the Hamiltonian dynamics of system (4.4) (with zero damping and forcing terms). Specifically, we mathematically study certain aspects of NNMs on the regular and singular backbone curves, and explain analytically the multiplicity (fine structure) of the family of singular backbone

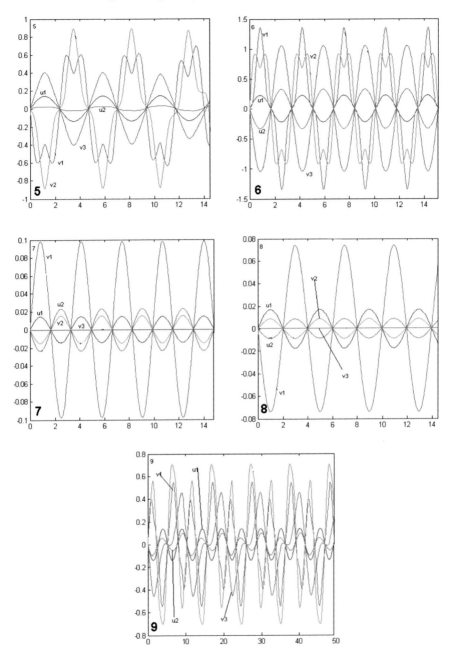

Fig. 4.18 Continued.

branches in the FEP. In the following analysis we consider in detail only the undamped and unforced System I, and show that the results can be extended to System II by an appropriate time transformation.

To this end we consider the system of coupled oscillators (Tsakirtzis, 2006; Tsakirtzis et al., 2007),

$$\ddot{u}_1 + (\omega_0^2 + \alpha)u_1 - \alpha u_2 = 0$$
$$\ddot{u}_2 + (\omega_0^2 + \alpha + \varepsilon)u_2 - \alpha u_1 - \varepsilon v_1 = 0$$
$$\mu \ddot{v}_1 + C_1(v_1 - v_2)^3 + \varepsilon(v_1 - u_2) = 0$$
$$\mu \ddot{v}_2 + C_1(v_2 - v_1)^3 + \varepsilon^2 C_2(v_2 - v_3)^3 = 0$$
$$\mu \ddot{v}_3 + \varepsilon^2 C_2(v_3 - v_2)^3 = 0 \qquad (4.13)$$

and assume that all parameters other than ε are $O(1)$ scalars. The main goal of the analysis is to study the periodic motions (NNMs) of this system that possess a dominant frequency ω away from the natural frequencies of the limiting linear system that results as $\varepsilon \to 0$. First, only *non-resonant motions* are considered. Under the condition of absence of linear resonances, and assuming that the system executes a periodic oscillation with frequency ω, the approximations $\ddot{u}_1 \approx -\omega^2 u_1$ and $\ddot{u}_2 \approx -\omega^2 u_2$ are introduced, which approximately reduce the two leading differential equations of (4.13) to the following algebraic relations (taking $\alpha = \omega_0^2 = 1$ for simplicity):

$$u_1 \approx \frac{\varepsilon v_1}{(1-\omega^2)(3-\omega^2) + \varepsilon(2-\omega^2)} = O(\varepsilon)$$

(ω away from roots of denominator)

$$u_2 \approx \frac{\varepsilon v_1 (2-\omega^2)}{(1-\omega^2)(3-\omega^2) + \varepsilon(2-\omega^2)} = O(\varepsilon) \qquad (4.14)$$

These approximate algebraic relations replace (and thus simplify) two of the ordinary differential equations of system (4.13). The rationale behind this approximation is that away from their resonances the two linear oscillators vibrate approximately in a harmonic fashion with common frequency ω.

It follows that in the absence of resonance the Hamiltonian dynamics is governed mainly by the MDOF NES, as the response of the linear system is approximately computed by (4.14). Moreover, for frequencies ω away from the roots of the denominator of (4.14) (i.e., the linearized natural frequencies of the limiting system as $\varepsilon \to 0$), the periodic orbits of (4.13) are mainly localized to the MDOF NES, and governed approximately by the following reduced system:

$$\ddot{v}_1 + C_1(v_1 - v_2)^3 + \varepsilon v_1 = 0$$
$$\ddot{v}_2 + C_1(v_2 - v_1)^3 + \varepsilon^2 C_2(v_2 - v_3)^3 = 0$$
$$\ddot{v}_3 + \varepsilon^2 C_2(v_3 - v_2)^3 = 0 \qquad (4.15)$$

where the rescaling of time $t \to \sqrt{\mu}t$ was introduced. Finally, with the change of variables, $3z = v_1 + v_2 + v_3$, $q_1 = v_1 - v_2$ and $q_2 = v_2 - v_3$, the reduced system (4.15) is expressed as:

$$\ddot{z} + (\varepsilon/3)[z + (2q_1/3) + (q_2/3)] = 0$$
$$\ddot{q}_1 + (\varepsilon/3)[z + (2q_1/3) + (q_2/3)] + 2C_1 q_1^3 - \varepsilon^2 C_2 q_2^3 = 0$$
$$\ddot{q}_2 + 2C_2 \varepsilon^2 q_2^3 - C_1 q_1^3 = 0 \qquad (4.16)$$

The variable z describes the (slow time scale) oscillation of the center of mass of the MDOF NES, whereas the variables q_1 and q_2 are the relative oscillations between the NES masses (which occur at a faster time scale). As a result, the reduced system can be further decomposed into a 'slowly varying' component, i.e., the z-oscillator, and two 'fast varying' components, namely the coupled oscillators governing q_1 and q_2.

The reduced system (4.16) is the starting point for the pertubation analysis that follows. Before proceeding further we show that the dynamics of System II (possessing small NES masses) can be reduced also to the form (4.16) by a transformation of the time variable. Indeed, considering the undamped and unforced System II – equations (4.12) – the time transformation $\tau = \varepsilon \sqrt{\mu}$ is introduced. Assuming that the dominant frequency ω of the periodic orbit is away from the linear resonances, it can be shown that the responses of the linear subsystem can be expressed approximately as, $u_1 \approx u_2/(2\varepsilon^2 \mu - \omega^2)$ and $u_2 \approx \varepsilon v_1 [\varepsilon + 2\varepsilon^2 \mu - \omega^2 - 2/(2\varepsilon^2 \mu - \omega^2)]$, so the system reduces again to system (4.16). Hence, the following analytical results derived for System I also apply to System II for the rescaled time variable $\tau = \varepsilon \sqrt{\mu}$.

4.2.3.1 The Low-Frequency Limit

Assuming that $\omega \ll \sqrt{\varepsilon/3}$, i.e., that the dominant frequency of the response is much less than the linearized natural frequency of the first equation of the set (4.16), we may approximately neglect the second derivative $\ddot{z}$ from the first equation, and derive the following approximate algebraic expression for z:

$$z \approx -(2q_1/3) - (q_2/3) \qquad (4.17)$$

This approximation is valid only in the *low-frequency limit*, since only for sufficiently small frequencies the inertia term in the linear oscillator in (4.16) is of much smaller magnitude that the stiffness terms. Hence, we can reduce further system (4.16) to a system of two essentially nonlinear coupled oscillators:

$$\ddot{q}_1 + 2C_1 q_1^3 - \varepsilon^2 C_2 q_2^3 = 0$$
$$\ddot{q}_2 + 2C_2 \varepsilon^2 q_2^3 - C_1 q_1^3 = 0 \qquad \text{(Low frequency limit)} \qquad (4.18)$$

This is a *symmetric* system in the terminology of Rosenberg (1966), and its periodic solutions are *similar nonlinear normal modes* (NNMs) satisfying linear modal relationships of the form $q_2 = kq_1$, where k is the modal constant (see Section 2.1). In addition, these are synchronous periodic motions of system (4.18) where both coordinates oscillate in-unison, reaching their extreme values of the same instant of time, so that the resulting motion is represented by a straight line in the configuration plane (q_1, q_2) of the reduced system. Substituting the relation $q_2 = kq_1$ into (4.18), and imposing the requirement that both equations produce identical periodic solutions, we derive the following equation for determining the modal constant k, possessing two real roots:

$$\varepsilon^2 (C_2/C_1) k^4 + 2(C_2/C_1) \varepsilon^2 k^3 - 2k - 1 = 0 \Rightarrow$$

$$k_1 = -\frac{1}{2} - \frac{3\varepsilon^2 C_2}{32 C_1} + O(\varepsilon^2) \qquad \text{(Regular root)}$$

$$k_2 = \left(\frac{2C_1}{C_2}\right)^{2/3} \varepsilon^{-2/3} - \frac{1}{2} + O(\varepsilon^{2/3}) \qquad \text{(Singular root)} \quad (4.19)$$

The characterization of the two roots as 'regular' and 'singular' is related to the analysis that follows below.

Summarizing, at the low frequency (and low energy) limit the system possesses two branches of periodic solutions. These are precisely the two *low regular* and *low singular backbones* shown in the FEP of Figure 4.13 of System I and of Figure 4.17 for System II. The periodic solutions (NNMs) of the system on these low frequency branches are computed through integration by quadratures of either one of equations (4.18) after the modal relation $q_2 = kq_1$ is imposed:

$$\ddot{q}_1 + (2C_1 - \varepsilon^2 C_2 k_{1,2}) q_1^3 = 0$$
$$q_2 = k_{1,2} q_1, \quad z \approx -(2q_1/3) - (q_2/3), \qquad \text{(Low frequency limit)} \quad (4.20)$$

The solutions of the reduced system (4.20) can expressed analytically in terms of elliptic functions. These periodic solutions represent the low-frequency/low-energy asymptotic limits of the branches of NNMs of System I (and also of System II through the time transformation discussed previously).

4.2.3.2 The Case of Finite $O(1)$ Frequencies

The other limiting case is when the basic frequency ω of the periodic orbit is of $O(1)$, but away from the linear resonances. In this case the term $\varepsilon z/3$ in the first equation of system (4.16) is small compared to the second derivative $\ddot{z}$, so we may neglect it and express approximately the (slow) oscillation of the center of mass of the MDOF NES as follows:

$$z \approx (\varepsilon/3\omega^2)[(2q_1/3) + (q_2/3)] + O(\varepsilon^2) = (\varepsilon/9\omega^2)(2q_1 + q_2) + O(\varepsilon^2) \quad (4.21)$$

It follows that in this case we may reduce the system (4.16) to a two-DOF system, similarly to the low-frequency case [see equation (4.18)]:

$$\ddot{q}_1 + \frac{\varepsilon}{9}\left(1 + \frac{1}{\omega^2}\right)(2q_1 + q_2) + 2C_1 q_1^3 - \varepsilon^2 C_2 q_2^3 = 0 \quad (O(1) \text{ frequency})$$

$$\ddot{q}_2 + 2C_2 \varepsilon^2 q_2^3 - C_1 q_1^3 = 0 \tag{4.22}$$

System (4.22) represents a perturbed system of two coupled nonlinear oscillators, with ε being the perturbation parameter. This may be regarded as a non-symmetric perturbation of the symmetric system (4.18) derived for the low-frequency limit; however the added non-symmetric term $(\varepsilon/9)(1 + 1/\omega^2)(2q_1 + q_2)$ may produce non-trivial perturbations to the dynamics (since it represents a perturbation of an already degenerate-symmetric system), and requires careful consideration in the asymptotic analysis. Indeed, in what follows we prove that there exist two general classes of periodic solutions of (4.22): *regular solutions* based on regular perturbation analysis of the reduced set; and *singular solutions* based on singular asymptotic expansions of that set. These two types of asymptotic solutions correspond to the two types of backbone curves identified in previous sections in the FEP of System I, namely, regular and singular backbone branches with each type possessing distinct topological features and dynamical characteristics.

Starting with regular perturbation analysis, and omitting terms that depend on ε from (4.22) the following generating symmetric system is obtained:

$$\ddot{q}_{10} + 2C_1 q_{10}^3 = 0$$

$$\ddot{q}_{20} - C_1 q_{10}^3 = 0 \tag{4.23}$$

whose periodic solutions may be exactly computed by quadratures (and expressed in terms of elliptic functions). The solutions are similar NNMs in the terminology of Rosenberg (1966), since they satisfy the linear modal relationship $q_2 = (-1/2)q_1$. Recalling the previous analysis of the lower limiting case, we infer that solutions of (4.22) that are expressed as perturbations of the generating solutions obtained from system (4.23) can be regarded as finite-frequency analogs of the ones lying on the low regular backbone branch corresponding to the regular root $k_1 = -1/2 + O(\varepsilon^2)$ in relations (4.19).

The perturbed solutions for $q_1(t)$ and $q_2(t)$ are expressed as *regular perturbations* of the generating solutions of (4.23):

$$q_1(t) = q_{10}(t) + \varepsilon q_{11}(t) + \varepsilon^2 q_{12}(t) + \cdots$$

$$q_2(t) = q_{20}(t) + \varepsilon q_{21}(t) + \varepsilon^2 q_{22}(t) + \cdots \tag{4.24}$$

Substituting (4.24) into (4.22) an hierarchy of problems is derived (in increasing powers of ε) that govern the higher-order corrections to the high-frequency periodic solutions. These regular perturbation solutions lie on a single backbone branch of the FEP of System I, which is the high-frequency (and high-energy) limit of the regular backbone curve. Based on this approximation it is confirmed that the regular backbone in System I consists of a single branch and does not possess the fine structure of the family of singular backbone branches (see Figures 4.13–4.15). Moreover, this analytic approximation leads to the following estimate for the oscillation of the right end mass of the NES,

$$v_3 \approx \frac{\varepsilon q_1}{6\omega^2} + O(\varepsilon^2), \quad \omega = O(1) \qquad (4.25)$$

which in the high-frequency limit is negligible. This fully confirms the numerical results reported in Section 4.2.1.

We now consider periodic solutions of (4.22) that may be regarded as finite frequency continuations of periodic solutions on the low singular backbone, and correspond to the singular root $k_2 = (2C_1/C_2)^{2/3}\varepsilon^{-2/3}$ in (4.19). Based on the numerical findings of Figures 4.13–4.15, we deduce that for increasing frequency (and energy) there occurs a series of bifurcations giving rise to additional singular backbones containing solutions of increasingly higher frequency content. For finite [i.e., $O(1)$] frequencies we obtain an entire family of singular backbone branches that is densely packed in energy. The following analysis aims to analytically study this family of singular backbones for $O(1)$ frequencies (but away from linear resonances).

In this case we approximate the solutions through *singular asymptotic analysis*, and introduce the transformations $(q_1, q_2) \rightarrow (Q_2 = \varepsilon q_2, \eta = q_2 - k_2 q_1)$. Substituting these transformations into (4.22) the following rescaled equations are obtained, which govern periodic solutions (NNMs) on the family of singular backbone branches:

$$\ddot{Q}_2 + (3/2)C_2 Q_2^3 + (3\varepsilon/2)C_2 Q_2^2 \eta = 0$$

$$\varepsilon^{2/3}\ddot{\eta} + 6\left(\frac{C_2^2}{2C_1}\right)^{1/3} Q_2^2 \eta = \frac{1}{9}\left(\frac{2C_1}{C_2}\right)^{1/3}\left(1 + \frac{1}{\omega^2}\right) Q_2 \qquad (4.26)$$

The first equation represents an $O(\varepsilon)$ parametric perturbation of a strongly nonlinear oscillator. The second equation is singular, as noted from the small coefficient of the derivative term. It is a quasi-linear equation with combined parametric and external excitations. It is well known that this type of excitation produces families of periodic solutions of increasingly higher frequency content (in the case of pure parametric excitation these periodic solutions lie on stability-instability boundaries according to Floquet theory). Hence, from the model (4.26) we may *indirectly infer the existence of countable infinities of periodic solutions (due to combined parametric/external resonances) with increasingly higher frequency contents*. These correspond to the family of periodic solutions realized on the family of singular backbone curves in the FEP; moreover, the previous analytical arguments indicate that the numerically observed fine structure of singular backbones of Figure 4.13 consists of a countable infinity of distinct branches. Apart from the common basic frequency ω, different members of the family of singular backbones possess increasingly higher harmonics at frequencies $n\omega$, $n = 2, 3, \ldots$, which are generated by the previously described combined parametric and external resonances in the second of equations (4.26).

The fine structure of the family of singular backbone branches (interestingly enough, it resembles *quantization at closely spaced discrete values of energy*) is analytically studied by defining the averaged energy of oscillation, E, of a periodic orbit (NNM):

$$\langle E \rangle_t = \left\langle \frac{\dot{v}_1^2}{2} + \frac{\dot{v}_2^2}{2} + \frac{\dot{v}_3^2}{2} + \frac{\varepsilon v_1^2}{2} + \frac{C_1 q_1^4}{4} + \frac{\varepsilon^2 C_2 q_2^4}{4} \right\rangle_t = \quad (4.27)$$

$$= \left\langle \frac{3\dot{z}^2}{2} + \frac{1}{3}\left(\dot{q}_1^2 + \dot{q}_1\dot{q}_2 + \dot{q}_2^2\right) + \frac{\varepsilon}{2}\left(z + \frac{2q_1}{3} + \frac{q_2}{3}\right) + \frac{C_1 q_1^4}{4} + \frac{\varepsilon^2 C_2 q_2^4}{4} \right\rangle_t$$

We claim that the averaged value of the potential energy between NNMs on distinct singular branches is almost unaffected by the perturbation due to the fine structure of the family. This claim is based in the following reasoning. As mentioned previously, the fine structure is formed due to parametric and external resonances in the second of equations (4.26), which, in addition to oscillations at the basic frequency ω, produce high-frequency harmonics in η possessing similar amplitudes but increasingly higher frequency components $n\omega$, $n = 2, 3, \ldots$. Actually, it holds that $|\eta| \sim |q_2|/k_2 \sim \varepsilon^{2/3}|q_2|$. Hence, *the corrections to the potential energies due to singular perturbations will be insignificant, and, as a result, the fine structure of the family of singular backbones will be determined mainly by fluctuations of the averaged kinetic energy T*.

The fluctuations of the kinetic energy between different branches of the family of singular backbones is evaluated as follows:

$$\langle T \rangle_t = \left\langle \frac{3\dot{z}^2}{2} + \frac{1}{3}\left(\dot{q}_1^2 + \dot{q}_1\dot{q}_2 + \dot{q}_2^2\right) \right\rangle_t$$

$$\sim \left\langle \frac{3\dot{z}^2}{2} + \frac{1}{3}\left(k_2^{-2}(\dot{q}_2 - \dot{\eta})^2 + \dot{q}_2 k_2^{-1}(\dot{q}_2 - \dot{\eta}) + \dot{q}_2^2\right) \right\rangle_t$$

$$\sim \left\langle \frac{3\dot{z}^2}{2} + \frac{\dot{q}_2^2}{3}\left(1 + \frac{1}{k_2} + \frac{1}{k_2^2}\right) + \frac{\dot{\eta}^2}{3k_2^2} \right\rangle_t \sim T_0 + \frac{1}{3k_2^2}\langle \dot{\eta}^2 \rangle_t \quad (4.28)$$

where T_0 is the average value of the kinetic energy, and we have taken into account that since $\dot{q}_2$ and $\dot{\eta}$ have different dominant frequencies they average out from the final expression in (4.28). Now, taking into account that at high frequencies it holds that, $Q_2 \sim \omega \Rightarrow q_2 \sim \varepsilon^{-1}\omega$, and that $|\eta| \sim |q_2|/k_2 \sim \varepsilon^{2/3}|q_2| \sim \varepsilon^{-1/3}\omega$, it follows that $\dot{\eta}^2 \sim n^2\omega^2|\eta|^2 \sim \varepsilon^{-2/3}n^2\omega^4$. From (4.28) it is concluded that in the high frequency limit the averaged kinetic energy behaves according to

$$\langle T \rangle_t = T_0 + \frac{\varepsilon^{2/3}}{3} C_0 \omega^4 n^2 \Rightarrow$$

$$\langle E \rangle_t = E_0 + \frac{\varepsilon^{2/3}}{3} C_0 \omega^4 n^2 \sim \omega^4 (\varepsilon^{-2} + D_0 n^2 \omega^2 \varepsilon^{2/3}) \qquad (4.29)$$

since $E_0 \sim \varepsilon^{-2} \omega^4$.

Hence, *the splitting distances between members of the family of singular backbones of System I is of $O(n^2 \omega^2 \varepsilon^{2/3})$*, where n is the order of parametric resonance of the periodic solution for η (or equivalently, the high-frequency harmonic component in η). On the logarithmic scale used to depict energy in the numerical FEP of Figure 4.13, the splitting distance is scaled according to $\ln(\langle E \rangle_t) \sim \ln(\varepsilon^{-2} + D_0 n^2 \omega^2 \varepsilon^{2/3}) \sim n^2 \omega^2 \varepsilon^{8/3}$. This analytical result is in agreement with the numerical results depicted in the FEP.

The previous analysis directly applies also to System II. Indeed, taking into account the previous rescaling of time that relates Systems I and II, we only need to apply the frequency rescaling, $\omega \to \omega/(\varepsilon\sqrt{\mu})$, to extend the previous analytical findings to System II. The resulting scaling in the frequency-energy plot of System II is $\ln(\langle E \rangle_t) \sim n^2 \omega^2 \varepsilon^{2/3}$, correctly predicting the 'spreading out' of the fine structure of the family of singular backbone branches.

4.2.4 Analysis of Resonant Periodic Orbits

We now consider resonant nonlinear responses and transient resonance captures (TRCs) of system (4.4) by reducing the dynamics of system to a single integro-differential equation; we then discuss methodologies for the analytical treatment of the reduced system.

First, we focus on the resonant motions of system (4.4). Specifically, we study the nonlinear undamped and damped dynamics in the neighborhoods of the linear natural frequencies of the system, and discuss methods to analyze the resonant nonlinear interactions between the linear primary system and the MDOF NES. Contrary to the non-resonant analysis of Section 4.2.3, during resonance the components of the linear subsystem oscillate with finite amplitudes, and strong energy exchanges with the NES take place. It is precisely these motions close to resonances that lead to TET phenomena when damping is introduced.

First, we study analytically the periodic orbits (NNMs) of the undamped and unforced system (4.4) that result from resonance interactions, i.e., that possess dominant frequency components close to the $O(1)$ natural frequencies of the linear limiting system (4.10). To this end, we introduce again the coordinate transformation

$$R = \frac{v_1 + v_2 + v_3}{3}, \quad X_1 = v_2 - v_1, \quad X_2 = v_3 - v_2 \qquad (4.30)$$

where X_1, X_2 and R denote the two relative displacements, and the displacement of the center of mass of the MDOF NES, respectively. Substituting (4.30) into (4.4), and omitting damping terms for the moment, the undamped equations of motion take the form:

$$\ddot{u}_1 + (\omega_0^2 + \alpha)u_1 - \alpha u_2 = 0$$

$$\ddot{u}_2 + (\omega_0^2 + \alpha + \varepsilon)u_2 - \alpha u_1 - \varepsilon R = -\frac{\varepsilon}{3}(2X_1 + X_2)$$

$$3\mu \ddot{R} + \varepsilon R - \varepsilon u_2 = \frac{\varepsilon}{3}(2X_1 + X_2)$$

$$\mu \ddot{X}_1 + 2C_1 X_1^3 - C_2 X_2^3 - \varepsilon(R - \frac{(2X_1 + X_2)}{3} - u_2) = 0$$

$$\mu \ddot{X}_2 + 2C_2 X_2^3 - C_1 X_1^3 = 0 \qquad (4.31)$$

In the following analysis, unless explicitly noted, the system parameters are assumed to be $O(1)$ quantities. Considering the transformed set of equations (4.31), it is noted that the motion of the center of mass of the NES also executes a linear (but slow) motion which results as weak perturbation of the rigid body mode $\ddot{R} = 0$.

The next step of the analysis involves a linear coordinate transformation that brings the leading three linear equations of system (4.31) into Jordan canonical form (note that the last two equations are perturbations of essentially nonlinear, i.e., non-linearizable, equations). To this end, we introduce the linear modal transformation

$$\begin{Bmatrix} u_1 \\ u_2 \\ R \end{Bmatrix} = \begin{bmatrix} 1 & 0 & 0 \\ 0 & 1 & 0 \\ 0 & 0 & 1/\sqrt{3\mu} \end{bmatrix} \begin{bmatrix} T_{1,1} & T_{2,1} & T_{3,1} \\ T_{1,2} & T_{2,2} & T_{3,2} \\ T_{1,3} & T_{2,3} & T_{3,3} \end{bmatrix} \begin{Bmatrix} Q_1 \\ Q_2 \\ Q_3 \end{Bmatrix} \qquad (4.32)$$

where T_{ij} denotes the j-th component of the i-th eigenvector of the following symmetric matrix:

$$\Omega = \begin{bmatrix} (\omega_0^2 + \alpha) & -\alpha & 0 \\ -\alpha & (\omega_0^2 + \alpha + \varepsilon) & -\varepsilon/\sqrt{3\mu} \\ 0 & -\varepsilon/\sqrt{3\mu} & \varepsilon/3\mu \end{bmatrix} \qquad (4.33)$$

Substituting the transformation (4.32) into (4.31), the following alternative set of equations of motion is obtained:

$$\ddot{Q}_1 + \hat{\omega}_1^2 Q_1 = \frac{\varepsilon}{3}(2X_1 + X_2)(T_{1,3}/\sqrt{3\mu} - T_{1,2})$$

$$\ddot{Q}_2 + \hat{\omega}_2^2 Q_2 = \frac{\varepsilon}{3}(2X_1 + X_2)(T_{2,3}/\sqrt{3\mu} - T_{2,2})$$

$$\ddot{Q}_3 + \hat{\omega}_3^2 Q_3 = \frac{\varepsilon}{3}(2X_1 + X_2)(T_{3,3}/\sqrt{3\mu} - T_{3,2})$$

$$\mu \ddot{X}_1 + 2C_1 X_1^3 - C_2 X_2^3 + \varepsilon \frac{(2X_1 + X_2)}{3}$$

$$= \varepsilon \left((T_{1,3}/\sqrt{3\mu} - T_{1,2})Q_1 + (T_{2,3}/\sqrt{3\mu} - T_{2,2})Q_2 + (T_{3,3}/\sqrt{3\mu} - T_{3,2})Q_3 \right)$$

$$\mu \ddot{X}_2 + 2C_2 X_2^3 - C_1 X_1^3 = 0 \qquad (4.34)$$

where the linearized natural frequencies are defined as follows:

$$\hat{\omega}_1^2 = \omega_1^2 + \varepsilon/2 + O(\varepsilon^2), \quad \hat{\omega}_2^2 = \omega_2^2 + \varepsilon/2 + O(\varepsilon^2), \quad \hat{\omega}_3^2 = \varepsilon/3\mu + O(\varepsilon^2) \quad (4.35)$$

and $\omega_1 > \omega_2$ are the two natural frequencies of the uncoupled linear system corresponding to $\varepsilon = 0$. The elements T_{ij} in (4.34) are defined as follows:

$$T_{1,1} = -1/\sqrt{2} + O(\varepsilon), \quad T_{1,2} = +1/\sqrt{2} + O(\varepsilon), \quad T_{1,3} = 0 + O(\varepsilon),$$
$$T_{2,1} = +1/\sqrt{2} + O(\varepsilon), \quad T_{2,2} = +1/\sqrt{2} + O(\varepsilon), \quad T_{2,3} = 0 + O(\varepsilon),$$
$$T_{3,1} = 0 + O(\varepsilon), \quad T_{3,2} = 0 + O(\varepsilon), \quad T_{3,3} = 1 + O(\varepsilon)$$

Physically, the variables $Q_1(t)$ and $Q_2(t)$ are modal coordinates of the out-of-phase and the in-phase modes, respectively, of the uncoupled linear primary system; whereas $Q_3(t)$ is the coordinate describing the (slow) motion of the center of mass of the MDOF NES. It is noted that the following relations hold between the linearized frequencies $\hat{\omega}_i$ and the natural frequencies f_i of the linear limiting (4.10) (these are defined in Section 4.2.1):

$$\hat{\omega}_1 = f_1 + O(\varepsilon), \quad \hat{\omega}_2 = f_2 + O(\varepsilon), \quad \hat{\omega}_3 = f_3 + O(\sqrt{\varepsilon}) = O(\sqrt{\varepsilon})$$

Considering the system of equations (4.34), we partition it into two subsets: a set of three linear uncoupled oscillators that are weakly 'forced' by terms that depend linearly on the NES relative displacements; and a set of two coupled, essentially nonlinear oscillators that govern the relative displacements within the MDOF NES. This partition is very useful in the following analysis in order to perform a reduction of the dynamics to a single integro-differential equation.

Finally, motivated again by the numerical results of the previous section, we introduce the additional assumption that the stiffness characteristic C_2 of the NES is small; this is imposed by introducing the rescaling $C_2 \to \varepsilon^2 C_2 = O(\varepsilon^2)$. Under these assumptions, and assuming that $0 < \varepsilon \ll 1$, the first subset of three uncoupled linear equations of the system (4.34) can be solved explicitly as follows:

$$Q_1(t) = Q_1(0)\cos\hat{\omega}_1 t + \frac{\dot{Q}_1(0)}{\hat{\omega}_1}\sin\hat{\omega}_1 t$$
$$+ \frac{\varepsilon\left(-T_{1,2} + T_{1,3}/\sqrt{3\mu}\right)}{3\hat{\omega}_1} \int_0^t [2X_1(\tau) + X_2(\tau)]\sin\hat{\omega}_1(t-\tau)d\tau$$

$$Q_2(t) = Q_2(0)\cos\hat{\omega}_2 t + \frac{\dot{Q}_2(0)}{\hat{\omega}_2}\sin\hat{\omega}_2 t$$
$$+ \frac{\varepsilon\left(-T_{2,2} + T_{2,3}/\sqrt{3\mu}\right)}{3\hat{\omega}_2} \int_0^t [2X_1(\tau) + X_2(\tau)]\sin\hat{\omega}_2(t-\tau)d\tau$$

$$Q_3(t) = Q_3(0)\cos\hat{\omega}_3 t + \frac{\dot{Q}_3(0)}{\hat{\omega}_3}\sin\hat{\omega}_3 t$$

$$+ \frac{\varepsilon\left(-T_{3,2}+T_{3,3}/\sqrt{3\mu}\right)}{3} \int_0^t [2X_1(\tau)+X_2(\tau)]\sin\hat{\omega}_3(t-\tau)d\tau \tag{4.36}$$

Hence, the modal coordinates of the linear subsystem and the displacement of the center of mass of the MDOF NES are expressed (in exact form) in terms of the relative displacements $X_1(t)$ and $X_2(t)$ between the particles of the NES. Note, however, that since $\hat{\omega}_3 = O(\sqrt{\varepsilon})$, *the center of mass of the NES executes a slow oscillation*; this was anticipated previously by the observation that this motion is the weak perturbation of the rigid body motion $\ddot{Q}_3 = 0$.

Considering now the last of equations (4.34), and taking into account the previous rescaling $C_2 \to \varepsilon^2 C_2 = O(\varepsilon^2)$, the following analytic approximation for the variable $X_2(t)$ is obtained:

$$\mu\ddot{X}_2 = C_1 X_1^3 - 2\varepsilon^2 C_2 X_2^3 \Rightarrow$$

$$X_2(t) = \mu^{-1} \int_0^t \int_0^\tau C_1 X_1^3(s) ds\, d\tau + O(\varepsilon^2) \tag{4.37}$$

where we have taken into account that the MDOF NES is initially at rest [so that the initial conditions $X_2(0) = \dot{X}_2(0) = 0$ were imposed in (4.37)]. As a result, the relative displacement $X_2(t)$ is approximately expressed in terms of the relative displacement $X_1(t)$. Finally, substituting the previous results into the fourth of equations (4.34) *the full dynamics is approximately reduced to a single, essentially nonlinear integro-differential equation in terms of the dependent variable* $X_1(t)$:

$$\ddot{X}_1 + (2C_1/\mu) X_1^3 + (2\varepsilon/3\mu) X_1$$

$$= -(\varepsilon/3\mu^2) \int_0^t \int_0^\tau C_1 X_1^3(s) ds\, d\tau + \varepsilon^2 \hat{C}_2 \left[\mu^{-1} \int_0^t \int_0^\tau C_1 X_1^3(s) ds\, d\tau\right]^3$$

$$+ (\varepsilon/\mu)\left[(T_{1,3}/\sqrt{3\mu} - T_{1,2})\left(Q_1(0)\cos\hat{\omega}_1 t + \frac{\dot{Q}_1(0)}{\hat{\omega}_1}\sin\hat{\omega}_1 t\right)\right.$$

$$+ \frac{\varepsilon(-T_{1,2} + T_{1,3}/\sqrt{3\mu})}{3\hat{\omega}_1} \int_0^t$$

$$\times \left[2X_1(\tau) + \mu^{-1}\int_0^\tau \int_0^w C_1 X_1^3(s) ds\, dw\right] \sin\hat{\omega}_1(t-\tau)d\tau\right)$$

$$+ (T_{2,3}/\sqrt{3\mu} - T_{2,2})\left(Q_2(0)\cos\hat{\omega}_2 t + \frac{\dot{Q}_2(0)}{\hat{\omega}_2}\sin\hat{\omega}_2 t\right.$$

$$+ \frac{\varepsilon(-T_{2,2} + T_{2,3}/\sqrt{3\mu}}{3\hat{\omega}_2} \int_0^t$$

$$\times \left[2X_1(\tau) + \mu^{-1}\int_0^\tau \int_0^w C_1 X_1^3(s) ds\, dw\right] \sin\hat{\omega}_2(t-\tau)d\tau\right)$$

$$+ (T_{3,3}/\sqrt{3\mu} - T_{3,2})\left(Q_3(0)\cos\hat{\omega}_3 t + \frac{\dot{Q}_3(0)}{\hat{\omega}_3}\sin\hat{\omega}_3 t\right.$$

$$+ \frac{\varepsilon(-T_{3,2} + T_{3,3}/\sqrt{3\mu})}{3}\int_0^t$$

$$\times \left[2X_1(\tau) + \mu^{-1}\int_0^\tau\int_0^w C_1 X_1^3(s)ds\,dw\right]\sin\hat{\omega}_3(t-\tau)d\tau\bigg)\bigg] + O(\varepsilon^3)$$

$$\equiv \varepsilon f_1(X_1;\varepsilon) + \varepsilon^2 f_2(X_1;\varepsilon) + O(\varepsilon^3) \quad (4.38)$$

Strongly nonlinear dynamical systems with similar structure to (4.38) were analyzed asymptotically in Vakakis et al. (2004) and Panagopoulos et al. (2004). Solutions that possess a dominant (fast frequency) harmonic component, may be portioned into slow and fast components by imposing the following *ansatz*:

$$X_1(t) \approx A(t)\cos\theta(t) \quad (4.39)$$

where $A(t)$ and $\theta(t)$ represent the slowly-varying amplitude and phase of the response, respectively. Hence, by expressing the solution of (4.38) in the form (4.39) the solution is expressed as a fast oscillation modulated by slowly varying envelope. Clearly the (slow) variation of the envelope represents the important (essential) dynamics that govern the resonance interactions between the primary system and the MDOF NES. Substituting (4.39) into (4.38) we obtain the following approximate modulation equations that govern the slow evolution of the amplitude and phase,

$$\frac{dA(t)}{dt} \approx \varepsilon g_1(A(t),\theta(t),\varepsilon^{1/2}t,\hat{\omega}_1 t,\hat{\omega}_2 t) + \varepsilon^2 g_2(A(t),\theta(t),\varepsilon^{1/2}t,\hat{\omega}_1 t,\hat{\omega}_2 t)$$

$$\frac{d\theta(t)}{dt} \approx \Omega(t) + \varepsilon h_1(A(t),\theta(t),\varepsilon^{1/2}t,\hat{\omega}_1 t,\hat{\omega}_2 t)$$

$$+ \varepsilon^2 h_2(A(t),\theta(t),\varepsilon^{1/2}t,\hat{\omega}_1 t,\hat{\omega}_2 t)$$

$$\Omega(t) = \frac{\pi A(t)\sqrt{2C_1/\mu}}{2K(1/\sqrt{2})} \quad (4.40)$$

where $\Omega(t) = O(1)$ is the instantaneous frequency of the fast oscillation, and $K(1/\sqrt{2})$ is the complete elliptic integral of the first kind. The functions g_i and h_i, $i = 1, 2$ in (4.40) are 2π-periodic in the slow angle θ and the slow time $\varepsilon^{1/2}t$, but their dependences on the other time scales $\hat{\omega}_1 t$ and $\hat{\omega}_2 t$ depend on the specific values of the linearized natural frequencies $\hat{\omega}_1$ and $\hat{\omega}_2$. This means that the terms on the right-hand sides of relations (4.40) might be either periodic or quasi-periodic functions in terms of the fast time t, depending on if the frequency ratio $\hat{\omega}_1/\hat{\omega}_2$ is a rational or irrational number, respectively. We note that these terms also depend on the slow time $\varepsilon^{1/2}t$.

Equations (4.40) are *modulation equations* and apply for arbitrary values of the basic fast frequency of the solution. For further analysis we need to impose additional restrictions on the fast frequency $\Omega(t)$, and confine the analysis locally in frequency; this will introduce an additional slow independent variable in the modulation equations that will enable us to analyze resonant periodic orbits of the Hamiltonian system with frequencies close to the natural frequencies of f_1 and f_2 of the linear subsystem (4.10) [or equivalently – correct to $O(\varepsilon)$ – to the frequencies $\hat{\omega}_1$ and $\hat{\omega}_2$].

To provide an example of such a local analysis we restrict the fast frequency $\Omega(t)$ to be approximately equal to $\hat{\omega}_1$, and aim to study resonant periodic motions of the Hamiltonian system with dominant frequencies close to the higher natural frequency of the linear subsystem. To this end, we define the amplitude of oscillation, R, by the following frequency relation:

$$\hat{\omega}_1 = \frac{\pi R \sqrt{2C_1/\mu}}{2K(1/\sqrt{2})} \quad (4.41)$$

and introduce two new variables, namely, a slow angle variable $\chi(t)$ and an amplitude perturbation $\alpha(t)$:

$$\chi(t) = \theta(t) - \hat{\omega}_1 t, \quad A(t) = R + \sqrt{\varepsilon}\alpha(t) \quad (4.42)$$

By considering the relations (4.42) into (4.40) we study periodic motions in an $O(\sqrt{\varepsilon})$-neighborhood of the 1-1 resonance manifold in the phase space of the system, defined by the resonance condition (4.41). Hence, we aim to reduce the general modulation equation (4.40) to a *local system* valid in the $O(\sqrt{\varepsilon})$-neighborhood of this 1-1 resonance manifold.

Substituting (4.42) into the general modulation equations (4.40) the following *local modulation equations* are obtained:

$$\frac{d\alpha(t)}{dt} \approx \varepsilon^{1/2} G\big(\alpha(t), \chi(t) + \hat{\omega}_1 t, \hat{\omega}_1 t, \hat{\omega}_2 t, \varepsilon^{1/2} t; \varepsilon\big) \quad (4.43)$$

$$\frac{d\chi(t)}{dt} \approx \varepsilon^{1/2} \frac{\pi \alpha(t) \sqrt{2C_1/\mu}}{2K(1/\sqrt{2})} + \varepsilon H\big(\alpha(t), \chi(t) + \hat{\omega}_1 t, \hat{\omega}_1 t, \hat{\omega}_2 t, \varepsilon^{1/2} t; \varepsilon\big)$$

where G and H represent appropriately defined functions with the arguments shown above. Further analysis of the reduced modulation equations (4.43) can be performed by applying perturbation techniques, for example by applying the method of averaging [indeed, equations (4.43) are in standard form for applying averaging over the 'fast' time variable t] or the method of multiple scales [as performed in Panagopoulos et al. (2004)]. The analysis will yield approximate asymptotic expressions for the periodic orbits and their frequencies. In addition, the dynamical flow in the approximate slow phase plane of the modulation equations (4.43) can be derived. It is clear that the analysis (and the dynamics of the local model) will depend among other factors on the nature of the ratio of the linearized natural frequencies $\hat{\omega}_1/\hat{\omega}_2$. For example, if this ratio is rational the functions $\varepsilon^{1/2}G$ and εH in (4.43) become periodic functions in the fast time t (so, for example, simple averaging can be applied with respect to the fast time scale in order to analyze the local dynamics); whereas, if the frequency ratio is irrational the same functions become quasi-periodic in the fast time scale. These observations will dictate the type of asymptotic analysis that should be applied to study the dynamics of the local undamped system (4.43).

We note that the above reduction into modulation equations governing the slow flow dynamics can be applied also to analytically study transient resonance captures (TRCs) in the weakly damped dynamics (for example, the dynamics depicted in Figure 4.12). Indeed, considering the weakly damped system (4.4), and applying the previous reduction process, the five equations of motion can be reduced to the following single reduced integro-differential equation:

$$\ddot{X}_1 + (2C_1/\mu)X_1^3 + \varepsilon\hat{\lambda}\dot{X}_1 + (2\varepsilon/3\mu)X_1 = \varepsilon\hat{f}_1(X_1) + \varepsilon^2\hat{f}_2(X_1) + O(\varepsilon^3) \quad (4.44)$$

where $\varepsilon\hat{\lambda}$ denotes a weak damping coefficient, and $\varepsilon\hat{f}_1$, $\varepsilon^2\hat{f}_2$ are integro-differential operators analogous to the operators εf_1, $\varepsilon^2 f_2$ in (4.38), respectively, but modified to account for the additional weak damping terms. The analysis follows the general steps discussed previously, and can be applied to study local TRCs in neighborhoods of resonance manifolds defined by frequency relations similar to (4.41) (Panagopoulos et al., 2004).

Perhaps a disadvantage of the described approach for studying resonant motions is that the resulting integro-differential equations are quite complicated, which makes their analytical treatment cumbersome. To address this limitation, in the remainder of this section we formulate an alternative approach for analyzing the global structure of the resonant periodic orbits (NNMs) of the undamped and unforced system (4.4), based on complexification and averaging (CX-A). This approach is similar to the analytical approach introduced in Chapter 3, and in the context of the present analysis, it is applied only to study the resonant periodic orbits that are connected to the regular backbone branch [where all particles of system (4.4) oscillate with identical dominant frequencies]; however, similar analysis can be applied to develop analytic approximations for solutions on the family of singular backbone branches and on the local subharmonic tongues. This can be performed by selecting in each case the appropriate ansatz to replace the one utilized in the following analysis.

The alternative method for analyzing resonant motions in system (4.4) relies on complexification of the dynamics, followed by slow / fast partition of the response (see Section 2.4). The analysis is performed under the assumption that the resonant response possesses a single 'fast' frequency (satisfying a rational relation with a linear eigenvalues of the primary system), that is modulated by a 'slowly' varying envelop containing the important (essential) dynamics that we wish to study. The following procedure outlines the formulation of a *slow flow problem*, e.g., the derivation of the set of slow modulation equations governing the essential dynamics. As discussed in Lee et al. (2006) and demonstrated in Section 3.3.2 this procedure can be extended to study periodic or quasi-periodic motions possessing more than one 'fast' frequencies.

The first step of the alternative analytical method based on CX-A is to introduce the following set of complex dependent variables, each of which contains as real part the velocity of a particle of the system and as imaginary part the corresponding displacement multiplied by the (single) fast frequency:

$$\psi_1 = \dot{u}_1 + j\omega u_1, \quad \psi_2 = \dot{u}_2 + j\omega u_2,$$

$$\psi_3 = \dot{v}_1 + j\omega v_1, \quad \psi_4 = \dot{v}_2 + j\omega v_2 \text{ and } \psi_5 = \dot{v}_3 + j\omega v_3$$

where $j = (-1)^{1/2}$, and ω is the dominant (fast) frequency of the periodic resonant motion that we wish to study. Then, the displacements and accelerations can be expressed in terms of the new complex variables and their complex conjugates; for example, considering the velocity and acceleration of the first mass of the primary system we obtain, $\dot{u}_1 = (\psi_1 - \psi_1^*)/2j\omega$ and $\ddot{u}_1 = \dot{\psi}_1 - (j\omega/2)(\psi_1 + \psi_1^*)$, where (*) denotes complex conjugate. Moreover, since we seek approximately monochromatic periodic solutions in the fast time scale (i.e., solutions that possess a single common fast frequency), the previous complex variables may be expressed in polar form as

$$\psi_1 = \phi_1 e^{j\omega t}, \quad \psi_2 = \phi_2 e^{j\omega t},$$

$$\psi_3 = \phi_3 e^{j\omega t}, \quad \psi_4 = \phi_4 e^{j\omega t}, \quad \psi_5 = \phi_5 e^{j\omega t} \quad (4.45)$$

where the complex, time-varying amplitudes $\phi_i(t)$, $i = 1, \ldots, 5$, are slowly-varying amplitude modulations of the 'fast' oscillations $e^{j\omega t}$.

Employing the *ansatz* (4.45) it is possible to perform a partition of the resonant response of the system into slow and fast components, and to derive the approximate set of modulation equations governing the slow flow dynamics. This is performed by expressing the undamped and unforced equations (4.4) in terms of the complex variables (ψ_i and then) ϕ_i, and averaging the transformed equations over the fast variable ωt to retain only terms of fast frequency ω. In essence, this averaging process amounts to disregarding terms in the nonlinear equations of motion that possess fast components possessing frequencies higher than ω; the resulting approximate set of averaged equations is expected to be valid only in neighborhoods of the FEP close to the fast frequency ω.

Adopting the previously described averaging procedure we derive the following approximate set of first-order complex equations governing the amplitudes ϕ_i:

$$\dot{\phi}_1 + \phi_1 \left(\frac{j\omega}{2} - \frac{j\omega_0^2}{2\omega} \right) - \frac{j\alpha}{2\omega}(\phi_1 - \phi_2) = 0$$

$$\dot{\phi}_2 + \phi_2 \left(\frac{j\omega}{2} - \frac{j\omega_0^2}{2\omega} \right) - \frac{j\alpha}{2\omega}(\phi_2 - \phi_1) - \frac{j\varepsilon}{2\omega}(\phi_2 - \phi_3) = 0$$

$$\mu \left(\dot{\phi}_3 + \frac{j\omega}{2}\phi_3 \right) - \frac{j\varepsilon}{2\omega}(\phi_3 - \phi_2) + \frac{jC_1}{8\omega^3}\left[-3\,|\phi_3 - \phi_4|^2 (\phi_3 - \phi_4) \right] = 0$$

$$\mu\left(\dot{\phi}_4 + \phi_4\frac{j\omega}{2}\right) + \frac{jC_1}{8\omega^3}\left[-3|\phi_4 - \phi_3|^2(\phi_4 - \phi_3)\right]$$

$$+ \frac{jC_2}{8\omega^3}\left[-3|\phi_4 - \phi_5|^2(\phi_4 - \phi_5)\right] = 0$$

$$\mu\left(\dot{\phi}_5 + \phi_5\frac{j\omega}{2}\right) + \frac{jC_2}{8\omega^3}\left[-3|\phi_5 - \phi_4|^2(\phi_5 - \phi_4)\right] = 0 \quad (4.46)$$

This represents the (approximate) slow flow of the undamped and unforced dynamical system (4.4) under the specific assumptions made. In a final step we introduce the following polar transformations:

$$\phi_1 = A_1 e^{ja_1}, \quad \phi_2 = A_2 e^{ja_2}, \quad \phi_3 = A_3 e^{ja_3}, \quad \phi_4 = A_4 e^{ja_4}, \quad \phi_5 = A_5 e^{ja_5}$$

which, when substituted into (4.46), and upon setting separately the real and imaginary parts equal to zero, yield the following set of ten real modulation equations governing the (real) amplitudes A_i and phases a_i:

$$\dot{A}_1 - A_2 \frac{\alpha}{2\omega} \sin(a_2 - a_1) = 0$$

$$A_1 \dot{a}_1 + A_1 \left(\frac{\omega}{2} - \frac{\omega_0^2}{2}\right) - \frac{\alpha}{2\omega}(A_1 - A_2 \cos(a_2 - a_1)) = 0$$

$$\dot{A}_2 - A_1 \frac{\alpha}{2\omega} \sin(a_2 - a_1) - A_3 \frac{\varepsilon}{2\omega} \sin(a_3 - a_2) = 0$$

$$A_2 \dot{a}_2 + A_2 \left(\frac{\omega}{2} - \frac{\omega_0^2}{2}\right) - \frac{\alpha}{2\omega}(A_1 \cos(a_2 - a_1) - A_2)$$

$$- \frac{\varepsilon}{2\omega}(A_2 - A_3 \cos(a_3 - a_2)) = 0$$

$$\mu \dot{A}_3 - \frac{\varepsilon A_2}{2\omega} \sin(a_3 - a_2) - A_4 \frac{(A_3^2 + A_4^2)C_1}{8\omega^3} \sin(a_4 - a_3) = 0$$

$$\mu A_3 \dot{a}_3 + \mu A_3 \frac{\omega}{2} - \frac{\varepsilon}{2\omega}(A_3 - A_2 \cos(a_3 - a_2))$$

$$- 3\frac{(A_3^2 + A_4^2)C_1}{8\omega^3}(A_3 - A_4 \cos(a_4 - a_3)) = 0$$

$$\mu \dot{A}_4 + 3\frac{(A_3^2 + A_4^2)C_1 A_3}{8\omega^3} \sin(a_4 - a_3) - 3\frac{(A_4^2 + A_5^2)C_2 A_5}{8\omega^3} \sin(a_5 - a_4) = 0$$

$$\mu A_4 \dot{a}_4 + \mu A_4 \frac{\omega}{2} - 3\frac{(A_4^2 + A_3^2)C_1}{8\omega^3}(A_4 - A_3 \cos(a_4 - a_3))$$

$$- 3\frac{(A_4^2 + A_5^2)C_2}{8\omega^3}(A_4 - A_5 \cos(a_5 - a_4)) = 0$$

$$\mu\dot{A}_5 + 3\frac{(A_5^2 + A_4^2)A_4}{8\omega^3}\sin(a_5 - a_4) = 0$$

$$\mu A_5\dot{a}_4 + \mu A_5\frac{\omega}{2} - 3\frac{(A_5^2 + A_4^2)C_2}{8\omega^3}(A_5 - A_4\cos(a_5 - a_4)) = 0 \qquad (4.47)$$

An inspection of (4.47) verifies that the steady state amplitudes satisfy the algebraic relationship $A_1^2 + A_2^2 + \mu(A_3^2 + A_4^2 + A_5^2) = N^2$, which may be regarded as an energy-like expression indicating conservation of total energy of the resonant periodic motion of the unforced and undamped system (4.4). Alternatively, this represents a first integral of the slow flow (4.47).

To compute periodic resonant solutions of the system, we impose two stationarity requirements in (4.47), namely, that, (i) the phase differences are trivial, $a_1 = a_2 = a_3 = a_4 = a_5 = a$, where a is arbitrary; and (ii) the derivatives of the amplitudes are equal to zero, $\dot{A}_i = 0$. The first condition can hold since the system is undamped. By imposing these stationarity conditions we obtain the following set of nonlinear algebraic equations:

$$A_1\left(\frac{\omega}{2} - \frac{\omega_0^2}{2\omega}\right) - \frac{\alpha}{2\omega}(A_1 - A_2) = 0$$

$$A_2\left(\frac{\omega}{2} - \frac{\omega_0^2}{2\omega}\right) - \frac{\alpha}{2\omega}(A_2 - A_1) - \frac{\varepsilon}{2\omega}(A_2 - A_3) = 0$$

$$\mu A_3\frac{\omega}{2} - \frac{\varepsilon}{2\omega}(A_3 - A_2) - \frac{3C_1}{8\omega^3}(A_3 - A_4)^3 = 0$$

$$\mu A_4\frac{\omega}{2} - \frac{3C_1}{8\omega^3}(A_4 - A_3)^3 - \frac{3C_2}{8\omega^3}(A_4 - A_5)^3 = 0$$

$$\mu A_5\frac{\omega}{2} - \frac{3C_2}{8\omega^3}(A_5 - A_4)^3 = 0 \qquad (4.48)$$

which governs the steady state amplitude of the resonant motions with fast frequency ω. By numerically solving it for varying frequency ω we obtain an approximation for the main backbone branch of the system (based on the assumption that the averaging operation is valid). Once the state amplitudes are numerically computed, the analytical approximation for the corresponding periodic orbit (NNM) of the system is given by

$$u_1 = \frac{A_1}{\omega}\sin(\omega t + a), \quad u_2 = \frac{A_2}{\omega}\sin(\omega t + a)$$

$$w_1 = \frac{A_3}{\omega}\sin(\omega t + a), \quad w_2 = \frac{A_4}{\omega}\sin(\omega t + a), \quad w_3 = \frac{A_5}{\omega}\sin(\omega t + a)$$

$$(4.49)$$

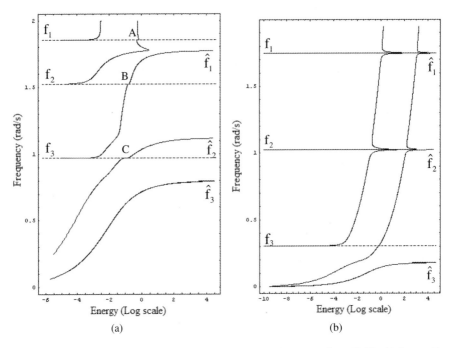

Fig. 4.19 Approximate regular backbone branches obtained from equations (4.48): (a) System I, (b) System II.

In Figure 4.19 we depict the approximate main backbone branches in the FEP for Systems I and II, resulting from the numerical solution of the set of steady state equations (4.48). The analytical results are in agreement with the numerical FEPs depicted in Figures 4.13 and 4.17; this validates the outlined analytical complexification/averaging method.

As mentioned previously, by modifying appropriately the ansatz (4.45) the previous analysis can be extended to approximate other types of periodic solutions in the FEPs of Systems I and II. Depending on the dominant fast frequencies of the motions of the particles of the system, one should define appropriate complex variables ψ_i, $i = 1, \ldots, 5$, and select suitable slow/fast partitions of the dynamics. Moreover, the complexification / averaging analysis can be applied to study damped transient responses of the full system (4.4), in similarity to the analysis performed in Chapter 3.

These results conclude the study of the FEP of periodic orbits of the underlying Hamiltonian system which results by neglecting the damping and forcing terms from (4.4). In the following section we present a study of damped transitions and TET in system (4.4) by adding weak damping and considering impulses applied to the linear primary system. We will show that the weakly damped transitions (and TET) of the impulsively forced system can be studied in terms of the underlying Hamiltonian dynamics.

4.3 TRCs and TET in the Damped and Forced System

The topological portraits of the FEPs of the Hamiltonian Systems I and II provide a clear indication of the complex topology of the periodic orbits of the undamped and unforced dynamics. In this section we show that this rich topological structure of periodic orbits of the underlying Hamiltonian systems leads to complicated transient dynamics of the forced and damped systems, including multi-frequency transitions between different branches of solutions, isolated TRCs and resonance capture cascades.

The study of transitions in the damped dynamics is performed by superimposing the wavelet transform (WT) spectra of the transient responses to the FEPs of the underlying Hamiltonian systems (Tsakirtzis, 2006; Tsakirtzis et al., 2007). In that way, and while supposing that the effect of weak damping is purely parasitic (as it cannot generate 'new dynamics,' but rather acts as perturbation of the underlying Hamiltonian response), the transient responses occur in neighborhoods of branches of periodic (or quasi-periodic) solutions of the corresponding Hamiltonian systems. Once this is recognized, the interpretation of the damped dynamics is possible, and an understanding of the resulting multi-frequency transitions can be gained.

4.3.1 Numerical Wavelet Transforms

The transient dynamics of the damped and forced system is processed by numerical wavelet transforms (WTs). The results are presented in terms of WT spectra, which are contour plots depicting the amplitude of the WT as function of frequency (vertical axis) and time (horizontal axis). Heavy shaded areas correspond to regions where the amplitude of the WT is high whereas lightly shaded regions correspond to low amplitudes. Both Morlet and Cauchy WTs were considered, but these two mother wavelets provided similar results when applied to the signals considered herein.

Representative WT spectra of the transient nonlinear responses of system (4.4) are presented in Figures 4.20–4.25. Specifically, we reconsider the responses of System II for $\alpha = 1.0$ and impulsive forcing condition (IFC) I3, studied previously in Figures 4.8, 4.11 and 4.12. Referring to the plot depicted in Figure 4.8 (with $\alpha = 1.0$), a peculiar behavior of the efficiency of targeted energy transfer (TET) from the primary linear system to the MDOF NES was noted. In particular, when the primary system was excited by a pair of out-of-phase impulses of magnitude Y, strong TET to the NES occurs at low energy levels (i.e., for weak applied impulses), with values of EDM reaching levels of 90% for $Y = 0.1$ (point C in Figure 4.8). By increasing the magnitude of the applied impulse the eventual energy transfer to the NES first decreases (with EDM reaching nearly 50% for $Y = 1.0$ – point A in Figure 4.8), before increasing again to higher levels (with EDM being nearly equal to 90% for $Y = 1.5$ – point B in Figure 4.8); further increase of Y decreases the portion of input energy eventually dissipated by the NES.

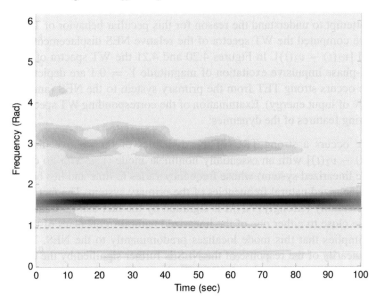

Fig. 4.22 WT spectrum of the relative NES displacement ($v_1 - v_2$) of System II for out-of-phase impulse magnitude Y; the linear natural frequencies of the uncoupled ($\varepsilon = 0$) primary system are indicated by dashed lines.

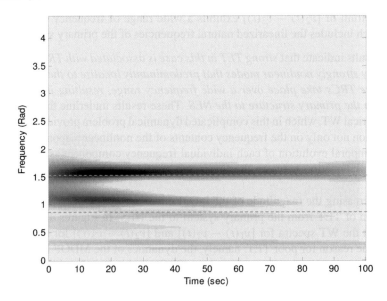

Fig. 4.23 WT spectrum of the relative NES displacement ($v_2 - v_3$) of System II for out-of-phase impulse magnitude $Y = 1.0$; the linear natural frequencies of the uncoupled ($\varepsilon = 0$) primary system are indicated by dashed lines.

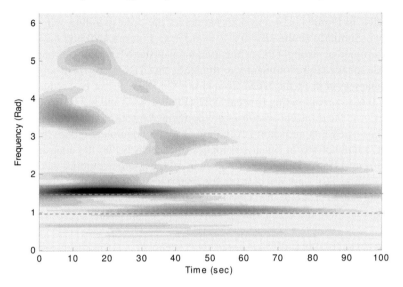

Fig. 4.24 WT spectrum of the relative NES displacement ($v_1 - v_2$) of System II for out-of-phase impulse magnitude $Y = 1.5$; the linear natural frequencies of the uncoupled ($\varepsilon = 0$) primary system are indicated by dashed lines.

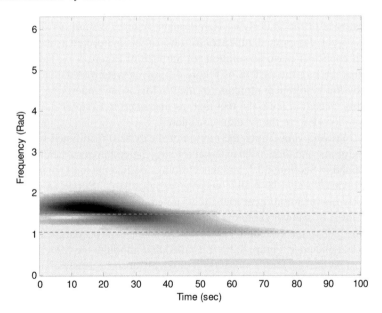

Fig. 4.25 WT spectrum of the relative NES displacement ($v_2 - v_3$) of System II for out-of-phase impulse magnitude $Y = 1.5$; the linear natural frequencies of the uncoupled ($\varepsilon = 0$) primary system are indicated by dashed lines.

Finally, in Figures 4.24 and 4.25 we depict the corresponding WT spectra for $Y = 1.5$. Similarly to the case for $Y = 0.5$ (see Figures 4.20 and 4.21), we note the occurrence of strong TRC of the dynamics of the NES with a strongly nonlinear mode localized predominantly to the NES; this TRC leads to strong TET from the primary system to the NES. Comparing the WT spectra of Figures 4.22 and 4.23 to those corresponding to weak TET (Figures 4.22 and 4.23), we note that in the later case the transient responses are dominated by sustained frequency components (i.e., by SRCs), indicating excitation of weakly nonlinear modes that are mere analytic continuations of linearized modes of System II. On the contrary, in cases where strong TET is realized, the frequencies of the nonlinear modes involved in the corresponding TRCs are not close to the linearized natural frequencies ω_1 and ω_2, indicating the presence in the response of strongly nonlinear modes with no linear counterparts; these modes localize predominantly to the NES.

4.3.2 Damped Transitions on the Hamiltonian FEP

Starting with System I, we perform a series of numerical simulations to study the transient dynamics of system (4.4) with weak damping, in an effort to demonstrate that complicated transitions in the dynamics of the weakly damped system closely follow branches of the underlying Hamiltonian system. We aim to show that, for sufficiently weak damping, damped transitions can be interpreted as jumps between different branches of periodic solutions of the FEP of Figure 4.14. Hence, we aim to show that TET in the system of Figure 4.2 [or in system (4.4)] is governed, in essence, by the topological structure of the NNMs of the underlying Hamiltonian system; this, occurs in spite the fact that, as discussed in Chapter 3, damping is a prerequisite for TET for the systems considered.

In the following simulations the motion of the system is initiated with different initial conditions, and there is no external forcing; the system parameters for System I were defined in Section 4.2.1, and the damping coefficients in (4.4) were assigned the (small) values $\varepsilon\lambda_1 = 8 \times 10^{-3}$ and $\varepsilon\lambda_2 = 1.6 \times 10^{-3}$. Hence, in what follows only weakly damped nonlinear transitions are examined. First, the motion is initiated at point A of a lower subharmonic tongue emanating from the main backbone curve of the FEP of the system in Figure 4.13, and the resulting damped transient responses are depicted in Figure 4.26. It is noted that although the MDOF NES starts with almost no energy, after $t = 1500$ s it passively absorbs nearly all of the energy of the (initially excited) linear primary system in an irreversible fashion.

Moreover, TET from the linear primary system to the NES coincides with the transition from a subharmonic tongue to the main backbone curve with decreasing energy (due to damping dissipation) as evidenced from the plots of Figure 4.26c; these plots depict the superposition of the FEP of Figure 4.13 to the WT spectra of the transient responses $[v_1(t) - v_2(t)]$ and $[v_2(t) - v_3(t)]$. These plots should be viewed from a purely phenomenological point of view, as they superpose weakly damped (the WT spectra) to undamped (the branches of periodic orbits on the FEP)

Passive Nonlinear Targeted Energy Transfer in Mechanical and Structural Systems 353

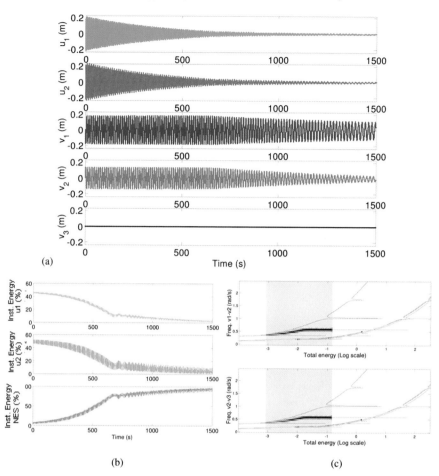

Fig. 4.26 Transient response of the weakly damped System I for initial conditions at point A of the FEP of Figure 4.13: (a) Time series, (b) partition of instantaneous energy of the system, and (c) WT spectra depicted in the FEP of the underlying Hamiltonian System I.

responses, and they should be used only for descriptive purposes. Nevertheless this type of superpositions help us interpret transitions that occur in the damped responses in terms of the topological portrait of the periodic orbits of the underlying Hamiltonian system; in this particular case, the only transition in the dynamics takes place from the subharmonic branch where the motion is initiated, to the main backbone branch, and there are no other transitions or jumps between branches of solutions (i.e., the transition is smooth with decreasing energy – see Figure 4.26c). Concerning the damped responses of Figure 4.26a, we note the nearly complete absence of motion of the third particle of the MDOF NES, in accordance to our previous discussion regarding the periodic motions (NNMs) on the regular backbone branch of System I.

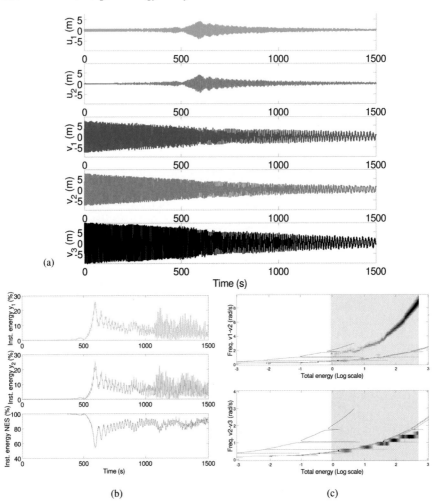

Fig. 4.27 Transient response of the weakly damped System I for initial conditions at point B of the FEP of Figure 4.13: (a) Time series, (b) partition of instantaneous energy of the system, and (c) WT spectra depicted in the FEP of the underlying Hamiltonian System I.

Next, the motion is initiated at point B on a branch of the family of singular backbones of the FEP of Figure 4.13, namely, on branch S161. The results of this simulation are depicted in Figure 4.27, and some major qualitative differences are observed compared to the previous simulation. In this case the last mass of the NES executes large-amplitude oscillations, and the dominant frequency components of the WT spectra of $[v_1(t) - v_2(t)]$ and $[v_2(t) - v_3(t)]$ differ (in contrast to motions on the regular backbone curve that are nearly monochromatic); finally, the motion is nearly localized to the MDOF NES. Indeed, the WT spectrum of the relative displacement $[v_2(t) - v_3(t)]$ follows a singular backbone branch, engaging at $t \approx 550$ s

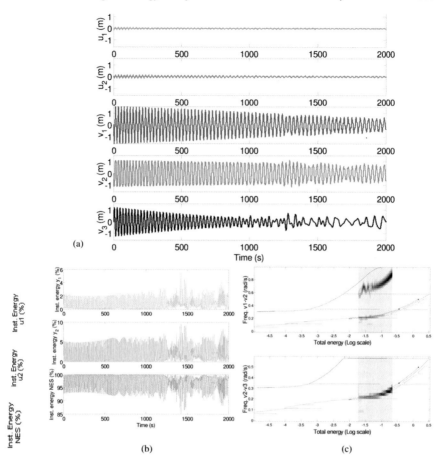

Fig. 4.28 Transient response of the weakly damped System I for initial conditions at point C of the FEP of Figure 4.13: (a) Time series, (b) partition of instantaneous energy of the system, and (c) WT spectra superimposed to the FEP of the underlying Hamiltonian System I.

in 1:1 TRC with the in-phase linearized mode at the natural frequency f_2 of system (4.10). On the other hand, the WT spectrum of $[v_1(t) - v_2(t)]$ does not generally follow the same singular backbone branch since its dominant harmonic component is six times the dominant harmonic component of $[v_2(t) - v_3(t)]$. When the dominant frequency of $[v_1(t) - v_2(t)]$ gets close to the neighborhood of the regular backbone branch, it is possible that TRCs occur involving the regular backbone $S111$ and the singular backbone $S161$.

An additional simulation is depicted in Figure 4.28, with the motion initiated on point C of $S131c$ (see Figure 4.13) not far from the coalescence point of this branch with $S131d$ (see Figure 4.15). Once the motion reaches the coalescence point for diminishing energy, a bifurcation occurs, which is clearly evidenced by the envelopes of the relative displacements of the NES. In addition, we note the occurrence of an interesting resonance capture at the final stage of the motion when the dominant harmonic component of the relative displacement $[v_1(t) - v_2(t)]$ (which is three times the dominant harmonic component of $[v_2(t) - v_3(t)]$) appears to engage in resonance capture with one of the lower tongues emanating from the regular backbone curve. This is precisely the type of resonance capture conjectured previously, leading to strong energy exchanges between the particles of the NES. As in the previous simulation, throughout the motion almost all of the energy of vibration is localized to the MDOF NES.

It is interesting to note that in general *weak TET occurs in System I*. This is concluded by performing a series of numerical simulations with initial forcing conditions similar to those considered in Section 4.1.2 (with IFCs I1-I3), and computing the portion of total impulsive energy eventually dissipated by the MDOF NES. In all cases it was found that only a small portion of input energy is eventually transferred to (and locally dissipated by) the NES. A representative result of weak TET is depicted in Figure 4.29 for the case of single impulsive excitation with magnitude $Y = 1.5$ applied to the left mass of the linear subsystem (corresponding to impulsive forcing condition I1).

We now consider the transient damped dynamics leading to TET in System II, corresponding to weak nonlinear stiffness C_2 and small NES masses. We will show that by decreasing the masses of the MDOF NES the complexity of the dynamics increases, and the capacity for TET significantly improves compared to System I. In the following simulations the motion of the system is initiated with different initial conditions, and no external forcing is considered; the system parameters for System II were defined in Section 4.2.2, and the damping coefficients in (4.4) were assigned the values $\varepsilon\lambda_1 = \varepsilon\lambda_2 = \varepsilon\lambda = 0.01$. So, again, only weakly damped nonlinear transitions are considered in what follows.

Revisiting an earlier result, we wish to reconsider and study in more detail the damped transitions associated with the peculiar behavior of the TET plot of System II for $\alpha = 1.0$ and IFC I3 depicted in Figure 4.8. More specifically, in Section 4.1.2 it was numerically shown that when the linear system is excited by a pair of out-of-phase impulses of magnitude Y, strong TET from the linear primary system to the NES occurs even at low values of the impulse (with EDM as high as 90% for $Y = 0.1$); by increasing the magnitude of the impulse, initially TET deteriorates (with EDM reaching nearly 50% for $Y = 1.0$), before improving back to high levels (with EDM increasing up to nearly 90% for $Y = 1.5$). Further increase of Y decreases the portion of input energy that is eventually dissipated by the NES, so that TET deteriorates.

The WT spectra of the responses of the particles of the NES for System II were depicted in Figures 4.20–4.25, and it was postulated that strong TET is associated with transient resonance captures (TRCs) of the transient dynamics by strongly nonlinear modes predominantly localized to the NES; whereas, weak TET is associated with sustained resonance captures (SRCs) of the dynamics by weakly nonlinear modes predominantly localized to the linear system. We wish to confirm these results by studying the WT spectra of the NES responses superimposed to the FEP of System II (depicted in Figure 4.17); by doing so we wish to observe directly

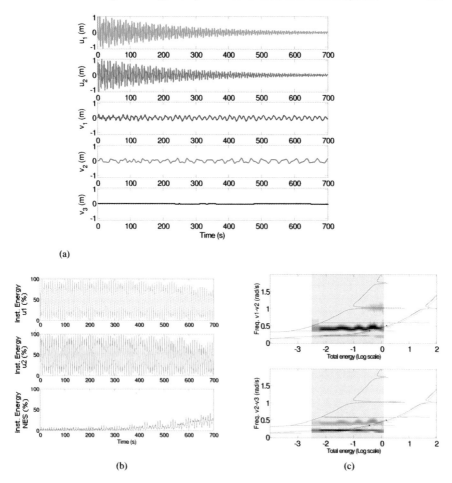

Fig. 4.29 Weak TET in System I for IFC I1 of magnitude $Y = 1.5$: (a) Time series, (b) partition of instantaneous energy of the system, and (c) WR spectra superimposed to the FEP of the underlying Hamiltonian System I.

the resulting TRCs and transitions between branches of periodic solutions. The WT spectra superimposed to the FEP for System II are depicted in Figures 4.30–4.32.

In Figure 4.30 the damped responses corresponding to IFC I3 and $Y = 0.1$ [i.e., impulses $F_1(t) = -F_2(t) = Y\delta(t)$ and zero ICs in system (4.4)] are presented. These responses correspond to point C of the TET diagram of Figure 4.8. In this case both relative displacements $[v_1(t) - v_2(t)]$ and $[v_2(t) - v_3(t)]$ of the NES follow regular backbone branches in the FEP as energy decreases due to damping dissipation. The relative displacement $[v_1(t) - v_2(t)]$ has a dominant frequency component which approaches the linearized natural frequency f_2 of the limiting system (4.10) with decreasing energy; in contrast, $[v_2(t) - v_3(t)]$ has two strong harmonic components that approach the linearized natural frequencies f_2 and f_3 with de-

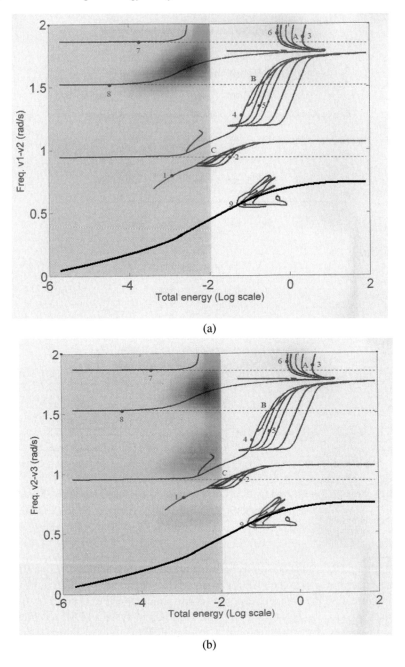

Fig. 4.30 Damped responses of System II for IFC I3 with $Y = 0.1$: (a, b) Cauchy WT spectra of the relative displacements $[v_1(t) - v_2(t)]$ and $[v_2(t) - v_3(t)]$ superimposed to the FEP of the Hamiltonian System II; these responses correspond to point C of the TET diagram of Figure 4.8.

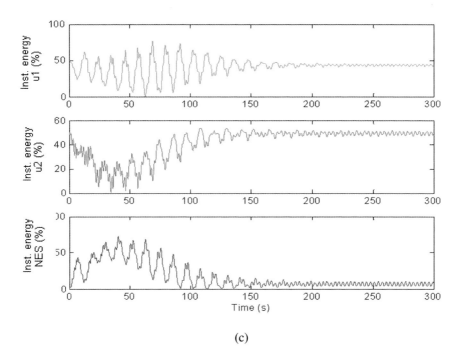

Fig. 4.30 Damped responses of System II for IFC I3 with $Y = 0.1$: (c) partition of instantaneous energy of the system; these responses correspond to point C of the TET diagram of Figure 4.8.

creasing energy; this indicates that *TET occurs simultaneously with two modes of the linear limiting system (4.10)*. Moreover, the same regular backbone branches are tracked by the response throughout the motion, and strong TET occurs right from the early stage of the dynamics. This explains the high value of EDM (~90%) that is realized even for this low level of impulsive excitation; clearly, this can not be realized through the use of SDOF NESs, as TET to this type of attachments takes place (is 'activated') only above a certain critical energy level. Hence, the described low-energy TET is a unique feature of the MDOF NES configuration.

By increasing the magnitude of the impulse to $Y = 1.0$ TET from the primary system to the MDOF NES significantly decreases. The damped response of System II in this case is depicted in Figure 4.31. Some major qualitative differences are observed compared to the lower-impulse simulation of Figure 4.30. Judging from the partition of the instantaneous energy among the linear and nonlinear systems, it is concluded that targeted energy transfer is significantly delayed, and, hence, occurs at lower energy levels; this explains the weak TET to the NES (EDM~50% in this case). This delay is explained when one studies the WT spectra of the NES relative responses superimposed to the FEP of Figure 4.31a. Noting that in the initial stage of the motion the dominant WT components of the NES relative displacements occur close to the linearized frequency f_1, we conclude that in the initial (high energy)

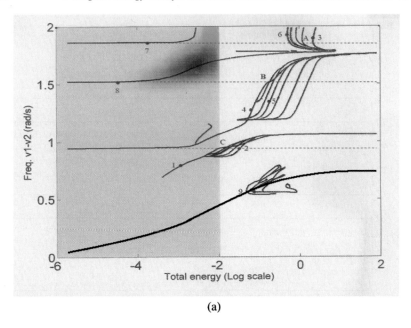

(a)

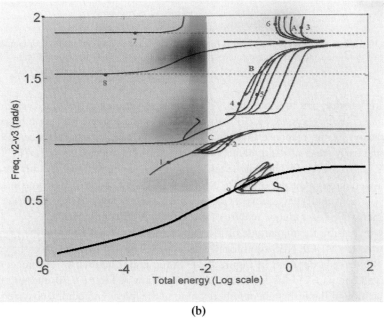

(b)

Fig. 4.31 Damped responses of System II for IFC I3 with $Y = 1.0$: (a, b) Cauchy WT spectra of the relative displacements $[v_1(t) - v_2(t)]$ and $[v_2(t) - v_3(t)]$ superimposed to the FEP of the Hamiltonian System II; these responses correspond to point A of the TET diagram of Figure 4.8.

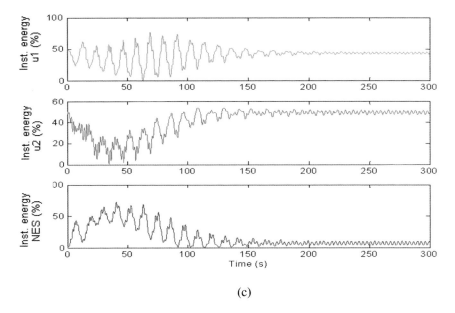

Fig. 4.31 Damped responses of System II for IFC I3 with $Y = 1.0$: (c) partition of instantaneous energy of the system; these responses correspond to point A of the TET diagram of Figure 4.8.

stage of the motion there occurs strong resonance capture of the damped motion by the linearized out-of-phase mode of the limiting system (4.10). This yields a motion mainly localized to the (directly excited) primary linear system, with only a small portion of energy 'spreading out' to the NES. As energy decreases due to damping dissipation, the damped motion 'escapes' from the initial out-of-phase resonance capture, and follows regular backbone branches; this results in TET (as in the simulations of Figure 4.30), which, however, occurs with a delay, at a stage where the energy of the system is small due to damping dissipation. Hence, no significant TET from the primary system to the NES takes place in this case.

By increasing the magnitude of the impulse to $Y = 1.5$, the dynamics escape from the strong initial out-of-phase resonance capture, yielding once again strong TET. This is depicted in Figure 4.32, showing that the NES relative responses possess multiple strong frequency components, indicating that strong TET takes place over multiple frequencies. Note in this case the early strong TET from the primary system to the NES, resulting in EDM of nearly 90%.

These results are in agreement with the conclusions drawn from the study of the WT spectra of the NES relative responses of the same system (see Figures 4.20–4.25 in Section 4.3.1). The superposition of the WT spectra to the FEP of the underlying Hamiltonian System II provides additional valuable insight to the sequences of resonance captures (transient or sustained) that facilitate or hinter TET from the primary system to the NES. This confirms the value of the FEP as a tool for interpreting the transient dynamics of the strongly nonlinear systems considered herein.

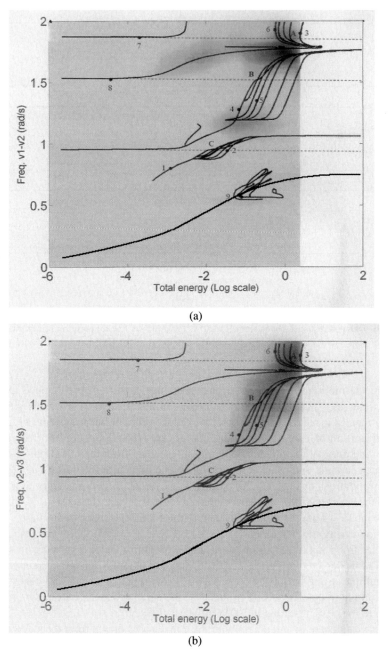

Fig. 4.32 Damped responses of System II for IFC I3 with $Y = 1.5$: (a, b) Cauchy WT spectra of the relative displacements $[v_1(t) - v_2(t)]$ and $[v_2(t) - v_3(t)]$ superimposed to the FEP of the Hamiltonian System II; these responses correspond to point B of the TET diagram of Figure 4.8.

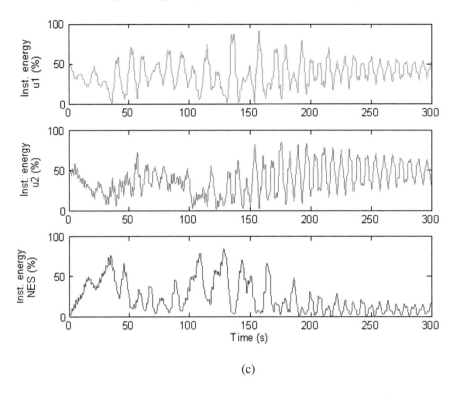

Fig. 4.32 Damped responses of System II for IFC I3 with $Y = 1.5$: (c) partition of instantaneous energy of the system; these responses correspond to point B of the TET diagram of Figure 4.8.

Similar results were obtained for alternative forcing excitations of the linear primary system, confirming the strong TET capacity of the NES in System II. A last example of strong TET is depicted in Figure 4.33, for the case of single impulse excitation of magnitude $Y = 1.5$ (IFC I1 – Figure 4.6, case $\alpha = 1.0$). Notice the strong multi-frequency content of the WT spectra of the internal displacements of the MDOF NES, proving that TET from the primary system to the NES takes place in a broadband fashion [i.e., simultaneously from the three linearized modes of the limiting subsystem (4.10)]; this results in nearly 85% of input energy being eventually transferred to, and dissipated by the MDOF NES. Compare this picture to the corresponding plot of Figure 4.29c for System I, where the NES dynamics is narrowband and weak TET occurs.

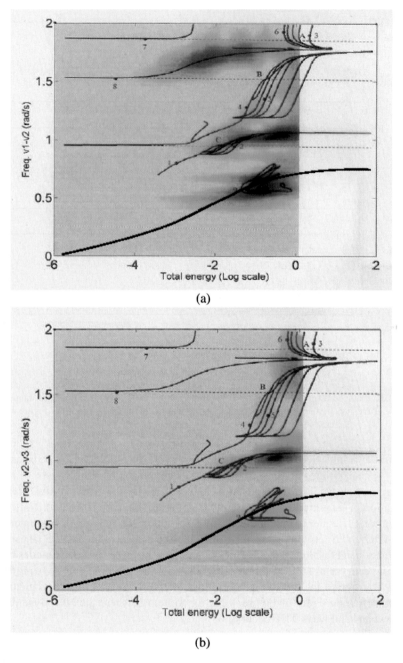

Fig. 4.33 Damped responses of System II for IFC I1 with $Y = 1.5$: (a, b) Cauchy WT spectra of the relative displacements $[v_1(t) - v_2(t)]$ and $[v_2(t) - v_3(t)]$ superimposed to the FEP of the Hamiltonian System II.

4.4 Concluding Remarks

The results presented in this chapter demonstrate that MDOF essentially nonlinear attachments (MDOF NESs) can be designed to be efficient and robust passive broadband absorbers of vibration or shock energy from the primary systems to which they are attached. Moreover, the extraction of vibration energy occurs in a multi-frequency fashion, through simultaneous dynamic interactions of multiple modes of the nonlinear attachments with multiple modes of the primary systems. This form of multi-frequency energy exchange is different than the resonance capture cascades encountered in the previous chapter, where TET to SDOF NESs occurs in a sequential manner, i.e., through resonance capture cascades.

The dynamical systems considered in this work possess complicated dynamics due to their degenerate structures. The considered MDOF NES has strong passive TET capacity, extracting in some cases as much as 90% of the vibration energy of the primary system to which it is attached. The capacity of the MDOF essentially nonlinear attachment to absorb broadband vibration energy was demonstrated numerically in this section, but it can also be analytically studied by a reduction process of the governing system of ordinary differential equations, and local slow/fast partition of the damped dynamics.

It was shown that MDOF essentially nonlinear attachments may be more efficient energy absorbers compared to SDOF ones, since they are capable of absorbing energy simultaneously from multiple structural modes, over wider frequency and energy ranges. Passive TET by the MDOF NES can be related to transient resonance captures (TRCs) of the damped dynamics, whereby orbits of the system in phase space are transiently captured in neighborhoods of resonance manifolds.

An interesting dynamical feature of the considered MDOF NES configurations is the existence of two classes of backbone branches in their frequency-energy planes: isolated regular backbone branch containing NNMs where all particles of the primary system and the NES oscillate with identical dominant frequencies; and additional families of densely packed singular backbone branches containing NNMs where particles oscillate with differing dominant harmonic components. It was proved that these families of singular backbones contain countable infinities of backbone branches, which are mainly generated by combined parametric and external resonances between the two relative displacements of the particles of the NES. It is conjectured that this interesting energy 'quantization' of the families of singular backbone branches may represent different modes of nonlinear interaction and energy exchange between the particles of the essentially nonlinear, MDOF attachment.

Finally, it was shown that complex transitions in the damped dynamics of the system with attached MDOF NES may be related to transitions or jumps between branches of NNMs of the underlying Hamiltonian system. In that context, TRCs leading to TET may be related to damped motions in neighborhoods of certain invariant manifolds of the underlying Hamiltonian system.

The methodologies and results presented in this chapter pave the way for applying lightweight MDOF essentially nonlinear attachments as shock and vibration absorbers of unwanted disturbances of structures. The proposed designs are modular and can be designed to be lightweight; hence they can be conveniently attached to existing structures with minimal structural modifications. Application of MDOF NESs for shock isolation of elastic continua is considered in the next chapter.

References

Georgiadis, F., Vakakis, A.F., McFarland, M., Bergman, L.A., Shock isolation through passive energy pumping caused by non-smooth nonlinearities, *Int. J. Bif. Chaos* (Special Issue on 'Non-Smooth Dynamical Systems: Recent Trends and Perspectives'), **15**(6), 1–13, 2005.

Gourdon, E., Lamarque, C.H., Energy pumping for a larger span of energy, *J. Sound Vib.* **285**, 711–720, 2005.

Gourdon, E., Coutel, S., Lamarque, C.H., Pernot, S. , Nonlinear energy pumping with strongly nonlinear coupling: Identification of resonance captures in numerical and experimental results, in *Proceedings of the 20th ASME Biennial Conference on Mechanical Vibration and Noise*, Long Beach, California, September 24–28, 2005.

Gourdon, E., Pernot, S, Lamarque, C.H., Energy pumping with multiple passive nonlinear absorbers, in *Proceedings of EUROMECH Colloquium 483 on Geometrically Nonlinear Vibrations of Structures*, FEUP, Porto, Portugal, July 9–11, 2007.

Guckenheimer, J., Holmes, P., *Nonlinear Oscillations, Dynamical System, and Bifurcation of Vector Fields*, Springer-Verlag, New York, 1983.

Kerschen, G., Lee, Y.S., Vakakis, A.F., McFarland, D.M., Bergman, L.A., Irreversible passive energy transfer in coupled oscillators with essential nonlinearity, *SIAM J. Appl. Math.* **66**, 648–679, 2006.

Lee, Y.S., *Passive Broadband Targeted Energy Transfers and Control of Self-Excited Vibrations*, PhD Thesis, Department of Mechanical Science and Engineering, University of Illinois at Urbana-Champaign, 2006.

Lee, Y.S., Kerschen, G., Vakakis, A.F., Panagopoulos, P.N., Bergman, L.A., McFarland, D.M., Complicated dynamics of a linear oscillator with a light, essentially nonlinear attachment, *Physica D* **204** (1–2), 41–69, 2005.

Ma, X., Vakakis, A.F., Bergman, L.A., Karhunen–Loeve analysis and order reduction of the transient dynamics of linear coupled oscillators with strongly nonlinear end attachments, *J. Sound Vib.* **309**, 569–587, 2008.

McFarland, D.M., Bergman, L.A., Vakakis, A.F., Experimental study of nonlinear energy pumping occurring at a single fast frequency, *Int. J. Nonlinear Mech.* **40**, 891–899, 2004.

Musienko, A.I., Lamarque, C.H., Manevitch, L.I., Design of mechanical energy pumping devices, *J. Vib. Control* **12**(4), 355–371, 2006.

Panagopoulos, P.N., Vakakis, A.F., Tsakirtzis, S., Transient resonant interactions of linear chains with essentially nonlinear end attachments leading to passive energy pumping, *Int. J. Solids Struct.* **41**(22–23), 6505–6528, 2004.

Rosenberg, R., On nonlinear vibrations of systems with many degrees of freedom, *Adv. Appl. Mech.* **9**, 155–242, 1966.

Tsakirtzis, S., *Passive Targeted Energy Transfers From Elastic Continua to Essentially Nonlinear Attachments for Suppressing Dynamical Disturbances*, PhD Thesis, National Technical University of Athens, Athens, Greece, 2006.

Tsakirtzis, S., Kerschen, G., Panagopoulos, P.N., Vakakis, A.F., Multi-frequency nonlinear energy transfer from linear oscillators to MDOF essentially nonlinear attachments, *J. Sound Vib.* **285**, 483–490, 2005.

Tsakirtzis, S., Panagopoulos, P.N., Kerschen, G., Gendelman, O., Vakakis, A.F., Bergman, L.A., Complex dynamics and targeted energy transfer in systems of linear oscillators coupled to multi-degree-of-freedom essentially nonlinear attachments, *Nonl. Dyn.* **48**, 285–318, 2007.

Vakakis, A.F., Manevitch, L.I., Mikhlin, Y.V., Pilipchuk, V.N., Zevin, A.A., *Normal Modes and Localization in Nonlinear Systems*, Wiley Interscience, New York, 1996.

Vakakis, A.F., Manevitch, L.I., Gendelman, O., Bergman, L.A., Dynamics of linear discrete systems connected to local essentially nonlinear attachments, *J. Sound Vib.* **264**, 559–577, 2003.

Vakakis, A.F., McFarland, D.M., Bergman, L.A., Manevitch, L.I., Gendelman, O., Isolated resonance captures and resonance capture cascades leading to single- or multi-mode passive energy pumping in damped coupled oscillators, *J. Vib. Acoust.* **126** (2), 235–244, 2004.

Wiggins, S., *Introduction to Applied Nonlinear Dynamical Systems and Chaos*, Springer-Verlag, New York, 1990.

Neimark–Sacker (NS) – II 382, 474, 477, 496, 525, 536, 549
neutral-saddle – II 494, 539, 540, 543-545, 547, 548, 562-564
pitchfork – I 19, 22; II 382, 384, 497
saddle-node (SN) – I 120, 123, 161, 163, 189, 203, 213, 218, 224, 233, 246, 249; II 164, 181, 266, 382, 493, 562–564, 626

Boundary value problem
linear (BVP) – II 102
nonlinear (NLBVP) – I 109-114; II 41, 101, 262, 266 Broadband I vii, 3, 5, 37, 72, 74, 86, 156, 170, 233, 237, 257, 258, 262, 264, 304, 313, 349, 363, 365; II 6, 9, 11, 12, 21, 22, 37, 109, 117, 120, 123, 126, 127, 131–133, 137, 158, 161, 202, 206, 226, 230, 231, 239, 241, 245, 254, 257, 312, 322, 323, 331, 353, 354, 452, 459, 461, 464, 501, 520, 530, 531, 533, 549, 552, 559, 615, 645, 646

Chaotic dynamics I 2, 8, 319; II 241
Clearance I 97; II 229, 230, 231, 233–236, 239–243, 245, 247, 249–251, 253, 256–260, 264, 271, 301, 572–574, 576–578, 590

Configuration
plane I 24–26, 112–114, 118, 119, 332, 121, 123, 130, 132, 137, 138; II 264, 269–271, 321
space I 16, 17, 239, 240, 244; II 67, 320

Complexification-averaging method (CX-A) I 10, 15, 54, 56, 58, 67, 69, 70, 93, 124, 125, 128, 130, 157, 173, 176, 212, 213, 238, 269, 276, 291, 293–295, 342, 343, 346; II 88, 91, 162, 175, 176, 216, 218, 344, 345, 359, 361, 398, 402, 409, 453, 483, 485

Confinement
energy – I 37
motion – I 16, 31, 37; II 351
passive – I 31, 32, 37

Continuation
analytic – I 17, 74, 239, 352
numerical – II 354, 357, 381, 382, 386, 394, 418, 447, 452, 471, 487, 493, 536, 559

Continuum
approximation I 272, 284
limit I 32, 272, 297, 298; II 98

Coupling
strong – II 100, 101, 105, 111, 113, 116, 127, 342, 536, 542
weak – I 20, 22, 28, 31, 41, 42, 270, 306, 311; II 67, 71, 74, 101, 113, 116, 230, 245, 259, 322, 331, 348, 536, 542, 546, 551

Cutting process, tool II 619, 620

Damping
inherent – I 81; II 645
viscous – I 5, 17; II 2, 13, 17, 18, 67, 88, 99, 100, 110, 134, 140, 155, 230, 256, 258, 260, 261, 274, 280, 285, 291, 297, 298, 300, 313, 314, 322, 331, 356, 503, 506, 509, 539, 544, 561, 573, 580, 585, 622, 631

Differential evolution II 576
Dissipative system I 4, 5, 7, 17, 18, 25, 45, 46, 48, 49, 51, 52, 129, 163, 165, 315
Duffing oscillator I 55, 151; II 60

Earthquake II 571, 572, 574, 576–582, 587–593, 595, 598-600, 604–615
Empirical mode decomposition (EMD) I 15, 70, 77–80; II 12, 15–19, 30–33, 35–37, 40, 56–66, 99, 107, 109, 110, 117, 118, 127, 132, 133, 141, 142, 147, 151, 153, 198, 282, 292, 293, 295, 301, 302, 331, 336, 338, 453, 463, 477, 478–484, 488, 520

Energy
absorber I 308, 311, 365; II 462, 559
dissipater I 289; II 21, 559
dissipation measure (EDM) I 99, 170, 201, 203, 205, 262, 263, 264, 310, 311, 312, 314, 318, 347, 356, 359, 361; II 4–7, 11, 16, 19-22, 24, 33, 109, 111, 112, 115, 117–119, 140, 142, 144, 147, 151, 153, 156, 230, 232, 233, 239, 244–251, 318, 325–330, 340, 342, 354, 453, 551
flow of – I 5, 272, 273; II 17, 24, 26, 141
harvesting I vii, 1, 3; II 646
threshold I 157, 163, 169, 170, 184, 208, 242, 246, 250, 256, 267, 282, 284, 313, 315; II 265, 271, 552, 599
transaction measure (ETM) II 16, 17, 25–27, 52, 53, 55, 141, 153–155
transfer
broadband – I 3, 37
multi-frequency – I 304

Equilibrium
stable – I 16, 44, 47, 52, 53, 161, 190, 191, 201; II 493, 625, 626, 628, 631, 632, 634, 637–640
unstable – I 44, 47, 52, 53, 161, 190; II 626, 634

Experimental
TET I 254
TRC II 333
Fast dynamics I 151, 174, 278; II 93
Fluid-structure instability II 418

166, 170, 172, 176, 184, 185, 213, 237, 238, 242, 245, 306, 307; II 61, 369, 398, 404, 432, 433, 438, 439, 516

Intrinsic mode function (IMF) I 71, 77–80; II 15, 18, 31–38, 40, 56–66, 99, 117, 127–132, 134, 146–148, 198, 200, 201, 293, 294, 301, 302, 336–339, 477–482, 488

Invariant
    manifold I 17, 20, 22, 53, 97, 114, 190, 191, 203, 204, 207, 224, 365; II 177, 179, 182, 183, 200, 280
        damped NNM – I 17, 18, 27, 52, 53, 171, 172, 190, 192, 193, 209, 213, 216, 247–249; II 50
        NNM – I 17, 18, 97, 171, 179; II 50
    torus I 8, 20–22, 39, 40

Isolation
    vibration – I vii, 10, 36, 81; II 161, 173, 202, 203, 206, 207–210, 212, 213, 218, 221–224, 226, 227, 242, 641, 645, 646
    seismic, base – II 571
    shock – I vii, 10, 70, 366; II 37, 109, 229, 230, 232, 233, 237, 239, 243, 251–256, 641, 646

KAM theorem, theory I 8, 21, 39
KAM tori I 122

Limit cycle oscillation (LCO)
    generation of – II 542, 549
    suppression of – II 462, 487, 496, 528
Localization I vii, 1–4, 10, 15, 16, 20, 25, 28–31, 33, 37, 68, 74, 105, 124, 130, 134, 137, 160, 161, 170–172, 179, 304, 349; II 285, 343, 344, 645

Map
    horseshoe – I 22
    one–dimensional – II 187, 188, 190, 193–196, 198, 220
    Poincaré – I 20, 21, 22, 40; II 187, 189, 197, 198, 199, 220, 271, 272

Mitigation
    LCO – II 428, 463
    seismic – I vii, 10, 70, 81, 85; II 157, 229, 259, 302, 305, 311, 571, 580
    shock – II 158
    ibration – II 223, 226, 620, 629, 635, 641, 642, 646

is I 27, 28, 251, 264, 265; II 313, 322, 348, 582–584
5, 24, 25
33–35

line I 16, 17
oscillator I 35; II 6, 65
relation I 33, 332, 333
response I 16, 250, 307, 308; II 432, 474
series I 260
TET I 267

Mode
    localized – I 45, 289; II 66, 67, 82, 84, 85
    nonlinear normal – (NNM) I 4, 9, 10, 15, 16, 94, 113, 166, 289, 306, 310, 332; II 50, 86, 109, 118, 123, 261, 265, 273, 280, 288, 320, 520, 559
    structural – I 3–5, 365; II 9, 11, 12, 24, 37, 158, 397, 400, 589, 590, 592, 628, 629

Modulation I 4, 55, 56, 63, 64, 67, 78, 80, 94, 102, 125, 126, 129, 143–145, 147, 148, 157, 158, 173, 175, 177, 180, 182, 187, 188, 192, 193, 195, 197, 198, 200, 208, 213–215, 219, 221, 222, 238, 276, 278–280, 283, 295, 297, 340–344; II 54, 93–97, 161, 163, 168, 176, 177, 179, 209, 216, 217, 315, 316, 336, 361, 364, 404, 410, 415, 417, 422, 439, 447, 461, 483, 484, 509

Multiple scales I 9, 21, 29, 44, 50, 54–59, 62, 64–66, 187, 188, 190, 193, 196, 341; II 180, 182, 415

Narrowband I vii, 5, 169, 237, 257, 363; II 9, 117, 161, 202, 209, 241, 245, 253
NATA II 503–514, 520
Non-integrability I 8, 122, 178; II 271
Nonlinear energy sink (NES)
    Configuration I (grounded) – I 96; II 311, 320, 330, 342, 344, 356
    Configuration II (ungrounded) – I 96; II 311, 356, 502, 505, 513
    MDOF – I 98, 303, 305, 307–311, 313–320, 324, 325, 330–332, 336, 338–340, 347, 349, 352–356, 359, 363, 365, 366; II 1, 99–101, 103–105, 109–117, 120, 123, 126–128, 132, 133, 140–142, 150–156, 226, 353, 397, 502, 520, 530–559
    piecewise–linear – II 230
    SDOF – I 97, 98, 103–105, 108, 165, 233, 237, 265, 269, 303, 305, 313, 316, 325, 359, 365; II 1, 2, 12, 14, 66, 99, 112, 132, 133, 140, 142, 145–157, 311, 320, 353, 394, 453, 455, 502, 520–533, 536, 541, 542, 546–550, 552, 559–561, 564
    vibro–impact (VI) – I 10, 85, 86; II 241–260, 264–268, 270, 274, 275, 277, 279–282, 284–305, 572–574, 576–581, 584–612

Index 373

Nonlinear modal interactions I 4, 15, 17, 18, 25, 41, 45, 53, 70, 165, 171, 173; II 12, 17, 27, 37, 101, 109, 117, 127, 130, 131, 133, 147, 153, 282, 418, 453, 462–464, 471, 474, 478, 520, 566
Nonlinearity
  essential – I 94, 96, 121, 134, 173, 185, 239, 265; II 4, 9, 20, 46, 78, 87, 105, 111, 213, 312, 313, 320, 344, 349, 526, 528, 530, 531, 536, 537, 540, 541, 542, 545, 546, 572, 573
  geometric – I 3, 81, 82; II 349
  non–smooth – I 10, 29, 311
  smooth – II 262, 269, 273, 285, 287, 288, 302
  vibro–impact (VI) – I 10, 86; II 242, 271
Nonlinear Schrödinger equation (NLS) I 2

Oil-drilling I 10
Optimal TET I 10, 93, 97, 163, 165, 170, 185, 207, 212, 213, 216–218, 220, 224, 232, 233, 264; II 112, 151, 273
Oscillation
  periodic – I 16, 17, 20, 131, 171, 280, 330; II 70, 80, 85, 88, 91, 104, 123, 167, 173, 220, 353, 382
  relaxation – II 161, 182, 184–191, 196, 197, 220, 647

Perturbation
  adiabatic – II 241
  regular – I 45, 111, 224, 333
  singular – I 8, 45, 335; II 179, 200
Pitch mode II 398, 399, 404–408, 418, 424, 427, 428, 432, 433, 438, 439, 452–456, 458, 460–463, 467, 468, 474, 478–480, 484, 487, 503, 515, 516, 528, 536, 551, 552, 561
Plate I 10, 264; II 132–158, 582
Propagation zone (PZ) I 269, 271–273, 275, 280, 289, 290, 294, 297; II 42, 66, 70, 80, 86–88, 90, 98, 103
Proper orthogonal decomposition (POD) I 304
Proper orthogonal mode (POM) I 264

Quasi-periodic
  beat I 147
  function I 340, 342
  LCO II 521, 523, 549, 562–564
  motion, response I 8, 39, 70, 148, 173, 342; II 210, 226, 260, 301, 471, 474, 477, 497
  orbit I 4, 7, 40, 44, 45, 97, 108, 135–137, 146, 148, 155, 156, 162, 241, 242; II 39, 269, 382, 471, 493, 625
  oscillation I 72, 349; II 174, 176

TET I 10; II 201, 203
Quasi-periodicity II 493, 525

Reduced order I 27, 264, 304; II 353, 439, 444, 447, 463, 468, 621
Resonance
  band I 22, 123, 172
  fundamental (1:1) – I 56, 62–65, 172; II 173, 198, 202, 203, 206, 207, 216, 218–221, 224, 225
  interaction I 4, 5, 45, 81, 254, 255, 267, 271, 275, 276, 278, 279, 303, 308, 336, 340; II 18, 30, 31, 33, 37, 58, 61, 86, 89, 90, 98, 114, 120, 123, 126, 127, 147, 152, 167, 213, 216, 292, 294, 336, 374, 432, 447, 468, 516, 530, 634, 640
  manifold I 40, 41, 43, 45–50, 52, 64, 65, 69, 94, 167, 172, 179, 184, 186, 187, 192, 197, 198, 208, 212–15, 218, 246, 247, 297, 304, 341, 365; II 39, 95, 280, 288, 336, 340, 369, 371, 408, 477, 478
  nonlinear – I 2, 5, 25, 45, 81, 172, 246, 254, 275; II 13, 18, 31, 37, 58, 147, 152, 162, 177, 264, 292, 293, 530, 533
  passage through – I 9, 48, 52
  scattering by – I 8
  subharmonic – I 64, 97, 103, 105, 179; II 366, 374, 468, 474, 483, 515, 516
  superharmonic – I 192; II 463, 474, 509
  transient – I 15, 37, 47, 48, 52, 55, 64, 65, 72, 80, 93, 94, 102, 166, 171, 173, 175, 197, 237, 255, 258, 303, 304, 307, 336, 342, 356, 365; II 9, 13, 18, 55, 115, 120, 128, 156, 274, 319, 365, 371, 394, 452, 501, 515, 519, 559, 565, 576, 578, 590, 599, 600, 629, 634, 639, 646
Resonance capture
  cascade (RCC) I 5, 233, 237, 246, 247, 255–262, 267–269, 298, 303–305, 347, 365; II 24, 30, 55, 98, 311, 453, 629
  escape, escape from – I 48, 173; II 9, 65, 316, 433, 438, 444, 452
  permanent – I 47; II 365
  sustained (SRC) – I 15, 37, 38, 46, 47, 52, 53, 64, 74, 172, 349, 352, 356; II 365, 399, 408, 428, 452, 515, 628, 640
  transient (TRC) – I 15, 37, 47, 48, 52, 55, 72, 80, 93, 94, 102, 166, 171, 173, 175, 197, 237, 303, 304, 307, 336, 342, 356, 365; II 9, 13, 18, 55, 115, 128, 274, 319, 341, 365, 371, 394, 399, 428, 452, 515, 559, 565, 576, 578, 590, 599, 600, 628, 629, 639, 640, 646
Robustness

- of LCO suppression II 353, 354, 357, 359, 391, 453, 467, 477, 487, 489, 493, 501, 502, 513, 516, 520, 521, 524–526, 528, 530, 533, 536, 541, 542, 546–549, 559, 561, 637
- of LCO elimination II 381, 382, 390, 391, 520
- of TET I 7, 9, 205, 303; II 11, 86, 157, 502

Rod I 10, 81; II 12–63, 66, 69–76, 78–82, 84–88, 90, 91, 93–120, 123, 126–128, 130, 132, 134, 155

Shock spectrum II 232, 238–241, 251, 253, 255–258
Sifting process I 77, 78
Signal processing I 15, 70, 106
Slow dynamics I 151, 198, 216–218; II 90, 183, 187, 362, 363, 366, 444, 447
Slow-fast partition I 56, 57, 67, 124, 141, 146, 155, 197, 276, 291; II 90–93, 402, 410, 177, 216
Slow flow I 44, 51–53, 57, 58, 66–68, 80, 97, 125, 126, 132, 143, 151, 154, 157, 159, 162, 163, 173, 175, 176, 177, 180–184, 187–189, 191, 197–198, 205–209, 212–232, 279, 282, 297, 342–345; II 15, 93, 94, 161, 163, 164, 168–170, 173–175, 177–180, 182, 184–189, 191–197, 203, 216, 218, 220, 295, 336, 359, 361, 365, 371, 383–385, 389, 398, 399, 402, 404, 408, 410–419, 422, 428, 432, 434–437, 439, 440, 444, 447, 452, 463, 478, 483–485
Slow invariant manifold (SIM) II 182–190, 192, 196, 201, 224
Soliton I 270; II 66, 67
Stability
  asymptotic – I 20; II 183, 626
  orbital – I 20
Steady state
  motion, response I 9, 55, 285; II 87, 161, 162, 173, 177, 178, 191, 201, 203, 205, 206, 208, 209, 212, 216, 343, 344, 351, 354, 357, 359, 386, 390, 391, 394, 396, 408, 422, 447, 455, 467, 474, 625, 630, 645
  TET I 54; II 167, 203, 212, 218, 342, 347, 351
Strongly modulated response (SMR) II 161, 177–180, 182, 183, 185–189, 191–194, 196–201, 203, 207, 209–214, 219–226, 342
Subharmonic
  motion I 122, 130, 323, 324, 326, 328; II 43, 48, 105, 120, 268, 269

TET I 97, 105, 171, 176, 178, 179, 181–186, 208, 246, 249, 250; II 276, 280, 301, 319, 412
tongue I 25, 121, 124, 125, 128, 131–133, 157, 171, 177, 179, 241, 243, 245, 250, 254, 322, 323, 326, 327, 342, 352; II 39, 46, 48, 49, 51, 52, 54, 55, 58, 104, 107, 120, 123, 126, 259, 268–270, 275, 276, 280, 285, 288–291, 320, 471
transient resonance capture (TRC) I 176, 249, 285; II 55, 201, 275, 276, 474, 480
Switch model II 625
System identification I 28, 72, 264; II 65, 293, 295, 312, 313, 331, 398, 508

Transversality I 47; II 185
Transform
  Fast Fourier (FFT) – I 71, 307, 324
  Hilbert – I 15, 38, 71, 77, 79, 80, 247, 258, 259; II 15, 78, 198, 200, 201, 294, 331, 336, 337, 463, 474, 477, 488
  wavelet (WT) – I 15, 70, 71, 95, 166, 247, 304, 305, 307, 309, 315–317, 347; II 12, 15–17, 19, 24, 27–31, 33, 37, 39, 40, 50, 51, 54–57, 59, 61–66, 105, 107, 109, 117–120, 122, 123, 125–127, 129, 130, 133, 141, 147, 148, 259, 260, 273–280, 282–285, 287, 288, 290, 291, 293, 294, 296, 297, 299, 302, 319, 335, 336, 342, 363, 365–373, 376, 378, 380, 398, 404, 406, 408, 432, 433, 453, 463, 468, 471, 474, 476, 483, 484, 514–519, 549, 552, 564, 633, 639–642
Transient TET II 177, 201, 462
Transition
  multi-frequency – II 39, 40, 126, 260
  multi-modal – II 259
Triggering II 319, 397, 398, 408, 417, 418, 422, 428, 433, 439, 444, 447, 452–454, 462, 468, 480, 484, 487, 509, 515, 516, 549
Tuned mass damper (TMD) I 5, 98, 103–107, 262–264; II 140, 153–158, 161, 202, 207, 226, 571, 615, 629
Twist II 399, 400

Van der Pol (VDP) oscillator II 353–357, 359 361, 364–366, 369, 371, 374, 390, 391, 394
Vibro-impact (VI)
  NES I 85; II 157, 241–260, 264–270, 274, 275, 277, 279–282, 284–305, 311, 572–574, 576–581, 585–604, 606–609, 611, 612
  oscillation II 241
  seismic mitigation I 86

Index 375

Vortex shedding I vii; II 400

Wave
  localized – I 33
  solitary – I 2, 35; II 66
  standing – I 270, 272, 275, 277, 278, 280, 284–287, 289, 290; II 42, 66, 70, 87, 98, 103
  traveling – I 36, 271, 273, 274, 280, 298; II 42, 66, 70, 72, 73, 86–88, 95, 97, 98, 103
Wavelet (WT) spectrum II 15, 17, 24, 27–31, 33, 37, 39, 40, 51, 54–57, 59, 61–63, 65, 66, 117–120, 122, 123, 125–127, 129, 130, 134, 147, 148, 259, 260, 273–280, 282–285, 287–291, 293, 294, 296, 297, 299–302, 319, 335, 365, 367–370, 373, 376, 378, 380, 404, 406, 407, 468, 471, 476, 483, 516, 552, 641, 642
Wing II 353, 354, 397–402, 404, 408, 409, 418, 428, 438, 447, 452–455, 458–464, 467, 468, 474, 477, 478, 483, 487, 497, 501, 502–509, 513, 516, 520, 521, 524–527, 529–533, 536, 541, 542, 546–554, 557–564